T0301797

# An Undergraduate Introduction to Financial Mathematics

## Fourth Edition

# An Undergraduate Introduction to Financial Mathematics

## Fourth Edition

**J Robert Buchanan**

Millersville University, USA

**World Scientific**

EW JERSEY · LONDON · SINGAPORE · BEIJING · SHANGHAI · HONG KONG · TAIPEI · CHENNAI · TOKYO

*Published by*

World Scientific Publishing Co. Pte. Ltd.

5 Toh Tuck Link, Singapore 596224

*USA office:* 27 Warren Street, Suite 401-402, Hackensack, NJ 07601

*UK office:* 57 Shelton Street, Covent Garden, London WC2H 9HE

**Library of Congress Cataloging-in-Publication Data**
Names: Buchanan, J. Robert, author.
Title: An undergraduate introduction to financial mathematics /
    J. Robert Buchanan, Millersville University, USA.
Description: Fourth edition. | New Jersey : World Scientific, [2023] |
    Includes bibliographical references and index.
Identifiers: LCCN 2022021015 | ISBN 9789811260308 (hardcover) |
    ISBN 9789811260315 (ebook for institutions) | ISBN 9789811260322 (ebook for individuals)
Subjects: LCSH: Business mathematics.
Classification: LCC HF5691 .B875 2023 | DDC 650.01/51--dc23/eng/20220603
LC record available at https://lccn.loc.gov/2022021015

**British Library Cataloguing-in-Publication Data**
A catalogue record for this book is available from the British Library.

For any available supplementary material, please visit
https://www.worldscientific.com/worldscibooks/10.1142/12964#t=suppl

Desk Editors: Jayanthi Muthuswamy/Lai Fun Kwong

Typeset by Stallion Press
Email: enquiries@stallionpress.com

Printed in Singapore

Dedication

For my wife, Monika.

# Preface

This is the fourth edition of the textbook, I created in 2006 to serve the educational needs of my undergraduate students who are interested in topics related to mathematical finance. The intended audience remains undergraduate students studying mathematics, economics, or finance. Anyone with an interest in learning about the mathematical modeling of prices of financial derivatives such as bonds, futures, and options can start with this book. The only mathematical prerequisite is multivariable calculus. The necessary theory of interest, statistical, stochastic, and differential equations material is developed in their respective chapters, with the goal of making this introductory text as self-contained as possible. Readers with more advanced preparation may skip chapters covering topics about which they feel they have sufficient background knowledge.

All of the chapters from the previous editions of the text have been extensively re-written and re-organized so that readers can get to the financial topics quickly while mathematical concepts are introduced as needed. Overall the number of pages of exposition has increased by 30% from the previous edition (not counting the exercise solutions). The material on American options has now been merged with similar topics related to European options in an effort to unify and streamline the presentation. The development of the normal and lognormal probability distributions from basic principles is more rigorous than in previous editions without sacrificing the accessibility of the material. The chapters on hedging portfolios and extensions of the Black–Scholes model have been expanded. The chapter from the previous editions on optimizing portfolios has been re-written to focus on the development of the Capital Asset Pricing Model (CAPM). Earlier editions of this text gave brief attention to a binomial lattice numerical technique for approximating option prices. This material has been expanded into a standalone chapter illustrating the wide-ranging utility of the binomial model for numerically estimating option prices. This new edition takes a different approach to solving the Black–Scholes partial differential equation, eliminating the need for the introduction of the Fourier transform. This edition also introduces a completely new chapter on the pricing of exotic options. The appendix now contains a chapter on linear algebra with sufficient background material to support a more rigorous development of the Arbitrage Theorem. To help students more fully engage in the learning process, by applying what they have read and filling in the details of the development of some topics, the number of exercises has been greatly expanded. The previous edition contained 295 exercises collected at the ends of the chapters. The fourth edition has 704 exercises arranged at the ends of sections within chapters. An instructor's solution manual is still available.

Chapter 1 introduces the theory of interest and the fundamentally important principle of present value. The second chapter provides background on discrete random variables, probability mass functions, and basic statistical measures. This chapter introduces the binomial random variable in particular. Chapter 3 contains an introduction to linear programming and the development of the arbitrage principle and risk-neutral probability. Readers without sufficient preparation in elementary linear algebra may read Appendix A before studying Chapter 3. Chapter 4 combines the notions of the theory of interest and statistics (specifically covariance and correlation) to present the CAPM. Chapters 5 and 6 present financial futures and options prior to the later material on pricing formulas. This is done so that properties of these financial instruments can be explored independently of how they are priced. Estimates of option prices for both European and American options are achieved using binomial lattice methods in Chapter 7. A one-semester course in the basics of option pricing could be based on Chapters 1 through 7. Such a course would not require the more advanced mathematical and statistical machinery used later in the textbook.

Chapter 8 broadens the discussion of random variables, probability, and statistics to the setting of the continuous. The ninth chapter uses the normal and lognormal random variables to develop the idea of Brownian motion which is assumed to be a model related to the returns on stocks and other assets. The chapter also introduces and justifies the stochastic integral and Itô's Lemma, fundamentally important tools in the rigorous development of option pricing formulas. Chapter 10 derives the Black–Scholes initial, boundary value problem for options with the European style of exercise and solves it. The final section of the chapter demonstrates that the limiting case of the binomial model converges to the partial differential equation solution. Chapter 11 demonstrates that the Black–Scholes option pricing models can be adapted to price options with different underlying assets such as futures, currency exchanges, and stocks which pay discrete dividends. The sensitivity of option pricing formulas to changes in independent variables and parameters is explored in Chapter 12. These ideas are used to investigate various hedging practices in Chapter 13. The final chapter rewards the dedicated reader and student with a rigorous development of the pricing formulas for many types of exotic options (options which do not satisfy the Black–Scholes equation). Chapter 14 presents not just the pricing formulas but also their derivations, parity relationships, and uses in portfolio management. There are many references providing the pricing formulas without the details of their development, but relatively few provide the formulas, their justifications, and some of their applications in one place. The material from Chapters 8 through 14 could form a second semester course in mathematical finance. Well-prepared and motivated students could cover all the of the material in the text in a single semester.

The fourth edition of the text expands and clarifies the earlier editions. While I hope this edition is error-free, should the readers find any errors please contact me with specifics at Robert.Buchanan@millersville.edu. A list of errata (if any) for this edition will be found on my homepage accessible through the millersville.edu website.

J Robert Buchanan
*Lancaster, PA, USA*
*March 31, 2022*

# About the Author

**J Robert Buchanan** received a BS in Physics from Davidson College (Davidson, North Carolina) and an MS and a PhD in Applied Mathematics from North Carolina State University (Raleigh, North Carolina). His mathematical interests include differential equations and financial mathematics. His other interests include art and travel. He is currently a Professor of Mathematics at Millersville University of Pennsylvania (Millersville, Pennsylvania).

# Contents

Chapter 1

# The Theory of Interest

One of the first types of investments that people learn about is some variation on the savings account. In exchange for the temporary use of an investor's money, a bank or other financial institution agrees to pay to the investor **interest**, which is related to the amount invested. There are many different schemes for paying interest. This chapter will describe some of the most common types of interest and contrast their differences. Along the way the reader will have the opportunity to renew their acquaintanceship with exponential functions and the geometric series. Since an amount of capital invested will earn interest and thus numerically increase in value in the future, the concept of **present value** will be introduced. Present value provides a way of comparing values of investments made at different times in the past, present, and future. As an application of present value, several examples of saving for retirement and calculation of mortgages will be presented. Sometimes investments pay the investor varying amounts of interest which change over time. The concept of **rate of return** can be used to convert these payments into effective interest rates, making comparison of investments easier. Present value and rate of return are two commonly used measures for comparing investments.

## 1.1 Simple Interest

In exchange for the use of a depositor's money, banks pay a fraction of the account balance back to the depositor. This fractional payment is known as **interest**. The money a bank uses to pay interest is generated by investments and loans that the bank makes with the depositor's money. Interest is paid in many cases at specified times of the year, but nearly always the fraction of the deposited amount used to calculate the interest is called the

**interest rate** and is expressed as a percentage paid per year. For example, a credit union may pay 6% annually on savings accounts. This means that if a savings account contains $100 now, then exactly 1 year from now the bank will pay the depositor $6 (which is 6% of $100) provided the depositor maintains an account balance of $100 for the entire year.

In this chapter and the chapters to follow, interest rates will be denoted symbolically by $r$. To simplify the formulas and mathematical calculations, when $r$ is used it will be converted to decimal form even though it may still be referred to as a percentage. The 6% annual interest rate mentioned above would be treated mathematically as $r = 0.06$ per year. The initially deposited amount which earns the interest will be called the **principal amount** and will be denoted $P$. The sum of the principal amount and any earned interest will be called the **amount due**. The symbol $A$ will be used to represent the amount due. The reader may even see the amount due referred to as the **compound amount**, though this use of the adjective "compound" is independent of its use in the term "compound interest" to be explored in Sec. 1.2. The relationship between $P$, $r$, and $A$ for a single year period is

$$A = P + Pr = P(1 + r).$$

In general, if the time period of the deposit is $t$ years then the amount due is expressed in the formula

$$A = P(1 + rt). \tag{1.1}$$

This implies the account balance for the period of the deposit is $P$ and when the balance is withdrawn (or the account is closed), the principal amount $P$ plus the interest earned $Prt$ is returned to the investor. No interest is credited to the account until the instant it is closed. This is known as the **simple interest** formula.

*Exercises*

**1.1.1** If a bank pays simple interest at a rate of 5% annually and an investor deposits $3,500 for 1 year, find the amount due at the end of the year.

**1.1.2** If a bank pays simple interest at a rate of 6.5% annually and an investor deposits $3,200 for 18 months, find the amount due at the end of the 18-month period.

**1.1.3** Find the principal amount deposited if a bank pays simple interest of 4% annually and the amount due on an investment after 1 year is $1,700.

**1.1.4** Find the principal amount deposited if a bank pays simple interest of 6% annually and the amount due on an investment after 3 and 1.5 years is $2,100.

**1.1.5** An investor deposits $1,500 for 2 years and earns $20 in interest. What simple interest rate (expressed as an annual percentage) is the bank paying this investor?

## 1.2 Compound Interest

Some financial institutions credit interest earned by the account balance at scheduled points in time. Banks and other financial institutions "pay" the depositor by adding the interest to the depositor's account. The interest, once paid to the depositor, is the depositor's to keep. Unless the depositor withdraws the interest or some part of the principal, the process begins again for another interest earning period. Whenever interest is allowed to earn interest itself, an investment is said to earn **compound interest**. In this situation, interest is paid to the depositor once or more times per year. Once paid, the interest begins earning interest. Consider the case of a bank which pays interest yearly on the anniversary of the initial deposit. If $P$ is initially invested, then after 1 year, the amount due according to Eq. (1.1) with $t = 1$ would be $P(1 + r)$. This amount can be thought of as the principal amount for the account at the beginning of the second year. Thus, 2 years after the initial deposit the amount due would be

$$A = P(1 + r) + P(1 + r)r = P(1 + r)^2.$$

Continuing in this way, $t$ years after the initial deposit of an amount $P$, the capital $A$ will grow to

$$A = P(1 + r)^t. \tag{1.2}$$

A mathematical "purist" may wish to establish Eq. (1.2) using the principle of induction.

Banks and other interest-paying financial institutions may pay interest more than one time per year. The yearly interest formula given in Eq. (1.2) must be modified to track the compound amount for interest periods of other than 1 year.

The typical interest bearing savings or checking account will be described by an investor as earning a **nominal annual interest rate** compounded some number of times per year. Investors will find interest compounded semi-annually, quarterly, monthly, weekly, or daily. This section will compare and contrast compound interest to the simple interest case of the previous section. Let $n$ denote the number of compounding periods per year. For example for interest "compounded monthly" $n = 12$. Only two small modifications to the interest formula in Eq. (1.2) are needed to calculate the amount due. First, it is now necessary to think of the interest rate per compounding period. If the annual interest rate is $r$, then the interest rate per compounding period is $r/n$. Second, the elapsed time should be thought of as some number of compounding periods rather than years. Thus, with $n$ compounding periods per year, the number of compounding periods in $t$ years is $nt$. Therefore, the formula for compound interest is

$$A = P\left(1 + \frac{r}{n}\right)^{nt}. \tag{1.3}$$

Equation (1.3) simplifies to the formula for the amount due given in Eq. (1.2) when $n = 1$.

**Example 1.1.** Suppose an account earns 5.75% annually compounded monthly. If the principal amount is \$3,104 find the amount due after three and 1.5 years.

**Solution.** The amount due will be

$$A = 3{,}104\left(1 + \frac{0.0575}{12}\right)^{(12)(3.5)} = \$3{,}794.15.$$

The reader should verify using Eq. (1.1) that if the principal in the previous example earned only *simple* interest at an annual rate of 5.75% then the amount due after 3.5 years would be only \$3,728.68. Thus happily for the depositor, compound interest builds value faster than simple interest. Frequently it is useful to compare a nominal interest rate compounded some number of times per year with an equivalent interest compounded annually, *i.e.*, to the annual interest rate which would generate the amount due as the nominal rate. This equivalent interest rate is called the **effective interest rate** and is denoted as $r_e$. For the rate mentioned in the previous example the effective interest rate can be found by solving the equation

$$\left(1 + \frac{0.0575}{12}\right)^{12} = 1 + r_e$$

$$r_e = 0.05904.$$

Thus, the nominal annual interest rate of 5.75% compounded monthly is equivalent to an effective annual rate of 5.904%.

In general, if the nominal annual rate $r$ is compounded $n$ times per year the equivalent effective annual rate $r_e$ is given by the formula:

$$r_e = \left(1 + \frac{r}{n}\right)^n - 1. \tag{1.4}$$

Intuitively it seems that more compounding periods per year implies a higher effective annual interest rate. The next section, will explore the limiting case of frequent compounding going beyond semiannually, quarterly, monthly, weekly, daily, hourly, *etc.* to continuous compounding.

## *Exercises*

**1.2.1** Suppose that \$3,659 is deposited in a savings account which earns 6.5% interest compounded semi-annually. What is the amount due after 5 years?

**1.2.2** Suppose that \$3,993 is deposited in an account which earns 4.3% interest. What is the amount due in 2 years if the interest is compounded monthly?

**1.2.3** Suppose \$3,750 is invested today. Find the amount due in 8 years if the interest rate is 0.375% 3-month interest compounded every 3 months.

**1.2.4** Find the effective annual interest rate which is equivalent to 8% interest compounded quarterly.

**1.2.5** Jordan is preparing to open a bank which will accept deposits and will pay interest compounded monthly. In order to be competitive Jordan must meet or exceed the interest paid by another bank which pays 5.25% compounded daily. What is the minimum interest rate Jordan can pay and remain competitive?

**1.2.6** Morgan deposits \$120 into a savings account which pays interest at an annual nominal rate of $r$, compounded semi-annually. Rory deposits \$180 into a different savings account at the same time which pays simple interest at an annual rate of $r$. Morgan and Rory earn the same amount of interest during the last 6 months of the 10th year. Find $r$.

## 1.3  Continuously Compounded Interest

Considering the effect on the amount due under more frequent compounding, requires a limit operation. In symbolic form amount due $A$ satisfies the equation

$$A = \lim_{n \to \infty} P\left(1 + \frac{r}{n}\right)^{nt}. \tag{1.5}$$

Fortunately, there is a simple expression for the value of the limit on the right-hand side of Eq. (1.5). Consider the limit

$$\lim_{n \to \infty} \left(1 + \frac{r}{n}\right)^n.$$

This limit is indeterminate of the form $1^\infty$. It is evaluated via a standard approach using the natural logarithm and l'Hôpital's Rule. The reader should consult an elementary calculus book such as [Smith and Minton (2002)] for more details. If $y = (1 + r/n)^n$, then

$$\ln y = \ln \left(1 + \frac{r}{n}\right)^n = n \ln \left(1 + \frac{r}{n}\right) = \frac{\ln(1 + r/n)}{1/n},$$

which is indeterminate of the form $0/0$ as $n \to \infty$. To apply l'Hôpital's Rule take the limit of the derivative of the numerator over the derivative of the denominator. Thus

$$\lim_{n \to \infty} \ln y = \lim_{n \to \infty} \frac{\frac{d}{dn} \left[\ln(1 + r/n)\right]}{\frac{d}{dn} \left[1/n\right]} = \lim_{n \to \infty} \frac{r}{1 + r/n} = r.$$

Therefore, $\lim_{n \to \infty} y = e^r$. Finally the formula for **continuously compounded interest**, can be expressed as

$$A = P e^{rt}. \tag{1.6}$$

This formula may seem familiar since it is often presented as the exponential growth formula in elementary algebra, precalculus, or calculus. The quantity $A$ has the property that $A$ changes with time at a rate proportional to $A$ itself.

**Example 1.2.** Suppose \$3,585 is deposited in an account which pays interest at an annual rate of 6.15% compounded continuously. Find the amount due after 2.5 years and the effective annual rate of interest.

**Solution.** After 2.5 years the principal plus earned interest will have grown to

$$A = 3{,}585 e^{(0.0615)(2.5)} \approx \$4{,}180.82.$$

The effective simple interest rate is the solution to the equation

$$e^{0.0615} = 1 + r_e,$$

which implies $r_e \approx 6.34\%$.

## Exercises

**1.3.1** Suppose that $3,995 is deposited in an account which earns 4.5% interest compounded continuously. What is the amount due after 3 years?

**1.3.2** Suppose a bank pays interest at the rate of 6% annually compounded continuously. What is the effective annual rate of interest?

**1.3.3** Suppose a bank pays interest at the rate of 7.5% per annum compounded continuously. How many years are required for the amount due to be twice the initial amount deposited?

**1.3.4** Angel deposits $1,000 into a bank account. The account earns interest at an annual nominal rate of interest of 5% compounded semi-annually. At the same time, Taylor deposits $1,000 into a separate account. Taylor's account earns interest at rate $r$ compounded continuously. After 6.5 years, the value of each account is the same. Find $r$.

**1.3.5** Suppose $1,000 will be deposited into each of two savings accounts. One account pays interest at an annual rate of 4.75% compounded daily, while the other pays interest at an annual rate of 4.75% compounded continuously. How long would it take for the amounts due to differ by $1?

**1.3.6** Finley deposits $500 in a savings account earning continuously compounded interest at an annual rate of 9%. After 4 years Finley deposits an additional amount $P$. The amount of interest earned between the end of the 4th year and the end of the 8th year is also $P$. Find $P$.

**1.3.7** Many textbooks determine the formula for continuously compounded interest through an argument which avoids the use of l'Hôpital's Rule (for example [Goldstein *et al.* (1999)]). Beginning with Eq. (1.5) let $h = r/n$. Then

$$P\left(1 + \frac{r}{n}\right)^{nt} = P(1 + h)^{(1/h)rt},$$

and focus on finding the $\lim_{h \to 0}(1 + h)^{1/h}$. Show that

$$(1 + h)^{1/h} = e^{(1/h)\ln(1+h)},$$

and take the limit of both sides as $h \to 0$. *Hint*: Use the definition of the derivative in the exponent on the right-hand side.

## 1.4 Present Value

One of the themes seen many times in the study of financial mathematics is the comparison of the value of a particular investment at the present time with the value of the investment at some point in the future. This is the comparison between the **present value** of an investment versus its

**future value.** Present and future value play central roles in planning for retirement and determining loan payments. Later in this book present and future values will help determine a fair price for financial derivatives.

The future value $t$ years from now of an invested amount $P$ subject to annual interest rate $r$ compounded continuously is

$$A = Pe^{rt}.$$

Thus, by comparison with Eq. (1.6), the future value of $P$ is just the amount due for $P$ monetary units invested in a savings account earning interest $r$ compounded continuously for $t$ years. By contrast, the present value of an amount due $A$ in an environment of interest rate $r$ compounded continuously for $t$ years is

$$P = Ae^{-rt}.$$

In other words, if an investor wishes to have $A$ monetary units in savings $t$ years from now and they can place money in a savings account earning interest at an annual rate $r$ compounded continuously, the investor should deposit $P$ monetary units now. There are also formulas for future and present value when interest is compounded at discrete intervals, not continuously. If the interest rate is $r$ annually with $n$ compounding periods per year then the future value of $P$ is

$$A = P\left(1 + \frac{r}{n}\right)^{nt}.$$

Compare this equation with Eq. (1.3). Simple algebra shows then the present value of $P$ earning interest at rate $r$ compounded $n$ times per year for $t$ years is

$$P = A\left(1 + \frac{r}{n}\right)^{-nt}.$$

The reader can also derive present and future value relationships for the case of simple interest.

**Example 1.3.** Suppose an investor will receive payments at the end of the next 6 years in the amounts shown in the table below.

| Year | 1 | 2 | 3 | 4 | 5 | 6 |
|---|---|---|---|---|---|---|
| **Payment** | 465 | 233 | 632 | 365 | 334 | 248 |

If the interest rate is 3.99% compounded monthly, what is the present value of the investments?

**Solution.** Let $a = (1 + 0.0399/12)^{-12}$, then assuming the first payment will arrive 1 year from now, the present value is the sum

$$465a + 233a^2 + 632a^3 + 365a^4 + 334a^5 + 248a^6 = \$2003.01.$$

Notice that the present value of the payments from the investment is different from the sum of the payments themselves (which is \$2,277).

Unless the reader is among the very fortunate few who can always pay cash for all purchases, the reader may some day apply for a loan from a bank or other financial institution. Loans are always made under the assumptions of a prevailing interest rate (with compounding), an amount to be borrowed, and the lifespan of the loan, *i.e.*, the time the borrower has to repay the loan. Usually portions of the loan must be repaid at regular intervals (for example, monthly). If the same amount is paid at regular intervals, this situation is sometimes called a loan with **level payments**. Now consider the relationship between the principal amount borrowed, the length of the loan, and the interest rate, and the level payment.

A helpful mathematical tool for answering questions regarding present and future values of level payment loans is the **geometric series**. Consider the sum

$$S = 1 + a + a^2 + \cdots + a^n, \tag{1.7}$$

where $n$ is a positive whole number. If both sides of Eq. (1.7) are multiplied by $a$ and then subtracted from Eq. (1.7), the difference is

$$S - aS = 1 + a + a^2 + \cdots + a^n - (a + a^2 + a^3 + \cdots + a^{n+1})$$
$$S(1 - a) = 1 - a^{n+1}$$
$$S = \frac{1 - a^{n+1}}{1 - a}, \tag{1.8}$$

provided $a \neq 1$.

The summation formula in Eq. (1.8) can be used to find the level payment amount of a loan. Suppose someone borrows $P$ to purchase a new car. The bank issuing the automobile loan charges interest at the annual rate of $r$ compounded $n$ times per year. The length of the loan will be $t$ years. The level payment can be calculated assuming the present value of all the payments made equals the principal amount borrowed. Suppose the level payment is $x$ and define the quantity $a = (1 + r/n)^{-1}$. If the first payment must be made at the end of the first compounding period, then the present value of all the payments is

$$xa + xa^2 + \cdots + xa^{nt} = xa(1 + a + \cdots a^{nt-1})$$

$$= xa \left( \frac{1 - a^{nt}}{1 - a} \right) = x \frac{1 - (1 + \frac{r}{n})^{-nt}}{\frac{r}{n}}.$$

Therefore, the relationship between the interest rate, the compounding frequency, the period of the loan, the principal amount borrowed, and the level payment is expressed in the following equation.

$$P = x \frac{n}{r} \left( 1 - \left[ 1 + \frac{r}{n} \right]^{-nt} \right). \tag{1.9}$$

**Example 1.4.** Suppose a person borrows \$25,000 for 5 years at an interest rate of 4.99% compounded monthly. Find the amount of the level monthly payments.

**Solution.** The payment amount will be

$$x = 25{,}000 \left( \frac{0.0499}{12} \right) \left( 1 - \left[ 1 + \frac{0.0499}{12} \right]^{-(12)(5)} \right)^{-1} \approx \$471.67.$$

Similar reasoning can be used when determining how much to save for retirement. Suppose a person is 25 years of age now and plans to retire at age 65. For the next 40 years they plan to invest a portion of their monthly income in securities which earn interest at the nominal annual rate of 10% compounded monthly. After retirement the person plans on receiving a monthly payment (an **annuity**) in the absolute amount of \$1,500 for 30 years. The amount of money the person should invest monthly while working can be determined by equating the present value of all their deposits with the present value of all their withdrawals. The first deposit will be made 1 month from now and the first withdrawal will be made 481 months from now. The last withdrawal will be made 840 months from now. The monthly deposit amount will be denoted by the symbol $x$. The present value of all the deposits made into the retirement fund is

$$x \sum_{i=1}^{480} \left( 1 + \frac{0.10}{12} \right)^{-i} = x \left( 1 + \frac{0.10}{12} \right)^{-1} \frac{1 - \left( 1 + \frac{0.10}{12} \right)^{-480}}{1 - \left( 1 + \frac{0.10}{12} \right)^{-1}}$$

$$\approx 117.765x.$$

Meanwhile, the present value of all the annuity payments is

$$1{,}500 \sum_{i=481}^{840} \left( 1 + \frac{0.10}{12} \right)^{-i} = 1{,}500 \left( 1 + \frac{0.10}{12} \right)^{-481} \frac{1 - \left( 1 + \frac{0.10}{12} \right)^{-360}}{1 - \left( 1 + \frac{0.10}{12} \right)^{-1}}$$

$$\approx 3{,}182.94.$$

Thus, $x \approx \$27.03$ per month. This seems like a small amount to invest, but such is the power of compound interest and starting to save early for retirement. If the person waits 10 years (*i.e.*, until age 35) to begin saving for retirement, but all other factors remain the same, then

$$x \sum_{i=1}^{360} \left(1 + \frac{0.10}{12}\right)^{-i} \approx 113.951x,$$

$$1{,}500 \sum_{i=361}^{720} \left(1 + \frac{0.10}{12}\right)^{-i} \approx 8{,}616.36,$$

which implies the person must invest $x \approx \$75.61$ monthly. Waiting 10 years to begin saving for retirement nearly triples the amount which the future retiree must set aside for retirement.

The initial amounts invested are of course invested for a longer period of time and thus contribute a proportionately greater amount to the future value of the retirement account.

**Example 1.5.** Suppose Karter and Ali will retire in 20 years. Karter begins saving $x$ per month for retirement but due to unforeseen circumstances must abandon their savings plan after 4 years. The amount Karter puts aside during those first 4 years remains invested, but no additional amounts are invested during the last 16 years of their working life. Ali waits 4 years before putting any money into a retirement savings account and saves $x$ per month for retirement only during the last 16 years of their working life. Assuming the nominal annual interest rate 5% compounded monthly find the present values of Karter's and Ali's retirement savings.

**Solution.** Upon retirement Karter has an account whose present value is

$$x \sum_{i=1}^{48} \left(1 + \frac{0.05}{12}\right)^{-i} \approx 43.423x.$$

The present value of Ali's retirement savings is

$$x \sum_{i=49}^{240} \left(1 + \frac{0.05}{12}\right)^{-i} \approx 108.102x.$$

Thus, the Ali retires with a larger amount of retirement savings; however, the ratio of Karter's retirement account value to Ali's is $43.423/108.102 \approx 0.40$. Karter saves, in only one fifth of the time, approximately 40% of what Ali saves.

The discussion of retirement savings makes no provision for rising prices. The economic concept of **inflation** is the phenomenon of the decrease in the purchasing power of a unit of money relative to a unit amount of goods or services. The rate of inflation (usually expressed as an annual percentage rate, similar to an interest rate) varies with time and is a function of many factors including political, economic, and international factors. While the causes of inflation can be many and complex, inflation is generally described as a condition which results from an increase in the amount of money in circulation without a commensurate increase in the amount of available goods. Thus, relative to the supply of goods, the value of the currency is decreased. This can happen when wages are arbitrarily increased without an equal increase in worker productivity.

Consider the effect that inflation may have on the worker planning to save for retirement. If the interest rate on savings is $r$ and the inflation rate is $i$, these quantities can be used to calculate the **inflation-adjusted rate** or as it is sometimes called, the **real rate of interest**. This derivation will test the reader's understanding of the concepts of present and future value discussed earlier in this chapter. Let the symbol $r_i$ denote the inflation-adjusted interest rate [Broverman (2004)]. Suppose at the current time one unit of currency will purchase one unit of goods. Invested in savings, that one unit of currency has a future value (in 1 year) of $1+r$. In 1 year the unit of goods will require $1 + i$ units of currency for purchase. The difference

$$(1 + r) - (1 + i) = r - i$$

will be the real rate of growth in the unit of currency invested now. However, this return on saving will not be earned until 1 year from now. This rate of growth is adjusted by finding its present value under the inflation rate. This produces the following formula for the inflation-adjusted interest rate.

$$r_i = \frac{r - i}{1 + i} \tag{1.10}$$

Note that when inflation is low ($i$ is small), $r_i \approx r - i$ and this latter approximation is sometimes used in place of the more accurate value expressed in Eq. (1.10).

Returning to the earlier example of the worker saving for retirement, consider the case in which $r = 0.10$, the worker will save for 40 years and live on a monthly annuity whose inflation adjusted value will be \$1,500 for 30 years, and the rate of inflation will be $i = 0.03$ for the entire lifespan of the worker/retiree. Thus $r_i \approx 0.0680$. Assuming the worker will make the first deposit in 1 month the present value of all deposits to be made is

$$x \sum_{i=1}^{480} \left(1 + \frac{0.068}{12}\right)^{-i} = x \left(1 + \frac{0.068}{12}\right)^{-1} \frac{1 - \left(1 + \frac{0.068}{12}\right)^{-480}}{1 - \left(1 + \frac{0.068}{12}\right)^{-1}}$$

$$\approx 164.756x.$$

The present value of all the annuity payments is given by

$$1500 \sum_{i=481}^{840} \left(1 + \frac{0.068}{12}\right)^{-i} = 1500 \left(1 + \frac{0.068}{12}\right)^{-481} \frac{1 - \left(1 + \frac{0.068}{12}\right)^{-360}}{1 - \left(1 + \frac{0.068}{12}\right)^{-1}}$$

$$\approx 15273.80.$$

Thus, the monthly deposit amount is approximately $92.71. This is roughly four times the monthly investment amount when inflation is ignored. However, since inflation does tend to take place over the long run, ignoring a 3% inflation rate over the lifetime of the individual would mean that the present purchasing power of the last annuity payment would be

$$1500 \left(1 + \frac{0.03}{12}\right)^{-840} \approx \$184.17.$$

This is not much money to live on for an entire month. Retirement planning should include provisions for inflation, varying interest rates, the period of retirement, the period of savings, and desired monthly annuity during retirement.

## Exercises

**1.4.1** Lyric borrows $15,000 for 6 years at a nominal annual interest rate of 3.5% compounded monthly. Find the level payment of the loan.

**1.4.2** Ellis purchases a 10-year annuity which pays $100 at the end of the first month, $105 at the end of the second month, $110 at the end of the third month and so on. The annual nominal interest rate is 6% compounded monthly. Find the present value of the annuity.

**1.4.3** An investor can receive one of the following two payment streams:

(i) $500 at $t = 0$, $1,000 in 5 years, and $1,500 in 10 years,
(ii) $3,500 in 8 years.

The present values of the two payment streams are equal when the interest rate is $r$ compounded annually. Find $r$.

**1.4.4** Which of the two investments described below is preferable? Assume the first payment will take place exactly 1 year from now and further payments are spaced 1 year apart. Assume the continually compounded annual interest rate is 2.75%.

| Year | 1 | 2 | 3 | 4 |
|---|---|---|---|---|
| **Investment A** | 200 | 211 | 198 | 205 |
| **Investment B** | 198 | 205 | 211 | 200 |

**1.4.5** Suppose Remi wishes to buy a house costing $200,000. Remi will put a down payment of 20% of the purchase price and borrow the rest from a bank for 30 years at a fixed interest rate $r$ compounded monthly. Remi wishes the monthly mortgage payment to be $1,500 or less. What is the maximum nominal annual interest rate for the mortgage loan?

**1.4.6** If the effective annual interest rate is 5.05% and the rate of inflation is 2.02%, find the nominal annual real rate of interest compounded quarterly.

**1.4.7** Suppose Aubrey puts $12,000 into a savings account that pays an effective annual interest of 3% compounded annually for 15 years. The interest is credited to her account at the end of each year. If Aubrey withdraws any money from the account during the first 10 years there will be a penalty of 5% of the withdrawal amount. To help pay for the education of a child, Aubrey withdraws $T$ from the account at the end of years 8, 9, 10, and 11. The balance of the account at the end of the 15th year is $12,000. Find the value of $T$.

**1.4.8** A homeowner receives a property tax bill on July 1 in the amount of $4,500. There are two schedules of payment described on the bill. The full amount minus 2% can be paid by August 31, or $1,500 can be paid on each of August 31, October 31, and December 31. If the homeowner can invest $4,500 in a savings account earning effective annual interest at rate $r$ compounded monthly, what is the minimum value of $r$ at which the homeowner would prefer the "three equal payments" plan?

**1.4.9** Gail has $1,500 to invest on July 1 and decides to invest in a Treasury Bill. From Gail's perspective a Treasury Bill is like a loan to the government that will be paid back in one lump sum (including principal and interest) at a specified time in the future. Gail has two options to consider:

(i)  buy a 6-month Treasury Bill which will pay $1,600 on December 31 and then invest that amount in a savings account earning simple interest at rate $r$ until June 30 of the following year,

(ii) buy for \$1,450 a 1-year Treasury Bill which will pay \$1,600 on June 30 and with the remaining \$50 open a savings account which will earn interest at rate $r$ compounded semiannually.

If the two options have the same present values, find the interest rate $r$.

**1.4.10** Teachers at a certain school may elect to receive their salary in 9 equal payments made at the end of month that school is in session, or in 12 equal payments made at the end of each month throughout the year. If a teacher's salary is $S$ and the interest rate is 2.5% compounded monthly, find the present value of the salary under the 9-pay plan and the 12-pay plan.

**1.4.11** Find the sum of the infinite series

$$\sum_{k=1}^{\infty} k\, x^{k-1} = k + 2x + 3x^2 + \cdots$$

when $0 < x < 1$. *Hint*: Use the sum of the geometric series $f(x) = \sum_{k=0}^{\infty} x^k$ for $0 < x < 1$ and differentiate.

## 1.5   Rate of Return

The present value of an investment is one way to determine the absolute worth of the investment and to compare its worth to that of other investments. Another way to judge the value of an investment which an investor may own or consider purchasing is known as the **rate of return**. If a person invests an amount $P$ now and receives an amount $A$ one time unit from now, the rate of return can be thought of as the interest rate per unit time that the invested amount would have to earn so that the present value of the payoff amount is equal to the invested amount. Since the rate of return is going to be thought of as an equivalent interest rate, it will be denoted by the symbol $r$. If there is a single amount invested and a single payment to the investor, then

$$P = A(1+r)^{-1} \text{ or equivalently } r = \frac{A}{P} - 1.$$

**Example 1.6.** If an investor loans \$100 today with the understanding that \$110 will repay the loan in 1 year's time, then the rate of return is $r = 0.10$ or 10%.

In a more general setting, a person may invest an amount $P$ now and receive a sequence of positive payoffs $\{A_1, A_2, \ldots, A_n\}$ at regular intervals. In this case the rate of return per repayment period is the interest rate such that the present value of the sequence of payoffs is equal to the amount invested. In this case

$$P = \sum_{i=1}^{n} A_i (1 + r)^{-i}.$$

It is not clear from this definition that $r$ has a unique value for all choices of $P$ and payoff sequences. However, if $P > 0$ and $A_i \geq 0$ for $i = 1, \ldots, n$, then $r$ is unique. Defining the function $f(r)$ to be

$$f(r) = -P + \sum_{i=1}^{n} A_i (1 + r)^{-i} \qquad (1.11)$$

then $f(r)$ is a continuous function of $r$ in the open interval $(-1, \infty)$. In the limit as $r$ approaches $-1$ from the right, the function values approach positive infinity. On the other hand as $r$ approaches positive infinity, the function values approach $-P < 0$ asymptotically. Thus by the Intermediate Value Theorem [Smith and Minton (2002), p. 108] there exists $r^*$ with $-1 < r^* < \infty$ such that $f(r^*) = 0$. The reader is encouraged to show that $r^*$ is unique in Exercise 1.5.7.

Rates of return can be either positive or negative. If $f(0) > 0$, *i.e.*, the sum of the payoffs is greater than the amount invested then $r^* > 0$ since $f(r)$ changes sign on the interval $[0, \infty)$. If the sum of the payoffs is less than the amount invested then $f(0) < 0$ and the rate of return is negative. In this case the function $f(r)$ changes sign on the interval $(-1, 0]$.

**Example 1.7.** Suppose Darby loans Emerson $100 with the agreement Emerson will pay Darby at the end of each year for the next 5 years the amounts $\{21, 22, 23, 24, 25\}$. Find the rate of return.

**Solution.** The rate of return per year is the solution to the equation,

$$-100 + \frac{21}{1+r} + \frac{22}{(1+r)^2} + \frac{23}{(1+r)^3} + \frac{24}{(1+r)^4} + \frac{25}{(1+r)^5} = 0.$$

Newton's Method [Smith and Minton (2002), Sec. 3.2]) can be used to approximate the solution as $r^* \approx 0.047$ or the rate of return is approximately 4.7%.

*Exercises*

**1.5.1** Suppose for an investment of $10,000 Dawson will receive payments at the end of each of the next 4 years in the amounts $\{2,000, 3,000, 4,000, 3,000\}$. What is Dawson's rate of return per year?

**1.5.2** Suppose Ellis has the choice of investing $1,000 in just one of two ways. Each investment will pay an amount listed in the table below at the end of each year for the next 5 years.

| Year | 1 | 2 | 3 | 4 | 5 |
|---|---|---|---|---|---|
| Investment A | 225 | 215 | 250 | 225 | 205 |
| Investment B | 220 | 225 | 250 | 250 | 210 |

(a) Using the present value of the investment to make the decision, which investment would Ellis choose? Assume the annual interest rate is 4.33%.
(b) Using the rate of return per year of the investment to make the decision, which investment would Ellis choose?

**1.5.3** An investor pays $50,000 today for a 4-year investment that pays $30,000 at the ends of years 3 and 4. Find the annual rate of return of the investment.

**1.5.4** A loan of $67,050 today and will repay $100,000 in 6 years. What is the continuously compounded rate of return?

**1.5.5** A loan of $50 today and repays $100 and has a continuously compounded rate of return of 18%. In how many years is the loan repaid?

**1.5.6** A stock selling today for $100 is worth $5 1 year later. What is the continuously compounded rate of return over the year?

**1.5.7** Use the Mean Value Theorem [Stewart (1999), p. 235] to show the rate of return defined by the root of the function in Eq. (1.11) is unique.

## 1.6 Time-Varying Interest Rates

All of the discussion so far has assumed that interest rates remain constant during the life of a loan or a deposit. However, interest rates change over time due to a variety of economic and political factors. This section will extend the ideas of present and future value to handle the case of time-varying interest rates.

Denote the continuously compounded interest rate as $r(t)$ (where the dependence on time $t$ is explicit). This is known as the **spot rate**. While the behavior of the spot rate can be quite complex, for the moment assume that it is a continuous function of time. Assuming the amount due on a deposit earning interest at the spot rate $r(t)$ at time $t$ is $A(t)$ then on the interval from $t$ to $t + \Delta t$ the interest rate remains near $r(t)$ and simple interest accrues. The amount due at $t + \Delta t$ is

$$A(t + \Delta t) \approx A(t)(1 + r(t)\Delta t).$$

Rearranging terms in this approximation produces

$$\frac{A(t + \Delta t) - A(t)}{\Delta t} \approx r(t)A(t)$$

which upon taking the limit of both sides as $\Delta t \to 0$ produces the equation

$$A'(t) = r(t)A(t). \tag{1.12}$$

This is an example of a first-order linear homogeneous differential equation. Many elementary calculus textbooks and most undergraduate-level texts on ordinary differential equations discuss solving this type of equation. For an extensive discussion the reader is referred to [Smith and Minton (2002)] or [Boyce and DiPrima (2001)]. The approach is to multiply both sides of Eq. (1.12) by an integrating factor and integrate with respect to $t$. Suppose the integrating factor is

$$\mu(t) = e^{- \int_0^t r(s)\,ds}$$

then, multiplying both sides of Eq. (1.12) by $\mu(t)$ results in

$$\mu(t)A'(t) = r(t)\mu(t)A(t)$$

$$e^{- \int_0^t r(s)\,ds} A'(t) - r(t)e^{- \int_0^t r(s)\,ds} A(t) = 0$$

$$\frac{d}{dt}\left[ e^{- \int_0^t r(s)\,ds} A(t) \right] = 0$$

Integrating both sides from 0 to $t$ produces the formula for the amount due at time $t$.

$$e^{- \int_0^t r(s)\,ds} A(t) - e^{- \int_0^0 r(s)\,ds} A(0) = 0$$

$$A(t) = A(0)e^{\int_0^t r(s)\,ds} \tag{1.13}$$

The present value of amount $A$ due at time $t$ under the time-varying schedule of interest rate $r(t)$ is

$$P(t) = Ae^{- \int_0^t r(s)\,ds}. \tag{1.14}$$

Closely associated with the definite integral of the spot rate is the average of the spot rate over the interval $[0, t]$. The average interest rate written as

$$\bar{r}(t) = \frac{1}{t} \int_0^t r(s)\,ds \tag{1.15}$$

is referred to as the **yield curve**. Thus, the formulas for amount due and present value can be written as

$$A(t) = A(0)e^{\bar{r}(t)\,t}$$

$$P(t) = Ae^{-\bar{r}(t)\,t}.$$

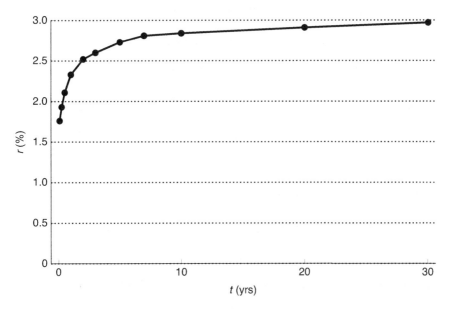

Fig. 1.1   This curve represents the interest rates associated with US Treasury bonds with maturities between 1 month and 30 years on June 29, 2018. The data for the yield curve for other dates can be found at www.treasury.gov.

If the spot rate is constant, these formulas simplify to the earlier forms.

The yield curve can also be observed in data kept by the United States Treasury. Daily the US Treasury publishes the yield rates on bonds sold by the US government. See Fig. 1.1 for a sample of rates for bonds with maturities between 1 month and 30 years.

**Example 1.8.** Suppose the spot rate is

$$r(t) = \frac{r_1}{1+t} + \frac{r_2 t}{1+t},$$

and find a formula for the yield curve and the present value of $1 due at time $t$.

**Solution.** By Eq. (1.15)

$$\bar{r}(t) = \frac{1}{t} \int_0^t \left( \frac{r_1}{1+s} + \frac{r_2 s}{1+s} \right) ds = r_2 + \frac{r_1 - r_2}{t} \ln(1+t).$$

Thus the present value of $1 is

$$P(t) = e^{-\bar{r}(t)\,t} = e^{-t\left(r_2 + \frac{r_1-r_2}{t}\ln(1+t)\right)} = (1+t)^{r_2 - r_1} e^{-r_2 t}.$$

## Exercises

**1.6.1** Confirm by differentiation that

$$\frac{d}{dt}\left[e^{-\int_0^t r(s)\,ds}A(t)\right] = e^{-\int_0^t r(s)\,ds}A'(t) - r(t)e^{-\int_0^t r(s)\,ds}A(t).$$

**1.6.2** An investor thinks that interest rates will rise over the next 5 years according to the function

$$r(t) = 0.04 + \frac{0.005t}{t+1}$$

for $0 \le t \le 5$.

(a) What is the average annual compound rate for $0 \le t \le 5$?
(b) What is the effective annual interest rate for the third year $2 \le t \le 3$?
(c) If the amount due at time $t = 5$ is \$1750, what is its present value at time $t = 1$?

**1.6.3** Find the future value at $t = 1$ of \$125 deposited at time $t = 0$ if interest rates are given by the function $r(t) = t^2/50$ for $t \ge 0$.

**1.6.4** An investor invests \$150 at time $t = 0$. Interest rates are given by the function $r(t) = t^2/50$ for $t \ge 0$. Find the amount of interest earned during the interval $2 \le t \le 3$.

**1.6.5** An investor deposits \$500 at time $t = 0$ and deposits $x$ at time $t = 2$. The amount of interest earned during the interval $2 \le t \le 3$ is also $x$. The interest rate is described by the function $r(t) = t^2/50$. Find $x$.

**1.6.6** The continuously compounded interest rate is expressed as

$$r(t) = \frac{t^2}{300 + 2t^3} \text{ for } t \ge 0.$$

If \$100 is invested at time $t = 0$, when will the amount due be \$1,000?

**1.6.7** Jan and Jesse each deposit \$100 in separate savings accounts at time $t = 0$. Jan's account earns interest at rate $r = k/20$ compounded annually. Jesse's account earns interest compounded continuously at rate $r(t) = 1/(k + t/5)$. After 5 years, the amounts due on the two accounts are the same. Find $k > 0$.

**1.6.8** A bank lends \$15,000 at time $t = 0$ and $x$ at time $t = 4$. The loan is repaid at $t = 8$ with a payment of \$55,000. The bank charges 6% interest compounded annually for the first 5 years and interest at rate $r(t) = 1/(1 + t)$ compounded continuously thereafter. Find $x$.

## 1.7  Continuous Income Streams

The treatment of interest, present value and future value has focused on discrete sums of money paid or received at distinct times spread throughout an interval. A large company may be receiving thousands or even hundreds of thousands of payments from customers each day. With income being received this frequently, it is preferable to think of the payments as a **continuous income stream** rather than as a sequence of distinct payments. Another situation which can be considered as a continuous income stream is ownership of an oil well. The well produces oil continuously and thus income is generated continuously. This section will develop the means to determine the present value and future value of continuous income streams.

Suppose the income received per unit time is the function $S(t)$. Over the short time interval from $t$ to $t + \Delta t$ assume that $S(t)$ is nearly constant and thus the income earned is approximately $S(t)\Delta t$. To determine the total income generated during an interval $[a, b]$ partition the interval as

$$a = t_0 < t_1 < \cdots < t_{n-1} < t_n = b,$$

and approximate the total income as

$$S_{\text{tot}} \approx \sum_{k=1}^{n} S(t_k)(t_k - t_{k-1}).$$

In elementary calculus this quantity is known as a **Riemann sum**. According to the definition of the definite integral, as $n \to \infty$ the total income is

$$S_{\text{tot}} = \lim_{n \to \infty} \sum_{k=1}^{n} S(t_k)(t_k - t_{k-1}) = \int_a^b S(t)\,dt.$$

The limit, and hence the definite integral, exists under fairly mild assumptions on $S(t)$. Here $S(t)$ is assumed to be piecewise continuous on $[0, T]$. A Riemann sum can be used to determine the present value of the income stream. Assuming that the continuously compounded interest rate is $r(t)$, the present value at time $t = 0$ of the income $S(t)\Delta t$ is $e^{-\bar{r}(t)t}S(t)\Delta t$. Therefore, the present value of the income stream $S(t)$ over the interval $[0, T]$ is

$$P = \int_0^T e^{-\bar{r}(t)t}S(t)\,dt = \int_0^T e^{-\int_0^t r(u)\,du}S(t)\,dt. \qquad (1.16)$$

Similarly, the future value at $t = T$ of the income stream is

$$A = Pe^{\bar{r}(T)T} = \int_0^T e^{\int_t^T r(u)\,du}S(t)\,dt. \qquad (1.17)$$

**Example 1.9.** Suppose the slot machine floor of a new casino is expected to bring in \$30,000 per day. What is the present value of the first year's slot machine revenue assuming the continuously compounded annual interest rate is 3.55%?

**Solution.** Using Eq. (1.16)

$$P = \int_0^1 (30{,}000)(365)e^{-0.0355t}\, dt = \left[\frac{(30{,}000)(365)}{-0.0355}e^{-0.0355t}\right]_{t=0}^{t=1}$$

$$\approx \$10{,}757{,}917.19.$$

The formulas for present value and future value in Eqs. (1.16) and (1.17) may require numerical integration depending on the functions $r(t)$ and $S(t)$.

*Exercises*

**1.7.1** Over the next three years an oil well will produce income at a rate of $50{,}000e^{-0.01t}$. If the continuous compounded interest rate is 4.25%, what is the present value of the income to be generated by the oil well?

**1.7.2** Suppose Ellis is savings for retirement by continuously depositing money in a savings account at the rate of \$18,000 per year for $0 \le t \le 40$. If the continuously compounded interest rate is 3.5% annually, what is the future value at $t = 40$ of the retirement account?

**1.7.3** In 6 years a company must pay a fine of \$1,000,000. The continuously compounded interest rate is 2.49%. At what continuous and constant rate must the company invest money so that the fine can be paid?

**1.7.4** Deposits are placed in a savings account at the continuous rate $S(t) = k(5 + t)$. The continuously compounded interest rate is $r = 0.05$. At time $t = 15$, the account has a balance of \$25,000. Find $k$.

**1.7.5** Deposits are placed in a savings account at the continuous rate $S(t) = k(5 + t)$. The continuously compounded interest rate is $r(t) = 1/(5 + t)$ for $t > 0$. At time $t = 15$, the account has a balance of \$25,000. Find $k$.

Chapter 2

# Discrete Probability

Since the number of forces and interactions driving the values of investments and prices in markets are so large and complex, development of a deterministic mathematical model of a market is likely to be impossible. In this book a probabilistic or stochastic model of a market will be developed instead. This chapter presents some elementary concepts of probability and statistics. Here the reader will find descriptions of discrete events and outcomes. A discrete outcome can take on only one value from a list of a finite (or countable) number of values. For example the outcome of a roll of a fair die can be only one of the six values in the set $\{1, 2, 3, 4, 5, 6\}$. No one ever rolls a die and discovers the outcome to be $\pi$ for example. Basic methods for determining the probabilities of outcomes will be presented. The concept of the **random variable**, a numerical quantity whose value is not known until an experiment is conducted, will be explained. There are many different kinds of discrete random variables, but one that frequently arises in financial mathematics is the **binomial random variable**. While statistics is a field of study unto itself, two important descriptive statistics will be introduced in this chapter, **expected value** (or mean) and **standard deviation**. The expected value provides a number which is representative of typical values of a random variable. The standard deviation is a number which provides a measure related to the width of an interval centered at the mean into which values of the random variable are likely to fall. As will be seen when discussing specific experiments, the standard deviation measures the degree to which values of the random variable are "spread out" around the mean. This chapter will also explore sums of random variables and properties of these sums.

## 2.1   Events and Probabilities

To the layman an event is something that happens. To the statistician, an **event** is an outcome or set of outcomes of an experiment. This brings up the question of what is an experiment? For our purposes an **experiment** will be any activity that generates an observable (generally unpredictable) outcome. The set of all possible outcomes of an experiment is called the **sample space** and is usually denoted $\Omega$.

**Example 2.1.** Some simple experiments include flipping a coin, rolling a pair of dice, and drawing cards from a deck. Describe the sample spaces of these experiments.

**Solution.** For the coin flip experiment the sample space is $\Omega = \{H, T\}$ (where $H$ denotes "heads" and $T$ denotes "tails"). For the experiment involving rolling a pair of dice the sample space is

$$\Omega = \left\{ \begin{array}{l} \boxdot\boxdot, \boxdot\boxdot, \boxdot\boxdot, \boxdot\boxdot, \boxdot\boxdot, \boxdot\boxdot, \boxdot\boxdot, \boxdot\boxdot, \boxdot\boxdot, \boxdot\boxdot, \boxdot\boxdot, \\ \boxdot\boxdot, \boxdot\boxdot, \boxdot\boxdot, \boxdot\boxdot, \boxdot\boxdot, \boxdot\boxdot, \boxdot\boxdot, \boxdot\boxdot, \boxdot\boxdot, \boxdot\boxdot, \boxdot\boxdot, \boxdot\boxdot, \\ \boxdot\boxdot, \boxdot\boxdot, \boxdot\boxdot, \boxdot\boxdot, \boxdot\boxdot, \boxdot\boxdot, \boxdot\boxdot, \boxdot\boxdot, \boxdot\boxdot, \boxdot\boxdot, \boxdot\boxdot, \boxdot\boxdot \end{array} \right\}.$$

For the experiment of drawing a card from a standard 52-card deck the sample space is

$$\Omega = \left\{ \begin{array}{l} 2\clubsuit, 3\clubsuit, 4\clubsuit, 5\clubsuit, 6\clubsuit, 7\clubsuit, 8\clubsuit, 9\clubsuit, 10\clubsuit, J\clubsuit, Q\clubsuit, K\clubsuit, A\clubsuit, \\ 2\diamondsuit, 3\diamondsuit, 4\diamondsuit, 5\diamondsuit, 6\diamondsuit, 7\diamondsuit, 8\diamondsuit, 9\diamondsuit, 10\diamondsuit, J\diamondsuit, Q\diamondsuit, K\diamondsuit, A\diamondsuit, \\ 2\heartsuit, 3\heartsuit, 4\heartsuit, 5\heartsuit, 6\heartsuit, 7\heartsuit, 8\heartsuit, 9\heartsuit, 10\heartsuit, J\heartsuit, Q\heartsuit, K\heartsuit, A\heartsuit, \\ 2\spadesuit, 3\spadesuit, 4\spadesuit, 5\spadesuit, 6\spadesuit, 7\spadesuit, 8\spadesuit, 9\spadesuit, 10\spadesuit, J\spadesuit, Q\spadesuit, K\spadesuit, A\spadesuit \end{array} \right\}.$$

Given an experiment and its sample space $\Omega$, the events are subsets of $\Omega$. An event for the coin flip experiment is $\{H\}$. An event for the dice rolling experiment is obtaining a sum of 7 through one of the outcomes in the set $\{\boxdot\boxdot, \boxdot\boxdot, \boxdot\boxdot, \boxdot\boxdot, \boxdot\boxdot, \boxdot\boxdot\}$. Lastly an event for the card drawing experiment is drawing a face card from the heart suit via an outcome in the set $\{J\heartsuit, Q\heartsuit, K\heartsuit, A\heartsuit\}$. The coin flip event can be thought of as "atomic" in the sense that it cannot be further broken down into non-empty subsets. Singleton events such as $\{\boxdot\boxdot\}$ are sometimes referred to as **outcomes**. For example, for the experiment of rolling a pair of dice, event $E$ may be "the sum of the dice is seven" while the outcome may be $\{\boxdot\boxdot\}$. An event can be a single outcome, the empty set, or even the entire sample space. For example suppose only the suit of a card drawn from a standard deck is observed. The events could then be labeled $\{\clubsuit, \diamondsuit, \heartsuit, \spadesuit\}$ where $\clubsuit$ is the subset,

$$\{2\clubsuit, 3\clubsuit, 4\clubsuit, 5\clubsuit, 6\clubsuit, 7\clubsuit, 8\clubsuit, 9\clubsuit, 10\clubsuit, J\clubsuit, Q\clubsuit, K\clubsuit, A\clubsuit\}.$$

Similar sets for $\diamondsuit$, $\heartsuit$, and $\spadesuit$ can also be enumerated. The event $\clubsuit$ consists of 13 outcomes.

The events associated with the experiment will be denoted with capital letters $(A, B, etc)$. The **probability** of event $A$ will be denoted $\mathbb{P}(A)$ and is a measure of the relative frequency of occurrence of $A$. Generally the proportion of the outcomes of the experiment in which event $A$ occurs can be thought of as $\mathbb{P}(A)$. More precisely probability is a function mapping the subsets of the sample space to the unit interval $[0, 1]$. Since the sample space is a set and events are subsets, the notation and operations of set theory will be found throughout this introduction to probability. To start. note that the collection $\{E_k\}_{k \geq 1}$ is mutually disjoint provided $E_i \cap E_j$ for all $i \neq j$. Every valid **probability measure** must satisfy the following three axioms.

**Axiom 1:** for all events $E$, $0 \leq \mathbb{P}(E) \leq 1$.
**Axiom 2:** $\mathbb{P}(\Omega) = 1$.
**Axiom 3:** for any countable collection of mutually disjoint sets $\{E_k\}_{k \geq 1}$,

$$\mathbb{P}\left(\bigcup_{k \geq 1} E_k\right) = \sum_{k \geq 1} \mathbb{P}(E_k).$$

If $E$ is an event for which $\mathbb{P}(E) = 0$, then $E$ is said to be an **impossible** event. If $\mathbb{P}(E) = 1$, then $E$ is said to be a **certain** event. Impossible events never occur, while certain events always occur. Events with probabilities closer to 1 are more likely to occur than events whose probabilities are closer to 0.

Set operations such as complement, union, and intersection yield insights into some probability rules and properties. Let $A$ and $B$ be events for an experiment then $A \cup B \subset \Omega$ is the event in which the outcomes of either event $A$ or event $B$ occur. The event $A \cap B \subset \Omega$ contains only the outcomes common to both events $A$ and $B$. The event $A^c \subset \Omega$ (referred to as the complement of event $A$) contains all the outcomes not in event $A$. These events are depicted in Fig. 2.1.

The language and symbolism of set theory enables the following theorem to be stated and proved.

**Theorem 2.1.** *Let $A$ and $B$ be events in sample space $\Omega$, then the following relationships hold:*

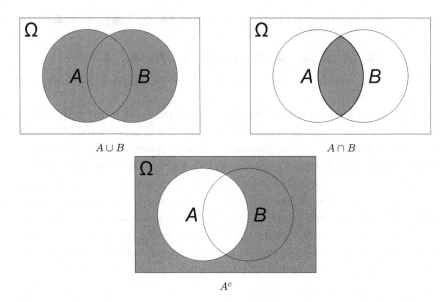

Fig. 2.1   Events $A \cup B$, $A \cap B$, and $A^c$ illustrated as shaded regions above represent the outcomes to occur either in event $A$ or $B$, the outcomes common to events $A$ and $B$, and the outcomes not in event $A$ respectively. A synergy exists between probability rules and set operations.

(i) *if $A \subset B \subset \Omega$, then* $\mathbb{P}(A) \leq \mathbb{P}(B)$,
(ii) $\mathbb{P}(A^c) = 1 - \mathbb{P}(A)$.

*Proof.* Suppose $A \subset B \subset \Omega$, then

$$\mathbb{P}(B) = \mathbb{P}(A \cup (A^c \cap B)) = \mathbb{P}(A) + \mathbb{P}(A^c \cap B) \geq \mathbb{P}(A).$$

Note that $A$ and $A^c \cap B$ are disjoint subsets of $\Omega$. Finally, note that the sets $A$ and $A^c$ are disjoint subsets of $\Omega$ with $A \cup A^c = \Omega$ and therefore

$$1 = \mathbb{P}(\Omega) = \mathbb{P}(A \cup A^c) = \mathbb{P}(A) + \mathbb{P}(A^c),$$

which is equivalent to the equation $\mathbb{P}(A^c) = 1 - \mathbb{P}(A)$.          □

As a consequence of Thm. 2.1 consider the trivial event $\emptyset$ containing no outcomes. Since $\Omega^c = \emptyset$ then $\mathbb{P}(\emptyset) = 1 - \mathbb{P}(\Omega) = 0$.

In this chapter, the outcomes of experiments will be thought of as **discrete** in the sense that the outcomes will be from a set whose members are isolated from each other by gaps. The discreteness of a coin flip, a roll of a pair of dice, and card draw are apparent due to the condition that there is no outcome between "heads" and "tails", or between 6 and 7, or between 2♣ and 3♣ respectively. Also in this chapter the number of different outcomes of an experiment will be either finite or countable (meaning that the outcomes can be put into one-to-one correspondence with a subset of the natural numbers).

## Exercises

**2.1.1** Suppose the four sides of a regular tetrahedron are labeled 1 through 4. If the tetrahedron is rolled like a die, what is the probability of it landing on 3?

**2.1.2** An experiment consists of rolling a pair of dice. List the outcomes in the event described by the sum of the dots showing on the top faces of the die is at least 10.

**2.1.3** Suppose that four DVDs are removed from their cases and then placed back into the empty cases in random order. What is the probability that at least one of the DVDs is in the correct case?

**2.1.4** An experiment is conducted in two parts. First a die is rolled. If the number of dots showing on the top face is prime, the die is rolled again. If the number of dots showing on the top face is not prime, a coin is tossed. List the outcomes in the sample space of this experiment.

**2.1.5** A pair of identical, fair dice is tossed. Find the probability that the outcome is a double, $\{⚁⚁, ⚀⚀, ⚂⚂, ⚃⚃, ⚄⚄, ⚅⚅\}$.

**2.1.6** One hundred people are attending a party at an embassy. Thirty guests speak english, 45 guests speak spanish, and 35 guests speak neither english nor spanish. How many party guests speak english and spanish?

**2.1.7** Suppose a card is drawn at random from a well-shuffled deck. Find the probability that the card is an ace or a queen.

**2.1.8** The concession stand at the baseball stadium sells drinks in sizes small, medium, and large. Records indicate that one-fourth as many medium-sized drinks are sold as small and large drinks combined. Three times as many large-sized drinks are sold as small-sized drinks. Find the probability that the next drink sold at the concession stand is a medium- or large-sized drink.

**2.1.9** Show that if $A_1, A_2, \ldots, A_n$ are pairwise disjoint subsets of sample space $\Omega$ with $\Omega = \bigcup_{i=1}^{n} A_i$ and $E$ is any event, then

$$\mathbb{P}(E) = \sum_{i=1}^{n} \mathbb{P}(E \cap A_i).$$

## 2.2   Random Variables

Consider an experiment in which a coin will be flipped ten times and the number of times that heads appears will be counted. Suppose the capital letter $X$ represents the number of times (out of ten) that heads results in this experiment. Since the result of an experiment is unknown until the experiment is conducted, the value of $X$ is unknown *a priori* but can be thought of as a function from the sample space $\Omega$ to the subset of the real numbers $\{0, 1, 2, \ldots, 10\}$. For example,

$$X\left(\{H, H, T, T, H, H, T, T, T, T\}\right) = 4.$$

A function from the sample space of an experiment to a subset of the real numbers is called a **random variable**. Since the result of an experiment is thought to be unpredictable then it is reasonable to ask for the probability that $X = x$, where $X$ is a random variable and $x$ is a real number. This is sometimes be denoted $\mathbb{P}\left(X = x\right)$.

**Example 2.2.** For the experiment involving the rolling of a pair of dice, describe the random variable mapping the events of the sample space to the sum of the number of dots appearing on the top faces of the dice.

**Solution.** The mapping can be expressed as follows.

$$\{⚀⚀\} = 2$$
$$\{⚀⚁, ⚁⚀\} = 3$$
$$\{⚀⚂, ⚁⚁, ⚂⚀\} = 4$$
$$\{⚀⚃, ⚁⚂, ⚂⚁, ⚃⚀\} = 5$$
$$\{⚀⚄, ⚁⚃, ⚂⚂, ⚃⚁, ⚄⚀\} = 6$$
$$\{⚀⚅, ⚁⚄, ⚂⚃, ⚃⚂, ⚄⚁, ⚅⚀\} = 7$$
$$\{⚁⚅, ⚂⚄, ⚃⚃, ⚄⚂, ⚅⚁\} = 8$$
$$\{⚂⚅, ⚃⚄, ⚄⚃, ⚅⚂\} = 9$$
$$\{⚃⚅, ⚄⚄, ⚅⚃\} = 10$$
$$\{⚄⚅, ⚅⚄\} = 11$$
$$\{⚅⚅\} = 12.$$

If $X$ is a random variable which takes on only finitely many distinct values then a **probability distribution** (or **probability mass function**) for $X$ is a function $f_X(x) = \mathbb{P}\left(X = x\right)$ satisfying the following two properties:

(1) $f_X(x) \geq 0$ for all $x \in \mathbb{R}$, and

(2) $\displaystyle\sum_{x \in \mathbb{R}} f_X(x) = 1.$

There are two approaches to assigning a probability to an event, the **classical** approach and the **empirical** approach. Adopting the empirical approach requires an investigator to conduct (or at least simulate) the experiment $N$ times (where $N$ is usually taken to be as large as practical). During the $N$ repetitions of the experiment the investigator counts the number of times that event $A$ occurred. Suppose this number is $x$. The $\mathbb{P}(A)$ is estimated to be $x/N$. The classical approach is a more theoretical exercise. The investigator must consider the experiment carefully and determine the sample space of atomic outcomes of the experiment, determine the likelihood of each atomic outcome, and then determine the subset of outcomes among the sample space in which event $A$ occurs. Suppose the cardinality of the sample space $\Omega$ is $M$ and that all the atomic outcomes in $\Omega$ are equally likely to occur. If the subset of outcomes in which event $A$ occurs is $y$, then the probability of event $A$ is then assigned the value $\mathbb{P}(A) = y/M$. When the assumption that each outcome is equally likely is true, the empirical and classical methods closely agree, especially when the number of repetitions of the experiment $N$ is very large.

**Example 2.3.** Consider the experiment of rolling a six-sided fair die. Describe the sample space, the probability distribution for the random variable $X$ which is the number of dots showing on the top face, and determine the probability that $X$ is a prime number.

**Solution.** The sample space of outcomes is $\Omega = \{\boxdot, \boxdot, \boxdot, \boxdot, \boxdot, \boxdot\}$. Since the die is assumed to be fair, the probability distribution is

$$f_X(x) = \begin{cases} 1/6 & \text{for } x = 1, 2, \ldots, 6 \\ 0 & \text{otherwise.} \end{cases}$$

The reader can quickly verify that the two properties of a probability distribution are satisfied:

$$0 \leq f_X(x) \leq 1 \text{ for all } x \in \mathbb{R} \text{ and } \sum_{x \in \mathbb{R}} f_X(x) = 1.$$

The probability that $X$ is a prime number is

$$\sum_{x \in \{2,3,5\}} f_X(x) = \sum_{x \in \{2,3,5\}} \frac{1}{6} = \frac{1}{6} + \frac{1}{6} + \frac{1}{6} = \frac{1}{2}.$$

## Exercises

**2.2.1** Four coins will be flipped simultaneously. Let $X$ be the random variable representing the number of coins landing with heads facing up. What is the range of this random variable?

**2.2.2** A fair die is thrown twice. Find the probability that the sum of the dice is 6.

**2.2.3** A pair of identical, fair dice is rolled. Let the random variable $X$ denote the larger of the numbers of dots showing on the top faces of the two dice. Find $\mathbb{P}\,(X = k)$ for $k = 1, 2, \ldots, 6$.

**2.2.4** A drawer contains four red socks and six black socks. Two socks are drawn randomly without replacement from the drawer. Let $X$ be the total number of red socks drawn. Find the probability distribution for $X$.

**2.2.5** A fair coin is flipped until the first time heads occurs. Let the random variable $X$ be the number of times the coin is flipped. Find $\mathbb{P}\,(X = n)$ for $n \in \mathbb{N}$.

**2.2.6** Suppose that a random variable $X$ has a probability distribution $f_X(x) = c/x$ for $x = 1, 2, \ldots, 10$ and is zero otherwise. Find the appropriate value of the constant $c$.

**2.2.7** Suppose $X$ is a discrete random variable taking on non-negative integer values. If $\mathbb{P}\,(X = 0) = \mathbb{P}\,(X = 1)$ and $\mathbb{P}\,(X = n + 1) = \frac{1}{n}\mathbb{P}\,(X = n)$ for $n \in \mathbb{N}$, then find $\mathbb{P}\,(X = 0)$.

## 2.3    Addition Rule

Suppose $A$ and $B$ are two events in the sample space of an experiment. An investigator may wish to know the probability that event $A$ or event $B$ occurs. Symbolically this would be represented as $\mathbb{P}\,(A \cup B)$. For example, an investigator rolling a pair of fair dice they may want to know the probability that a total of 2 or 12 occurs. Let event $A$ be the outcome of 2 and $B$ be the outcome of 12. Since $\mathbb{P}\,(A) = 1/36$ and $\mathbb{P}\,(B) = 1/36$ and the two events are **mutually exclusive**, that is, they cannot both simultaneously occur, $\mathbb{P}\,(A \cup B) = \mathbb{P}\,(A) + \mathbb{P}\,(B) = 1/18$. Events $A$ and $B$ are mutually exclusive if and only if sets $A$ and $B$ are disjoint, *i.e.*, have an empty intersection. Suppose instead that the investigator wants to know the probability that a total of less than 6 or an odd total results. Let event $A$ be the outcomes of a total less than 6 (that is, $A = \{2, 3, 4, 5\}$) and let event $B$ be the outcomes of an odd total (specifically $B = \{3, 5, 7, 9, 11\}$). In this case events $A$ and $B$ are not mutually exclusive, there are outcomes which are contained in both events, namely $A \cap B = \{3, 5\}$. To properly calculate

$\mathbb{P}(A \cup B)$ the sum $\mathbb{P}(\{2,3,4,5\}) + \mathbb{P}(\{3,5,7,9,11\})$ must be found, but this sum includes the $\mathbb{P}(\{3,5\})$ twice, once in $\mathbb{P}(A)$ and a second time in $\mathbb{P}(B)$. Thus the "extra" $\mathbb{P}(A \cap B)$ must be debited from $\mathbb{P}(A) + \mathbb{P}(B)$. Thus $\mathbb{P}(A \cup B) = \mathbb{P}(A) + \mathbb{P}(B) - \mathbb{P}(A \cap B)$ and therefore

$$\mathbb{P}(A \cup B) = \frac{5}{18} + \frac{1}{2} - \frac{1}{6} = \frac{11}{18}.$$

Thus the calculation of the probability of event $A$ or event $B$ occurring is different depending on whether $A$ and $B$ are mutually exclusive.

The concept outlined above is known as the **Addition Rule** for probabilities and can be stated in the form of a theorem.

**Theorem 2.2 (Addition Rule).** *For events $A$ and $B$, the probability of $A$ or $B$ occurring is*

$$\mathbb{P}(A \cup B) = \mathbb{P}(A) + \mathbb{P}(B) - \mathbb{P}(A \cap B). \tag{2.1}$$

*Proof.* Let $A$ and $B$ be any two events, then $A \cup B = A \cup (A^c \cap B)$. Since $A$ and $A^c \cap B$ are mutually exclusive events, then

$$\mathbb{P}(A \cup B) = \mathbb{P}(A) + \mathbb{P}(A^c \cap B). \tag{2.2}$$

Similarly event $B = (A \cap B) \cup (A^c \cap B)$ where $A \cap B$ and $A^c \cap B$ are mutually exclusive events, thus

$$\mathbb{P}(B) = \mathbb{P}(A \cap B) + \mathbb{P}(A^c \cap B). \tag{2.3}$$

Solving Eqs. (2.2) and (2.3) for $\mathbb{P}(A^c \cap B)$, equating the results, and solving for $\mathbb{P}(A \cup B)$ produce Eq. (2.1). $\qquad\square$

If $A$ and $B$ are mutually exclusive events then $\mathbb{P}(A \cap B) = 0$ and the Addition Rule simplifies to

$$\mathbb{P}(A \cup B) = \mathbb{P}(A) + \mathbb{P}(B),$$

which is sometimes referred to as the Addition Rule for disjoint events.

Determining the probability of the occurrence of $A$ or $B$ rests on determining the probability that *both* $A$ and $B$ occur. This topic is explored in the next section.

*Exercises*

**2.3.1** A fair die is thrown twice. Find the probability that the sum of the dice is divisible by 3.
**2.3.2** If the probability that a batter strikes out in the first inning of a baseball game is $1/3$ and the probability that the batter strikes out in the fifth inning is $1/4$, and the probability that the batter strikes out in both

innings is $1/10$, then what is the probability that the batter strikes out in either inning?

**2.3.3** Find the probability of drawing a jack or a heart a random from a standard deck of cards.

**2.3.4** There are two ATMs denoted $A$ and $B$ in the lobby of a bank. The probability that $A$ is in use is 0.25, the probability that $B$ is in use is 0.40, and the probability that both are in use is 0.10. What is the probability that neither ATM is in use?

**2.3.5** There are two ATMs denoted $A$ and $B$ in the lobby of a bank. The probability that $A$ is in use is 0.75, the probability that $B$ is in use is 0.40. What is the minimum probability that both ATMs are in use?

**2.3.6** Let $A$ and $B$ be two events in the sample space of an experiment. Suppose $\mathbb{P}(A) = 0.75$ and $\mathbb{P}(B) = 0.65$. Are events $A$ and $B$ mutually exclusive?

**2.3.7** Suppose $\mathbb{P}(A \cup B) = 0.75$ and $\mathbb{P}(A \cup B^c) = 0.85$. Find $\mathbb{P}(A)$.

**2.3.8** Let the sample space of events be $\Omega = \{X_1, X_2, \ldots, X_n\}$. Show that

$$\mathbb{P}\left(\bigcup_{k=1}^{n} X_k\right) \leq \sum_{k=1}^{n} \mathbb{P}(X_k).$$

**2.3.9** The random variable $X$ can take on any natural number value, 1, 2, 3, *etc.* The probability that $X = x$ is given by $f_X(x) = 1/2^x$ for $x \in \mathbb{N}$. Find the probability that $X$ is even.

## 2.4 Conditional Probability and Multiplication Rule

A very famous puzzle involving probability has come to be known as the "Monty Hall Problem". This paradox of probability was published in different but equivalent forms in Martin Gardner's "Mathematical Games" feature of *Scientific American* [Gardner (1959)] and in the *American Statistician* [Selvin (1975)]. It appeared in its present form in the "Ask Marilyn" column of *Parade Magazine* [vos Savant (1990)].

> A game show host hides a prize behind one of three doors. A contestant must guess which door hides the prize. First, the contestant announces the door they have chosen. The host will then open one of the two doors, not chosen, in order to reveal the prize is not behind it. The host then tells the contestant they may keep their original choice or switch to the other unopened door. Should the contestant switch doors?

At first glance when faced with two identical unopened doors, it may seem that there is no advantage to switching doors; however, if the contestant switches doors they win the prize with probability $2/3$. When the contestant makes the first choice they have a 1/3 chance of being correct and a 2/3 chance of being incorrect. When the host reveals the non-winning, unchosen door, the contestant's first choice still has a 1/3 chance of being correct, but now the unchosen, unopened door has a 2/3 probability of being correct, so the contestant should switch. A more detailed explanation of the reason for switching doors is given in [Barrow (2008), Chapter 30].

This example illustrates the concept known as **conditional probability**. Essentially the contestant is given additional information about the state of the experiment which allows the probabilities to be updated to reflect the new information. The probability that event $B$ occurs given that event $A$ has occurred is denoted $\mathbb{P}\left(B|A\right)$.

The reader should consider the classical experiment of selecting balls from an urn. Suppose the urn contains 20 balls, six of which are blue and the other 14 are green. Two balls will be drawn and the second will be drawn without replacing the first. What is the probability that the second ball is green, given that the first ball was green? The answer to this question will motivate a simple statement of the multiplication rule of probability. Consider the decision tree illustrated in Fig. 2.2. Prior to the first selection there are 20 balls in the urn (6 blue, 14 green) and thus the probabilities of selecting blue or green balls are respectively 3/10 and 7/10.

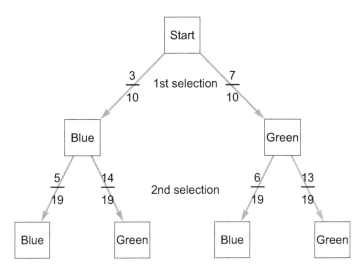

Fig. 2.2   A tree diagram illustrating the states of an urn originally containing 20 balls (6 blue and 14 green). Two balls are drawn at random without replacement. The probabilities associated with transitioning from the initial state of the experiment on the left to the final state of the experiment on the right are used as labels on the line segments connecting the outcomes.

The probabilities of selecting blue or green balls on the second selection are changed not only by the fact that there are only 19 remaining balls in the urn, but also by the outcome of the first selection. Given that the first selection was blue the probability of the second selection being blue is $5/19$ and the probability of the second selection being green is $14/19$. Given that the first selection was green the probability of the second selection being blue is $6/19$ and the probability of the second selection being green is $13/19$. Thus $\mathbb{P}$ (2nd green|1st green) $= 13/19$.

Another approach to answering the question involves determining the probability that when two balls are drawn without replacement they are both green. The probability that both selections are green would be the number of two green ball outcomes divided by the total number of outcomes. There are 20 candidates for the first ball selected and there are 19 candidates for the second ball selected. Thus the total number of outcomes is 380. Of those outcomes $(14)(13) = 182$ are both green balls. Thus the probability that both balls are green is $182/380$. Note the probability that both balls are green is the product of the probabilities of transitioning from "Begin" to "First Ball Green" and from "First Ball Green" to "Second Ball Green" as shown in Fig. 2.2. Let event $A$ be the event in which the first ball drawn is green, $B$ be the event in which the second ball drawn is green, and event $B|A$ be the event in which the second ball drawn is green given the first ball drawn was green.

$$\mathbb{P}\left(A \cap B\right) = \frac{182}{380} = \frac{13}{19} \cdot \frac{14}{20} = \mathbb{P}\left(B|A\right) \mathbb{P}\left(A\right).$$

Thus $\mathbb{P}\left(B|A\right) = \mathbb{P}\left(A \cap B\right) / \mathbb{P}\left(A\right)$.

The concept illustrated above is known as the **Multiplication Rule** for probability and can be stated in the form of a theorem.

**Theorem 2.3 (Multiplication Rule).** *For events $A$ and $B$, the probability of $A \cap B$ is*

$$\mathbb{P}\left(A \cap B\right) = \mathbb{P}\left(B|A\right) \mathbb{P}\left(A\right) = \mathbb{P}\left(A|B\right) \mathbb{P}\left(B\right). \tag{2.4}$$

Equation (2.4) can be used to find $\mathbb{P}\left(B|A\right)$ directly

$$\mathbb{P}\left(B|A\right) = \frac{\mathbb{P}\left(A \cap B\right)}{\mathbb{P}\left(A\right)}.$$

This expression is meaningful only when $\mathbb{P}\left(A\right) > 0$.

**Example 2.4.** One type of roulette wheel, known as the American type, has 38 potential outcomes represented by the integers 1 through 36 and two special outcomes 0 and 00. The positive integers are placed on alternating red and black backgrounds while 0 and 00 are on green backgrounds. What is the probability that the outcome is less than 10 and more than 3 given that the outcome is an even number?

**Solution.** Let event $A$ be the set of outcomes in which the number is even. $\mathbb{P}(A) = 10/19$ if 0 and 00 are treated as even numbers. Let $B$ be the set of outcomes in which the number is greater than 3 and less than 10. Then $\mathbb{P}(A \cap B) = 3/38$ and

$$\mathbb{P}(B|A) = \frac{3/38}{10/19} = 3/20.$$

To expand on the previous example, suppose the roulette wheel will be spun twice. One could ask what is the probability that both spins have a red outcome. If event $A$ is the outcome of red on the first spin and event $B$ is the outcome of red on the second spin, then we have as before $\mathbb{P}(A \cap B) = \mathbb{P}(A)\mathbb{P}(B|A)$. However there is no reason to believe that the wheel somehow "remembers" the outcome of the first spin while it is being spun the second time. The first outcome has no effect on the second outcome. In any experiment, if event $A$ has no effect on event $B$ then $A$ and $B$ are said to be **independent**. In this situation $\mathbb{P}(B|A) = \mathbb{P}(B)$. Thus for independent events the Multiplication Rule can be modified to

$$\mathbb{P}(A \cap B) = \mathbb{P}(A)\,\mathbb{P}(B).$$

Therefore the probability that both spins will have red outcomes is $\mathbb{P}(A \cap B) = (9/19)(9/19) = 81/361$.

Theorem 2.3 can be generalized for cases involving more than just a pair of events.

**Theorem 2.4 (General Multiplication Rule).** *For events $A$, $B_1$, $B_2$, $\ldots$, $B_n$ then*

$$\mathbb{P}(A \cap B_1 \cap B_2 \cdots \cap B_n)$$
$$= \mathbb{P}(A)\,\mathbb{P}(B_1|A)\,\mathbb{P}(B_2|A \cap B_1) \cdots \mathbb{P}(B_n|A \cap B_1 \cap \cdots \cap B_{n-1}). \quad (2.5)$$

*Proof.* Equation (2.5) can be proved by induction. The case of $n = 1$ has already been settled in Thm. 2.3. Suppose Eq. (2.5) holds for some $k \geq 1$, then

$$\mathbb{P}(A \cap B_1 \cap B_2 \cdots \cap B_k \cap B_{k+1})$$
$$= \mathbb{P}(B_{k+1}|A \cap B_1 \cap B_2 \cdots \cap B_k)\,\mathbb{P}(A \cap B_1 \cap B_2 \cdots \cap B_k)$$
$$= \mathbb{P}(A)\,\mathbb{P}(B_1|A)\,\mathbb{P}(B_2|A \cap B_1) \cdots \mathbb{P}(B_{k+1}|A \cap B_1 \cap \cdots \cap B_k),$$

by the induction hypothesis. Consequently Eq. (2.5) holds for $k+1$ and the proof is complete. $\qquad\square$

If $X$ and $Y$ are discrete random variables then the **joint probability distribution** of $X$ and $Y$ will be denoted as $f_{X,Y}(x,y)$ where

$$f_{X,Y}(x,y) = \mathbb{P}\left((X=x) \cap (Y=y)\right).$$

Some texts refer to $f_{X,Y}(x,y)$ as the **joint probability mass function** for random variables $X$ and $Y$. The **conditional probability distribution** of $X$ given that $Y = y$ is defined as

$$f_{X|Y}(x|y) = \mathbb{P}\left(X=x|Y=y\right).$$

Using the notion of conditional probability discussed earlier in this section, the conditional probability distribution can be calculated as

$$f_{X|Y}(x|y) = \frac{\mathbb{P}\left((X=x) \cap (Y=y)\right)}{\mathbb{P}\left(Y=y\right)} = \frac{f_{X,Y}(x,y)}{f_Y(y)}, \qquad (2.6)$$

provided $f_Y(y) > 0$. The conditional probability distribution of $Y$ given $X = x$ is defined in a similar way.

**Example 2.5.** Suppose a pair of fair dice are rolled. Let the random variables $X$ and $Y$ be the smaller and the larger of the number of dots on the top faces of the dice respectively. Find the joint probability distribution $f_{X,Y}(x,y)$ and the conditional probability distribution $f_{X|Y}(x,3)$.

**Solution.** Since the minimum and maximum values can occur on either die the joint probability distribution can be expressed as $f_{X,Y}(x,y) = 1/18$ if $1 \le x < y \le 6$ and $f_{X,Y}(x,y) = 1/36$ if $1 \le x = y \le 6$. The conditional probability distribution can be calculated as

$$f_{X|Y}(x,3) = \frac{\mathbb{P}\left((X=x) \cap (Y=3)\right)}{\mathbb{P}\left(Y=3\right)} = \frac{f_{X,Y}(x,3)}{\frac{5}{36}} = \begin{cases} 2/5 & \text{if } x=1, \\ 2/5 & \text{if } x=2, \\ 1/5 & \text{if } x=3, \\ 0 & \text{otherwise.} \end{cases}$$

Given a joint probability distribution for discrete random variables $X$ and $Y$ the probability distribution for $X$ can be obtained by summing over all the values $Y$ may take on.

$$f_X(x) = \sum_{y \in \mathbb{R}} f_{X,Y}(x,y). \qquad (2.7)$$

This is called the **marginal probability distribution** for discrete random variable $X$. The marginal probability distribution for $Y$ is defined in a similar way.

**Example 2.6.** Find the marginal distribution for the discrete random variable $X$ in Example 2.5.

**Solution.** Using the joint probability distribution found in the solution to Example 2.5 and Eq. (2.7)

$$f_X(x) = \begin{cases} 11/36 & \text{if } x = 1, \\ 1/4 & \text{if } x = 2, \\ 7/36 & \text{if } x = 3, \\ 5/36 & \text{if } x = 4, \\ 1/12 & \text{if } x = 5, \\ 1/36 & \text{if } x = 6. \end{cases}$$

Analogous to the statements made about independent events is the following theorem for random variables.

**Theorem 2.5.** *Random variables $X$ and $Y$ are independent if and only if $f_{X,Y}(x,y) = f_X(x)f_Y(y)$ for all $x, y \in \mathbb{R}$.*

*Proof.* Suppose $X$ and $Y$ are independent random variables, then

$$f_{X,Y}(x,y) = \mathbb{P}\left((X = x) \cap (Y = y)\right) = \mathbb{P}\left(X = x\right) \mathbb{P}\left(Y = y\right) = f_X(x)f_Y(y),$$

for all $x, y \in \mathbb{R}$. Now suppose the last equation holds and that $f_Y(y) > 0$, then using Eq. (2.6)

$$\mathbb{P}\left(X = x | Y = y\right) = f_{X|Y}(x|y) = \frac{f_X(x)f_Y(y)}{f_Y(y)} = \mathbb{P}\left(X = x\right),$$

which implies random variables $X$ and $Y$ are independent. ☐

*Exercises*

**2.4.1** On the last 100 spins of an American style roulette wheel, the outcome has been black. What is the probability of the outcome being black on the 101st spin?

**2.4.2** On the last 5,000 spins of an American style roulette wheel, the outcome has been 00. What is the probability of the outcome being 00 on the 5,001st spin?

**2.4.3** A fair die is thrown twice. Find the probability that:

(a) A five turns up twice.
(b) Both numbers are even.

**2.4.4** A fair coin is tossed repeatedly. What is the probability that on the $n$th toss:

(a) A tail appears for the first time.
(b) The numbers of heads and tails that have appeared are equal.
(c) Exactly two tails have appeared.
(d) At least two tails have appeared.

**2.4.5** A rare disease affects one person in 100,000. A test for the disease shows positive with probability 0.99 when the person has the disease. A test for the disease shows positive with probability 0.01 when the person does not have the disease (the so-called "false positive").

(a) What is $\mathbb{P}\left(+\,|\,\text{ill}\right)$?
(b) What is $\mathbb{P}\left(\text{ill}\right)$?
(c) What is $\mathbb{P}\left(+\,|\,\text{healthy}\right)$?
(d) What is $\mathbb{P}\left(\text{healthy}\right)$?
(e) What is $\mathbb{P}\left(\text{ill}\,|\,+\right)$?

**2.4.6** If event $X$ is independent of itself, prove that $\mathbb{P}\left(X\right) = 0$ or that $\mathbb{P}\left(X\right) = 1$.

**2.4.7** If $\mathbb{P}\left(X\right) = 0$ or if $\mathbb{P}\left(X\right) = 1$, show that event $X$ is independent of all other events.

**2.4.8** Suppose $n$ people are in a group. None of the people was born during a leap year. Show that the probability that at least two of the people share a birthday ("birthday buddies") is

$$p = 1 - \frac{365!}{(365 - n)!\,365^n}.$$

If enrollment in mathematics courses is holding steady at 30 students per class section, what is the probability of birthday buddies in such a class?

**2.4.9** Suppose cards will be drawn without replacement from a standard 52-card deck. What is the probability that the first two cards will be aces?

**2.4.10** Suppose cards will be drawn without replacement from a standard 52-card deck. What is the probability that the second card drawn will be an ace and that the first card was not an ace?

**2.4.11** Suppose cards will be drawn without replacement from a standard 52-card deck. What is the probability that the fourth card drawn will be an ace given that the first three cards drawn were all aces?

**2.4.12** Suppose cards will be drawn without replacement from a standard 52-card deck. On which draw is the card mostly likely to be the first ace drawn?

**2.4.13** Consider the following three random variables:

$X$ is a random variable which equals 1 with probability $p$ and which equals 0 with probability $1 - p$,
$Y = 1 - X$,
$Z = XY$.

(a) Find the sample space of outcomes of random variable $Y$.
(b) Find the sample space of outcomes of random variable $Z$.
(c) Find $\mathbb{P}\left((X = x) \cap (Y = y)\right)$ for all possible outcomes of $X$ and $Y$.
(d) Find $\mathbb{P}\left((X = x) \cap (Z = z)\right)$ for all possible outcomes of $X$ and $Z$.

**2.4.14** Suppose that a box contains 15 black balls and 5 white balls. Three balls will be selected without replacement from the box. Determine the probability distribution for $X$ the number of black balls selected.

**2.4.15** Let $A$ and $B$ be events with $\mathbb{P}(A) > 0$ and $\mathbb{P}(B) > 0$. Show that $A$ and $B$ are independent if and only if $\mathbb{P}(A \cap B) = \mathbb{P}(A)\,\mathbb{P}(B)$.

**2.4.16** A shipment of 144 smart phones contains 7 defective phones. If an inspector samples 2 of the smart phones without replacement, find the probabilities of the following events.

(a) Obtaining no defective smart phones in the sample.
(b) Obtaining two defective smart phones in the sample.
(c) Obtaining exactly one defective smart phone in the sample.

**2.4.17** The board of directors of XYZ Corporation consist of ten men and nine women. A committee of three is selected at random from the board of directors. Find the probability that the number of women on the committee exceeds the number of men.

**2.4.18** A poker hand consisting of five cards of the same suit is called a "flush". What is the probability of drawing a flush from a well-shuffled standard deck of 52 cards?

**2.4.19** Suppose a pair of fair dice is rolled. Find the probability that the sum of the number of dots on the top faces of the dice is 8 given that the numbers of dots showing on the top faces are different.

**2.4.20** Two cards are dealt from a standard 52-card deck. Let $X$ be the number of clubs dealt and $Y$ be the number of diamonds dealt. Find the joint probability distribution for $X$ and $Y$.

**2.4.21** A fair die is rolled once and random variable $X$ is the number of dots on the top face. A fair second die is rolled and random variable $Y$ is the minimum of the number the dots on the top face of the second die and the number of dots on the top face of the first die.

(a) Find the joint probability distribution of $X$ and $Y$.
(b) Find the marginal probability distribution of $Y$.

**2.4.22** A manufacturer of laptop computers uses three different suppliers for its keyboards. Supplier A provides 45% of the keyboards and has a defect rate of 2%. Supplier B provides 30% of the keyboards and has a 1% defect rate. Supplier C provides 25% of the keyboards and has a 3% defect rate. If randomly selected laptop has a defective keyboard, what is the probability the keyboard came from supplier C?

**2.4.23** Alex inspects keyboards made by supplier A and Brent inspects keyboards made by supplier B. The defect rate for keyboards from supplier A is 2% while the defect rate for keyboards from supplier B is 1%. Alex and Brent will inspect keyboards until each finds a defective keyboard. Once either of Alex and Brent find a defective keyboard they stop inspecting. Find the probability that Brent inspects fewer keyboards than Alex.

**2.4.24** One version of the table game called "craps" is played by having participants roll a pair of dice.

- If the player rolls 7 or 11, the player wins.
- If the player rolls 2, 3, or 12, the player loses.
- If the player rolls 4, 5, 6, 8, 9, or 10 then the player must keep rolling the dice until they roll their original number or 7. If the player rolls their original number before they roll 7, they win, otherwise they lose.

(a) What is the probability the player will roll a 7 or 11 on the first roll of the dice?
(b) What is the probability the player will roll a 2, 3, or 12 on the first roll of the dice?
(c) Suppose the player rolls a 4 on the first roll of the dice, what is the probability the player will roll another 4 before rolling a 7?
(d) What is the probability the player will roll a 4 on the first roll of the dice and then win the game?
(e) What is the probability the player will roll a 5 (or 6, 8, 9, 10) on the first roll of the dice and then win the game?
(f) What is the probability that a player will win at the game of craps?

## 2.5  Cumulative Distribution Functions

Earlier in this chapter the notion of a random variable was defined. Associated with each random variable $X$ is its probability distribution $f_X(x)$, a function which gives the probability that $X = x$ for each real number $x$. Another useful function is the **cumulative distribution function** which gives the probability that $X \leq a$ for each real number $a$. The cumulative distribution function will be denoted $F_X(a)$ and defined as

$$F_X(a) = \sum_{x \leq a} f_X(x). \tag{2.8}$$

If $X$ is a discrete random variable the graph of $F(x)$ will be a nondecreasing, piecewise constant curve.

**Example 2.7.** In Example 2.3 the probability distribution for the outcomes of rolling a fair die was found to be $f_X(x) = 1/6$ for $x \in \{1, 2, 3, 4, 5, 6\}$ and zero otherwise. Find the cumulative distribution function for this experiment and graph the function.

**Solution.** Using Eq. (2.8) the cumulative distribution function can be expressed as the following piecewise defined function.

$$F_X(x) = \begin{cases} 0 & \text{if } x < 1 \\ 1/6 & \text{if } 1 \leq x < 2 \\ 1/3 & \text{if } 2 \leq x < 3 \\ 1/2 & \text{if } 3 \leq x < 4 \\ 2/3 & \text{if } 4 \leq x < 5 \\ 5/6 & \text{if } 5 \leq x < 6 \\ 1 & \text{if } 6 \leq x. \end{cases}$$

The graph of this function is shown in Fig. 2.3.

If $X$ and $Y$ are both random variables then the **joint cumulative distribution function** is denoted $F_{X,Y}(x, y)$ and is defined as

$$F_{X,Y}(a, b) = \mathbb{P}\left((X \leq a) \cap (Y \leq b)\right) = \sum_{x \leq a} \sum_{y \leq b} f_{X,Y}(x, y). \tag{2.9}$$

The **marginal cumulative distribution function** for either random variable can be determined from the joint cumulative distribution function. For instance, $F_X(x) = \lim_{y \to \infty} F_{X,Y}(x, y)$ (if $Y$ is a discrete random variable

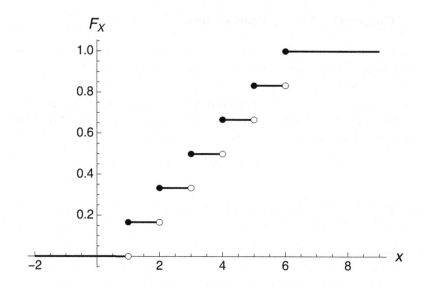

Fig. 2.3    The cumulative distribution function for the outcome of rolling a fair die.

then the limit is certain to exist). The marginal cumulative distribution for random variable $Y$ is defined in the same fashion.

**Example 2.8.** Suppose a card is drawn four times with replacement from a well-shuffled standard deck of 52 cards. Random variable $X$ will be the number of times a $\heartsuit$ appears in the first three draws and random variable $Y$ will be the number of times a $\heartsuit$ appears in the last three draws. Find the joint probability distribution and the joint cumulative probability distribution for $X$ and $Y$.

**Solution.** Each card drawn has a probability of $1/4$ of being a $\heartsuit$. The joint probability distribution of $X$ and $Y$ is presented in the table below.

|   |   | $Y$ | | | | |
|---|---|---|---|---|---|---|
|   |   | 0 | 1 | 2 | 3 | $f_X$ |
|   | 0 | 81/256 | 27/256 | 0 | 0 | 27/64 |
| $X$ | 1 | 27/256 | 63/256 | 9/128 | 0 | 27/64 |
|   | 2 | 0 | 9/128 | 15/256 | 3/256 | 9/64 |
|   | 3 | 0 | 0 | 3/256 | 1/256 | 1/64 |
|   | $f_Y$ | 27/64 | 27/64 | 9/64 | 1/64 | 1 |

The joint cumulative probability distribution of $X$ and $Y$ is presented next.

|   | $Y$ | | | |
|---|---|---|---|---|
| | 0 | 1 | 2 | 3 |
| 0 | 81/256 | 27/64 | 27/64 | 27/64 |
| 1 | 27/64 | 99/128 | 27/32 | 27/32 |
| 2 | 27/64 | 27/32 | 249/256 | 63/64 |
| 3 | 27/64 | 27/32 | 63/64 | 1 |

$X$ labels rows 0–3.

**Exercises**

**2.5.1** The probability density for discrete random variable $X$ is given in the following table. Find the cumulative distribution function for $X$.

| $x$ | 3 | 5 | 7 | 9 | 11 |
|---|---|---|---|---|---|
| $f_X(x)$ | 0.10 | 0.15 | 0.20 | 0.30 | 0.25 |

**2.5.2** Suppose random variable $X$ can take on only the discrete values $x_1 < x_2 < \cdots < x_n$ and let $F_X(x)$ be the cumulative distribution function for $X$. Show that $\mathbb{P}(X = x_1) = F_X(x_1)$ and that

$$\mathbb{P}(X = x_k) = F_X(x_k) - F_X(x_{k-1}) \text{ for } k = 2, 3, \ldots, n.$$

**2.5.3** Suppose the cumulative distribution function for discrete random variable $X$ is given by

$$F_X(x) = \begin{cases} 0 & \text{if } x < -1 \\ 1/6 & \text{if } -1 \leq x < 0 \\ 1/2 & \text{if } 0 \leq x < 2 \\ 3/4 & \text{if } 2 \leq x < 3 \\ 5/6 & \text{if } 3 \leq x < 4 \\ 1 & \text{if } 4 \leq x. \end{cases}$$

Find the probability distribution for $X$.

**2.5.4** Random variable $X$ has probability distribution $f_X(x) = \frac{3}{4}\left(\frac{1}{4}\right)^x$ for $x = 0, 1, 2, \ldots$. Find the cumulative distribution function for $X$.

**2.5.5** A biased coin has a 3/5 probability of landing on "heads". Suppose the coin is flipped five times. Find the probability density and cumulative density function for the number of heads that results.

**2.5.6** The probability distribution for jointly distributed discrete random variables $X$ and $Y$ is given in the following table.

<table>
<tr><td></td><td></td><td colspan="4" style="text-align:center">$Y$</td></tr>
<tr><td></td><td></td><td>0</td><td>1</td><td>2</td><td>$f_X$</td></tr>
<tr><td rowspan="4">$X$</td><td>0</td><td>2/27</td><td>4/27</td><td>2/27</td><td>8/27</td></tr>
<tr><td>1</td><td>1/9</td><td>2/9</td><td>1/9</td><td>4/9</td></tr>
<tr><td>2</td><td>1/18</td><td>1/9</td><td>1/18</td><td>2/9</td></tr>
<tr><td>3</td><td>1/108</td><td>1/54</td><td>1/108</td><td>1/27</td></tr>
<tr><td></td><td>$f_Y$</td><td>1/4</td><td>1/2</td><td>1/4</td><td>1</td></tr>
</table>

Find the following quantities and expressions.

(a) $\mathbb{P}\left((X = 1) \cap (Y = 2)\right)$,
(b) $\mathbb{P}\left(X = 1\right)$,
(c) $\mathbb{P}\left((X > 0) \cup (Y \leq 1)\right)$,
(d) a table showing the values of the cumulative distribution function $F_{X,Y}(x,y)$.

## 2.6   Binomial Random Variables

An important probability distribution for the study of financial mathematics is the lognormal distribution. Its name suggests a connection with the normal distribution. The normal distribution will be derived in Chapter 8 as a limiting case of the **binomial distribution** introduced in this section. The discussion of binomial random variables begins with a simple discrete random variable known as the **Bernoulli random variable** which takes on only one of two possible values, often thought of as true or false (or sometimes as success and failure). It is mathematically convenient to designate the outcomes as 0 and 1. The probability distribution of a Bernoulli random variable is $f_X(1) = \mathbb{P}\left(X = 1\right) = p$ where $0 \leq p \leq 1$ and $f_X(0) = 1 - p$.

A **binomial random variable** is the number of success outcomes occurring in a fixed number of repetitions of the Bernoulli experiment. More specifically $X$ is a binomial random variable if there is a fixed number of independent and identically distributed trials of a Bernoulli experiment. Thus the number of trials $n$ and the probability of success on an individual trial $p$ describe completely the binomial random variable. If $X$ is a binomial random variable with parameters $n$ and $p$ then $X$ is concisely described as $X \sim \mathcal{B}(n, p)$ which can be read as "random variable $X$ is binomially

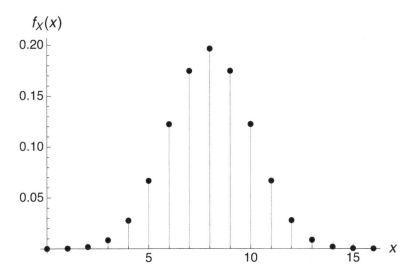

Fig. 2.4   A histogram of the probability distribution of a binomial random variable with parameters $n = 16$ and $p = 1/2$.

distributed with parameters $n$ and $p$". The range of $X$ is $\{0, 1, \ldots, n\}$ and the probability distribution is given by the formula,

$$f_X(x) = \binom{n}{x} p^x (1-p)^{n-x} = \frac{n!}{x!(n-x)!} p^x (1-p)^{n-x}. \tag{2.10}$$

The biased coin described in Exercise 2.5.5 is an example of a binomial random variable.

One interesting feature of the probability distribution of a binomial random variable is its symmetry. Consider a binomially distributed random variable $X$ with parameters $n = 16$ and $p = 1/2$. A histogram of the probability distribution of $X$ resembles that shown in Fig. 2.4. This section will conclude with more examples of binomial random variables.

**Example 2.9.** Consider a family with four children. Assume the probability of each child being male is $1/2$ and that the genders of the children are independent of one another. Find the probability distribution of the random variable $X$ representing the number of male children.

**Solution.** Using the parameter values $n = 4$ and $p = 1/2$ and Eq. (2.10), the probabilities are

$$f_X(0) = \binom{4}{0}\left(\frac{1}{2}\right)^0 \left(\frac{1}{2}\right)^4 = \frac{1}{16},$$

$$f_X(1) = \binom{4}{1}\left(\frac{1}{2}\right)^1 \left(\frac{1}{2}\right)^3 = \frac{1}{4},$$

$$f_X(2) = \binom{4}{2}\left(\frac{1}{2}\right)^2 \left(\frac{1}{2}\right)^2 = \frac{3}{8},$$

$$f_X(3) = \binom{4}{3}\left(\frac{1}{2}\right)^3 \left(\frac{1}{2}\right)^1 = \frac{1}{4},$$

$$f_X(4) = \binom{4}{4}\left(\frac{1}{2}\right)^4 \left(\frac{1}{2}\right)^0 = \frac{1}{16}.$$

**Example 2.10.** The probability that a computer memory chip is defective is 0.02. A SIMM (single in-line memory module) contains 16 chips for data storage and a 17th chip for error correction. The SIMM can operate correctly if one chip is defective, but not if two or more are defective. The probability that a SIMM will not function is

$$\mathbb{P}\left(X \geq 2\right) = 1 - F_x(1) = 1 - \binom{17}{0}(0.02)^0(0.98)^{17} - \binom{17}{1}(0.02)^1(0.98)^{16}$$

$$\approx 0.044578.$$

*Exercises*

**2.6.1** An apartment building has three elevators. The probability of an elevator being out of service is $1/3$ independent of the other elevators. Find the probability distribution for the number of elevators out of service.

**2.6.2** Suppose that a particular geographical area is subject to at most one hurricane per year with a probability of $3/20$ and that occurrences of hurricanes in different years are independent. What is the probability of five or more hurricanes in a 20-year period?

**2.6.3** Suppose Taylor will wager on red on 20 spins of the roulette wheel (see Example 2.4). What is the probability Taylor wins at least 10 times?

**2.6.4** Quality control for a manufacturer of integrated circuits is done by randomly selecting 25 chips from the previous days manufacturing run. Each of the 25 chips is tested. If two or more chips are faulty, then the entire run is discarded. Previously gathered evidence indicates that the defect rate for chips is 0.0016. What is the probability that a manufacturing run of chips will be discarded?

**2.6.5** Each of the 30 subjects in a medical study drop out before the completion of the study with probability 0.1 independently of each other. What is the probability that the study is completed with at least 20 subjects participating?

**2.6.6** The 30 participants in a medical study are divided into two groups of 15 subjects each. Subjects in the study drop out before the completion of the study with probability 0.1 independently of each other. What is the probability that both groups complete the study with at least 10 subjects each?

**2.6.7** Airlines find that each passenger who reserves a seat on a plane fails to turn up for the flight with probability 0.10 independent of all other passengers. The airline sells 10 tickets for airplanes with only 9 seats and 20 tickets for airplanes with only 18 seats. Which type of airplane is more likely to be oversold?

**2.6.8** A performer at the county fair claims to have the power of telekinesis (the ability to move objects with the mind). The audience throws 8 coins and 7 of them land on heads. The performer claims that he/she made this happen. If the performer has no supernatural powers, what is the probability of this event?

**2.6.9** A sample of size 5 is drawn with replacement from a drawer containing 8 red socks and 6 black socks. Find the probability that the fourth sock drawn was black given that the sample contains exactly four black socks.

**2.6.10** Suppose that $X$ is a binomial random variable representing the number of successful trials out of $m$ trials where the probability of success on a single trial is $0 < p < 1$. Likewise suppose that $Y$ is also a binomial random variable for which the probability of success on a single trial is also $p$ but the number of trials is $n$, where $m$ and $n$ can be different. Assume $X$ and $Y$ are independent and find a formula for

$$\mathbb{P}\left(X = x | X + Y = k\right).$$

## 2.7 Expected Value

When faced with experimental results, summary statistics are often useful for making sense of the data. In this context "statistics" refers to numbers which can be calculated from the data rather than the means and algorithms by which these numbers are calculated. In financial mathematics the statistical needs are somewhat more specialized than in a general purpose course in statistics. The focus is on the hypothetical question, "if an experiment was to be performed an infinite number of times, what would be the typical outcome?" This section introduces the concept of **expected value** used throughout the remainder of this text. A reader interested in a broader, deeper, and more rigorous background in statistics should consult one of the many textbooks devoted to the subject, for example [Ross (2006)].

To take an example, if a fair die was rolled an infinite number of times, what would be the typical result? In some ways expected value is synonymous with the mean or average of a list of numerical values; however, it can differ in at least two important ways. First, the expected value usually refers to the typical value of a random variable whose outcomes are not necessarily equally likely whereas the mean of a list of data treats each observation as equally likely. Second, the expected value of a random variable is the typical outcome of an experiment performed an infinite number of times whereas the statistical mean is calculated based on a finite collection of observations of the outcome of an experiment.

If $X$ is a discrete random variable with probability distribution $f_X(x)$ then the expected value of $X$ is denoted $\mathbb{E}(X)$ and defined as

$$\mathbb{E}(X) = \sum_{x \in \mathbb{R}} (x \cdot f_X(x)). \tag{2.11}$$

In the case that $X$ takes on only a finite number of values with non-zero probability, then this sum is well-defined. If $X$ may assume an infinite number of values with probabilities greater than zero, it is assumed that the sum converges. Since each value of $X$ is multiplied by its corresponding probability, the expected value of $X$ is a weighted average of the variable $X$. Returning to the question posed in the previous paragraph regarding the typical outcome of rolling a fair die an infinite number of times, since $X \in \{1, 2, 3, 4, 5, 6\}$ and $f_X(x) = 1/6$ for all possible values of $X$, then the expected value of $X$ is

$$\mathbb{E}(X) = \sum_{x=1}^{6} \frac{x}{6} = \frac{1}{6} \sum_{x=1}^{6} x = \left(\frac{1}{6}\right) \frac{(6)(7)}{2} = \frac{7}{2}.$$

Thus the average outcome of rolling a fair die is 3.5.

**Example 2.11.** Let random variable $X$ represent the number of female children in a family of four children. Assuming that births of males and females are equally likely and that all births are independent events, what is the $\mathbb{E}\,(X)$?

**Solution.** The sample space of $X$ is the set $\{0, 1, 2, 3, 4\}$. Using the binomial probability formula of Eq. (2.10) with $n = 4$ and $p = 1/2$,

$$\mathbb{E}\,(X) = \sum_{x=0}^{4} \left( x \cdot \binom{4}{x} \left(\frac{1}{2}\right)^x \left(\frac{1}{2}\right)^{4-x} \right) = 2.$$

In families having four children, typically there are two female children and consequently two male children.

The notion of the expected value of a random variable $X$ can be extended to the expected value of a function of $X$. Thus we say that if $g$ is a function applied to $X$, then

$$\mathbb{E}\,(g(X)) = \sum_{x \in \mathbb{R}} (g(x) \cdot f_X(x)).$$

When the function $G$ is merely multiplication by a constant then the expected value takes on a simple form.

**Theorem 2.6.** *If $X$ is a random variable and $a$ is a constant, then* $\mathbb{E}\,(a\,X) = a\,\mathbb{E}\,(X)$.

*Proof.* By the definition of expected value

$$\mathbb{E}\,(a\,X) = \sum_{x \in \mathbb{R}} ((a\,x) \cdot f_X(x)) = a \sum_{x \in \mathbb{R}} (x \cdot f_X(x)) = a\,\mathbb{E}\,(X). \qquad \square$$

Later in this work sums of random variables will become important. Thus some attention must be given to the expected value of the sum of random variables. Conveniently, the expected value of a sum of random variables is the sum of the expected values of the random variables. This notion is made more precise in the following theorem.

**Theorem 2.7.** *If $X_1, X_2, \ldots, X_k$ are random variables then*

$$\mathbb{E}\,(X_1 + X_2 + \cdots X_k) = \mathbb{E}\,(X_1) + \mathbb{E}\,(X_2) + \cdots + \mathbb{E}\,(X_k).$$

*Proof.* If $k = 1$ then the proposition is certainly true. If $k = 2$ then

$$
\begin{aligned}
\mathbb{E}\left(X_1 + X_2\right) &= \sum_{x_1 \in \mathbb{R}} \sum_{x_2 \in \mathbb{R}} \left((x_1 + x_2) f_{X_1, X_2}(x_1, x_2)\right) \\
&= \sum_{x_1 \in \mathbb{R}} \sum_{x_2 \in \mathbb{R}} x_1 f_{X_1, X_2}(x_1, x_2) + \sum_{x_2 \in \mathbb{R}} \sum_{x_1 \in \mathbb{R}} x_2 f_{X_1, X_2}(x_1, x_2) \\
&= \sum_{x_1 \in \mathbb{R}} x_1 \sum_{x_2 \in \mathbb{R}} f_{X_1, X_2}(x_1, x_2) + \sum_{x_2 \in \mathbb{R}} x_2 \sum_{x_1 \in \mathbb{R}} f_{X_1, X_2}(x_1, x_2) \\
&= \sum_{x_1 \in \mathbb{R}} x_1 f_{X_1}(x_1) + \sum_{x_2 \in \mathbb{R}} x_2 f_{X_2}(x_2) \\
&= \mathbb{E}\left(X_1\right) + \mathbb{E}\left(X_2\right).
\end{aligned}
$$

For a finite value of $k > 2$ the result is true by induction. Suppose the result is true for $n < k$ where $k > 2$, then

$$
\begin{aligned}
\mathbb{E}\left(X_1 + \cdots + X_{k-1} + X_k\right) &= \mathbb{E}\left(X_1 + \cdots + X_{k-1}\right) + \mathbb{E}\left(X_k\right) \\
&= \mathbb{E}\left(X_1\right) + \cdots + \mathbb{E}\left(X_{k-1}\right) + \mathbb{E}\left(X_k\right).
\end{aligned}
$$

The last step is true by the induction hypothesis. □

An application of Thm. 2.7 determines the expected value of a binomial random variable. Suppose $X$ is a binomial random variable with parameters $n$ and $p$. The random variable $X$ can be thought of as the sum of the results of $n$ Bernoulli experiments. The expected value of a single Bernoulli random variable $X_i$ (thought of as the outcome of the $i$th trial of the binomial experiment) is

$$
\mathbb{E}\left(X_i\right) = (1)p + (0)(1 - p) = p.
$$

Thus the binomial random variable $X = X_1 + X_2 + \cdots + X_n$ is

$$
\mathbb{E}\left(X\right) = \mathbb{E}\left(X_1 + \cdots + X_n\right) = \mathbb{E}\left(X_1\right) + \cdots + \mathbb{E}\left(X_n\right) = p + \cdots + p = n\,p.
$$

If functions are applied to random variables a corollary to Thm. 2.7 can be stated.

**Corollary 2.1.** *Let $X_1$, $X_2$, ..., $X_k$ be random variables and let $g_i$ be a function defined on $X_i$ for $i = 1, 2, \ldots, k$ then*

$$
\begin{aligned}
\mathbb{E}\left(g_1(X_1) + g_2(X_2) + \cdots + g_k(X_k)\right) \\
= \mathbb{E}\left(g_1(X_1)\right) + \mathbb{E}\left(g_2(X_2)\right) + \cdots + \mathbb{E}\left(g_k(X_k)\right).
\end{aligned}
$$

*Proof.* The assertion will be proved for the case when $k = 2$.

$$\mathbb{E}\left(g_1(X_1) + g_2(X_2)\right)$$
$$= \sum_{x_1 \in \mathbb{R}} g_1(x_1) \sum_{x_2 \in \mathbb{R}} f_{X_1, X_2}(x_1, x_2) + \sum_{x_2 \in \mathbb{R}} g_2(x_2) \sum_{x_1 \in \mathbb{R}} f_{X_1, X_2}(x_1, x_2)$$
$$= \sum_{x_1 \in \mathbb{R}} g_1(x_1) f_{X_1}(x_1) + \sum_{x_2 \in \mathbb{R}} g_2(x_2) f_{X_2}(x_2)$$
$$= \mathbb{E}\left(g_1(X_1)\right) + \mathbb{E}\left(g_2(X_2)\right).$$

The proof for the case of $k > 2$ is left to the reader. $\qquad\square$

The expected value of a product of random variables is also of interest, but a simple formula for the expected value of a product of random variables requires an assumption not needed when considering the expected value of a sum of random variables.

**Theorem 2.8.** *Let $X_1$, $X_2$, ..., $X_k$ be pairwise independent random variables, then*

$$\mathbb{E}\left(X_1 X_2 \cdots X_k\right) = \mathbb{E}\left(X_1\right) \mathbb{E}\left(X_2\right) \cdots \mathbb{E}\left(X_k\right).$$

*Proof.* When $k = 1$ the theorem is certainly true. If $X_1$ and $X_2$ are independent random variables with joint probability distribution $f_{X_1, X_2}(x_1, x_2) = f_{X_1}(x_1) f_{X_2}(x_2)$ then

$$\mathbb{E}\left(X_1 X_2\right) = \sum_{x_1 \in \mathbb{R}} \sum_{x_2 \in \mathbb{R}} x_1 x_2 f_{X_1, X_2}(x_1, x_2) = \sum_{x_1 \in \mathbb{R}} \sum_{x_2 \in \mathbb{R}} x_1 x_2 f_{X_1}(x_1) f_{X_2}(x_2)$$
$$= \sum_{x_1 \in \mathbb{R}} x_1 f_{X_1}(x_1) \sum_{x_2 \in \mathbb{R}} x_2 f_{X_2}(x_2) = \mathbb{E}\left(X_1\right) \mathbb{E}\left(X_2\right).$$

For a finite value of $k > 2$ the result is established by induction. Suppose the result is true for $n < k$ where $k > 2$, then

$$\mathbb{E}\left(X_1 \cdots X_{k-1} X_k\right) = \mathbb{E}\left(X_1 \cdots X_{k-1}\right) \mathbb{E}\left(X_k\right)$$
$$= \mathbb{E}\left(X_1\right) \cdots \mathbb{E}\left(X_{k-1}\right) \mathbb{E}\left(X_k\right).$$

The last step is true by the induction hypothesis. $\qquad\square$

A corollary to Thm. 2.8 holds for functions of pairwise independent random variables as well.

**Corollary 2.2.** *Let $X_1$, $X_2$, ..., $X_k$ be pairwise independent random variables and let $g_i$ be a function defined on $X_i$ for $i = 1, 2, \ldots, k$ then*

$$\mathbb{E}\left(g_1(X_1) g_2(X_2) \cdots g_k(X_k)\right) = \mathbb{E}\left(g_1(X_1)\right) \mathbb{E}\left(g_2(X_2)\right) \cdots \mathbb{E}\left(g_k(X_k)\right).$$

*Proof.* Once again the result will be proved for the case when $k = 2$.

$$\mathbb{E}\left(g_1(X_1)g_2(X_2)\right) = \sum_{x_1 \in \mathbb{R}} \sum_{x_2 \in \mathbb{R}} g_1(x_1)g_2(x_2)f_{X_1,X_2}(x_1,x_2)$$

$$= \sum_{x_1 \in \mathbb{R}} g_1(x_1)f_{X_1}(x_1) \sum_{x_2 \in \mathbb{R}} g_2(x_2)f_{X_2}(x_2)$$

$$= \mathbb{E}\left(g_1(X_1)\right)\mathbb{E}\left(g_2(X_2)\right).$$

The case when $k > 2$ is left to the reader.   □

If the reader is interested in more properties of the expected value of sum and products of random variables, consult a textbook on probability such as [Ross (2003)].

The notion of **conditional expected value** follows from an understanding of conditional probability. Since $\mathbb{P}\left(X|Y\right)$ is understood to mean the probability of event $X$ given event $Y$, then the notation $\mathbb{E}\left(X|Y = y\right)$ or the equivalent, but more compact notation $\mathbb{E}\left(X|y\right)$ will mean the expected value of random variable $X$ given that random variable $Y$ has the value $y$. The conditional expected value can be calculated according to the formula

$$\mathbb{E}\left(X|y\right) = \sum_{x \in \mathbb{R}} \left(x \cdot f_{X|Y}(x|y)\right). \tag{2.12}$$

For example if a pair of fair dice consists of one red die $R$ and one green die $G$, the expected value of their sum given that the red die shows "3" can be calculated as follows.

$$\mathbb{E}\left(G + R|R = 3\right) = \sum_{g=1}^{6} \left((g+r) \cdot f_{G|R}(g|3)\right) = \frac{1}{6}\sum_{g=1}^{6}(g+3) = \frac{13}{2}.$$

The conditional expected value $\mathbb{E}\left(X|Y = y\right)$ can also be thought of as a random variable itself.

$$\mathbb{E}\left(X|Y = y\right) = \sum_{x \in \mathbb{R}} \left(x \cdot f_{X|Y}(x|y)\right) = g(y).$$

The expression $g(Y)$ is itself a random variable.

**Example 2.12.** Suppose integer random variables $(X,Y)$ satisfy the inequality $-5 \le Y \le X \le 5$ (as shown in Fig. 2.5) and each ordered pair is equally likely. Find an expression for the random variable $\mathbb{E}\left(X|Y\right)$.

**Solution.** To find $\mathbb{E}\left(X|Y\right)$ first note that since there are 11 potential values for $Y \in \{-5, -4, \dots, 5\}$ and when $Y = y$ then there are $6 - y$ potential

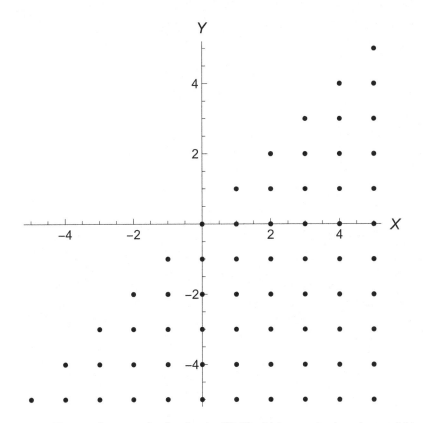

Fig. 2.5 The sample space of ordered pairs $(X, Y)$ of integer-valued random variables satisfying the inequality $-5 \leq Y \leq X \leq 5$.

values for $X$. Thus the set of outcomes of a random selection of $(X, Y)$ contains

$$\sum_{y=-5}^{5} (6 - y) = 66 \text{ elements.}$$

Therefore $f_{X,Y}(x, y) = 1/66$. The conditional probability $f_{X|Y}(x|y)$ can be found from the marginal probability for $Y$.

$$\mathbb{P}\left(X = x | Y = y\right) = \frac{\mathbb{P}\left((X = x) \cap (Y = y)\right)}{\mathbb{P}\left(Y = y\right)} = \frac{f_{X,Y}(x, y)}{\sum_{x=y}^{5} f_{X,Y}(x, y)}$$

$$= \frac{1/66}{\sum_{x=y}^{5} 1/66} = \frac{1}{6 - y}.$$

Applying the definition of conditional expectation in Eq. (2.12) yields

$$\mathbb{E}\left(X|Y=y\right) = \sum_{x=y}^{5}\left(x \cdot \frac{1}{6-y}\right) = \frac{1}{6-y}\sum_{x=y}^{5}x$$

$$= \frac{1}{6-y} \cdot \frac{30 - y^2 + y}{2} = \frac{y+5}{2}.$$

Conditional expected value also shares the linearity property with the original notion of expected value (see Exercise 2.7.6).

The expected value of a random variable specifies the average or typical outcome of an infinite number of repetitions of an experiment. In the next section the notions of variance and standard deviation are introduced. They specify measures of the spread of the outcomes from the expected value.

### Exercises

**2.7.1** A pair of fair dice are rolled. What is the expected value of the sum of the numbers of dots appearing on the top faces of the dice?

**2.7.2** The probabilities of a child being born male or female are not exactly equal to 1/2. Typically there are nearly 105 live male births per 100 live female births. Determine the expected number of female children in a family of 6 total children using these birth ratios and ignoring infant mortality.

**2.7.3** Show that for constants $a$ and $b$ and discrete random variable $X$ that $\mathbb{E}\left(aX + b\right) = a\,\mathbb{E}\left(X\right) + b$.

**2.7.4** Show that $\mathbb{E}\left(X(X-1)\right) + \mathbb{E}\left(X\right) = \mathbb{E}\left(X^2\right)$ for random variable $X$.

**2.7.5** Not every random variable has an expected value. Consider the discrete random variable $X$ whose probability density is $f_X(x) = \frac{2}{3}\left(\frac{1}{3}\right)^x$ for $x = 0, 1, 2, \ldots$. Show that the expected value of $Y = 3^X$ is undefined.

**2.7.6** Suppose that $X$, $Y$ and $Z$ are jointly distributed discrete random variables.

(a) Show that $\mathbb{E}\left(X + Y|Z = z\right) = \mathbb{E}\left(X|Z = z\right) + \mathbb{E}\left(Y|Z = z\right)$.
(b) Suppose $c$ is a constant and show that $\mathbb{E}\left(cX|Y = y\right) = c\,\mathbb{E}\left(X|Y = y\right)$.

**2.7.7** Suppose a standard deck of 52 cards is well shuffled and one card at a time will be drawn without replacement from the deck. What is the expected value of the position of the first ace drawn (in other words, of the first, second, third, *etc.* cards drawn, on average which will be the first ace drawn)?

**2.7.8** Kelsey plays a type of lottery game in which three numbers are chosen from the set $\{0, 1, 2, \ldots, 9\}$. If Kelsey's three numbers match the chosen three numbers in order, Kelsey wins $100, otherwise Kelsey loses the price of the lottery ticket which is $1. What are Kelsey's expected winnings from this game?

**2.7.9** A fair die is rolled repeatedly. Random variable $X$ denotes the number of rolls necessary to obtain a result of 1. Random variable $Y$ denotes the number of rolls necessary to obtain a result of 2. Find $\mathbb{E}\left(Y|X=2\right)$. *Hint*: The summation formula derived in Exercise 1.4.11 may be helpful.

## 2.8 Variance and Standard Deviation

The **variance** of a random variable is a measure of the spread of values of the random variable about the expected value of the random variable. The variance is defined as

$$\mathbb{V}\mathrm{ar}\left(X\right) = \mathbb{E}\left((X - \mathbb{E}\left(X\right))^2\right). \tag{2.13}$$

As the reader can see from Eq. (2.13), the variance is always non-negative. The expression $X - \mathbb{E}\left(X\right)$ is the signed deviation of $X$ from its expected value. Without the squaring operation the expected value of merely the deviation from the mean of $X$ would always be 0. The variance may be interpreted as the average of the squared deviation of a random variable from its expected value. An alternative formula for the variance is sometimes more convenient in calculations.

**Theorem 2.9.** *Let $X$ be a random variable, then the variance of $X$ is* $\mathbb{E}\left(X^2\right) - \mathbb{E}\left(X\right)^2$.

*Proof.* By definition,

$$\mathbb{V}\mathrm{ar}\left(X\right) = \mathbb{E}\left((X - \mathbb{E}\left(X\right))^2\right) = \mathbb{E}\left(X^2\right) - \mathbb{E}\left(2X\mathbb{E}\left(X\right)\right) + \mathbb{E}\left(\mathbb{E}\left(X\right)^2\right)$$

$$= \mathbb{E}\left(X^2\right) - 2\mathbb{E}\left(X\right)\mathbb{E}\left(X\right) + \mathbb{E}\left(X\right)^2 = \mathbb{E}\left(X^2\right) - \mathbb{E}\left(X\right)^2.$$

The third and fourth steps of this derivation made use of Theorems 2.7 and 2.6 respectively. $\square$

Returning to Example 2.11 describing a hypothetical family with four children, the variance in the number of female children can now be found. From Example 2.11 recall that $\mathbb{E}\left(X\right) = 2$. Using Thm. 2.9 then

$$\mathbb{V}\mathrm{ar}\left(X\right)$$

$$= (0^2)\left(\frac{1}{16}\right) + (1^2)\left(\frac{1}{4}\right) + (2^2)\left(\frac{3}{8}\right) + (3^2)\left(\frac{1}{4}\right) + (4^2)\left(\frac{1}{16}\right) - 2^2 = 1.$$

The variance of a binomial random variable will be important to later derivations. Developing a simple formula for the variance provides the opportunity to introduce some properties of the variance. Consider a Bernoulli random variable $X$. If the $\mathbb{P}(X = 1) = p$ with $0 \le p \le 1$ then according to Thm. 2.9,

$$\mathbb{V}\mathrm{ar}(X) = (1)^2 p + (0)^2(1-p) - [(1)p + (0)(1-p)]^2 = p(1-p).$$

The following theorem provides an convenient formula for calculating the variance of *independent* random variables.

**Theorem 2.10.** *Let $X_1$, $X_2$, $\ldots$, $X_k$ be pairwise independent random variables, then*

$$\mathbb{V}\mathrm{ar}(X_1 + X_2 + \cdots + X_k) = \mathbb{V}\mathrm{ar}(X_1) + \mathbb{V}\mathrm{ar}(X_2) + \cdots + \mathbb{V}\mathrm{ar}(X_k).$$

*Proof.* If $k = 1$ then the result is trivially true. Take the case when $k = 2$. By the definition of variance,

$$
\begin{aligned}
\mathbb{V}\mathrm{ar}(X_1 + X_2) &= \mathbb{E}\left(((X_1 + X_2) - \mathbb{E}(X_1 + X_2))^2\right) \\
&= \mathbb{E}\left(((X_1 - \mathbb{E}(X_1)) + (X_2 - \mathbb{E}(X_2)))^2\right) \\
&= \mathbb{E}\left((X_1 - \mathbb{E}(X_1))^2\right) + \mathbb{E}\left((X_2 - \mathbb{E}(X_2))^2\right) \\
&\quad + 2\mathbb{E}((X_1 - \mathbb{E}(X_1))(X_2 - \mathbb{E}(X_2))) \\
&= \mathbb{V}\mathrm{ar}(X_1) + \mathbb{V}\mathrm{ar}(X_2) + 2\mathbb{E}((X_1 - \mathbb{E}(X_1))(X_2 - \mathbb{E}(X_2))).
\end{aligned}
$$

Since random variables $X_1$ and $X_2$ are independent, then by Thm. 2.8

$$
\begin{aligned}
\mathbb{E}((X_1 - \mathbb{E}(X_1))(X_2 - \mathbb{E}(X_2))) &= \mathbb{E}(X_1 - \mathbb{E}(X_1))\mathbb{E}(X_2 - \mathbb{E}(X_2)) \\
&= (\mathbb{E}(X_1) - \mathbb{E}(X_1))(\mathbb{E}(X_2) - \mathbb{E}(X_2)) \\
&= 0,
\end{aligned}
$$

and thus

$$\mathbb{V}\mathrm{ar}(X_1 + X_2) = \mathbb{V}\mathrm{ar}(X_1) + \mathbb{V}\mathrm{ar}(X_2).$$

The result can be extended to any finite integer value of $k$ by induction. Suppose the result has been shown true for $n < k$ with $k > 2$. Then

$$
\begin{aligned}
\mathbb{V}\mathrm{ar}(X_1 + \cdots + X_{k-1} + X_k) &= \mathbb{V}\mathrm{ar}(X_1 + \cdots + X_{k-1}) + \mathbb{V}\mathrm{ar}(X_k) \\
&= \mathbb{V}\mathrm{ar}(X_1) + \cdots + \mathbb{V}\mathrm{ar}(X_{k-1}) + \mathbb{V}\mathrm{ar}(X_k),
\end{aligned}
$$

where the last equality is justified by the induction hypothesis. $\square$

Readers should think carefully about the validity of the claim that $X_1 - \mathbb{E}(X_1)$ and $X_2 - \mathbb{E}(X_2)$ are independent in light of the assumption that $X_1$ and $X_2$ are independent.

Now consider a binomial random variable $X$ whose probability distribution is parameterized by $n$ independent Bernoulli trials $(X_1, X_2, \ldots, X_n)$ for which the $\mathbb{P}(X_k = 1) = p$ with $0 \leq p \leq 1$ for $k = 1, 2, \ldots, n$. Binomial random variable $X$ denotes the total number of successes accrued over the $n$ trials.

$$\mathbb{V}\mathrm{ar}\,(X) = \mathbb{V}\mathrm{ar}\,\left(\sum_{j=1}^{n} X_j\right) = \sum_{j=1}^{n} \mathbb{V}\mathrm{ar}\,(X_j) = \sum_{j=1}^{n} p(1-p) = np(1-p).$$

The notion of **standard deviation** is closely related to that of variance. Standard deviation of a random variable $X$ is denoted by $\sigma_X$ and

$$\sigma_X = \sqrt{\mathbb{V}\mathrm{ar}\,(X)}.$$

There exist formulas for the variance of the product of random variables as well. The most general results for the variance of a product can be complicated, but the case of the product of pairwise independent random variables is tractable to state and prove.

**Theorem 2.11.** *Let $X_1$, $X_2$, $\ldots$, $X_k$ be pairwise independent random variables, then*

$$\mathbb{V}\mathrm{ar}\,(X_1 X_2 \cdots X_k) = \mathbb{E}\left(X_1^2\right)\mathbb{E}\left(X_2^2\right)\cdots\mathbb{E}\left(X_k^2\right) - (\mathbb{E}(X_1)\mathbb{E}(X_2)\cdots\mathbb{E}(X_k))^2.$$

*Proof.* The case when $k = 1$ follows from Thm. 2.9. Consider the case when $k > 1$.

$$
\begin{aligned}
\mathbb{V}\mathrm{ar}\,(X_1 X_2 \cdots X_k) &= \mathbb{E}\left((X_1 X_2 \cdots X_k)^2\right) - (\mathbb{E}(X_1 X_2 \cdots X_k))^2 \\
&= \mathbb{E}\left(X_1^2 X_2^2 \cdots X_k^2\right) - (\mathbb{E}(X_1)\mathbb{E}(X_2)\cdots\mathbb{E}(X_k))^2 \\
&= \mathbb{E}\left(X_1^2\right)\mathbb{E}\left(X_2^2\right)\cdots\mathbb{E}\left(X_k^2\right) - (\mathbb{E}(X_1)\mathbb{E}(X_2)\cdots\mathbb{E}(X_k))^2.
\end{aligned}
$$

The last equation holds as a result of Cor. 2.2. $\qquad\square$

*Exercises*

**2.8.1** Suppose random variable $X$ has range $\{-1, 1\}$ and $f_X(-1) = f_X(1) = 1/2$. Find the expected value and variance of $X$.

**2.8.2** Consider the experiment of rolling a fair die. Let the random variable $X$ represent the number of dots showing on the top face of the die. Find the variance of $X$.

**2.8.3** For the situation described in Exercise 2.7.7 determine the variance in the occurrence of the first ace drawn.

**2.8.4** Show that for constants $a$ and $b$ and discrete random variable $X$ that $\mathbb{V}\mathrm{ar}\,(aX + b) = a^2\,\mathbb{V}\mathrm{ar}\,(X)$.

**2.8.5** Suppose $X$ is a random variable with $\mathbb{E}\,(X) = 2$ and $\mathbb{V}\mathrm{ar}\,(X) = 7$. Suppose random variable $Y = 3X + 4$. Find $\mathbb{E}\,(Y)$ and $\mathbb{V}\mathrm{ar}\,(Y)$.

## 2.9   Covariance and Correlation

The concept of **covariance** is related to the degree to which two random variables tend to change in the same or opposite direction relative to one another. If $X$ and $Y$ are the random variables then mathematically the covariance, denoted $\mathbb{C}\mathrm{ov}\,(X, Y)$, is defined as

$$\mathbb{C}\mathrm{ov}\,(X, Y) = \mathbb{E}\,((X - \mathbb{E}\,(X))(Y - \mathbb{E}\,(Y))). \tag{2.14}$$

Making use of the definition and properties of expected value one can see that

$$\begin{aligned}
\mathbb{C}\mathrm{ov}\,(X, Y) &= \mathbb{E}\,(XY - Y\mathbb{E}\,(X) - X\mathbb{E}\,(Y) + \mathbb{E}\,(X)\mathbb{E}\,(Y)) \\
&= \mathbb{E}\,(XY) - \mathbb{E}\,(Y)\mathbb{E}\,(X) - \mathbb{E}\,(X)\mathbb{E}\,(Y) + \mathbb{E}\,(X)\mathbb{E}\,(Y) \\
&= \mathbb{E}\,(XY) - \mathbb{E}\,(X)\mathbb{E}\,(Y), \tag{2.15}
\end{aligned}$$

where Eq. (2.15) is generally more convenient to use than the expression on the right-hand side of Eq. (2.14).

**Example 2.13.** Table 2.1 lists the heights and arm spans of a sample of 20 children [Shodor (2007)]. Let $X$ represent height and $Y$ represent arm span for each child. Find the covariance of height and arm span.

Table 2.1   A sample of heights and arm spans for children.

| Child | Ht. (cm) | Span (cm) | Child | Ht. (cm) | Span (cm) |
|-------|----------|-----------|-------|----------|-----------|
| 1  | 142 | 138 | 11 | 150 | 147 |
| 2  | 148 | 144 | 12 | 152 | 141 |
| 3  | 152 | 148 | 13 | 148 | 144 |
| 4  | 150 | 145 | 14 | 152 | 148 |
| 5  | 141 | 136 | 15 | 144 | 140 |
| 6  | 142 | 139 | 16 | 148 | 143 |
| 7  | 149 | 144 | 17 | 150 | 146 |
| 8  | 151 | 145 | 18 | 138 | 134 |
| 9  | 147 | 144 | 19 | 145 | 142 |
| 10 | 152 | 148 | 20 | 142 | 138 |

**Solution.** The reader can readily calculate that $\mathbb{E}\,(X) = 147.15$ cm and $\mathbb{E}\,(Y) = 142.70$ cm. The expected value of the pairwise product of height and arm span is $\mathbb{E}\,(XY) = 21013.8$. Thus the covariance of height and arm span for this sample is $\mathbb{Cov}\,(X,Y) = 15.445$. Note that the covariance is positive, indicating that, in general, as height increases so does arm span.

From the definition of covariance several properties of this concept follow almost immediately. If $X$ and $Y$ are independent random variables then $\mathbb{E}\,(XY) = \mathbb{E}\,(X)\,\mathbb{E}\,(Y)$ by Thm. 2.8 and thus the covariance of independent random variables is zero. However, the converse is not true. Random variables can have zero covariance and not be independent. The following set of relationships can also be established.

**Theorem 2.12.** *Suppose $X$, $Y$, and $Z$ are random variables and $a$ is a constant, then the following statements are true:*

(i) $\mathbb{Cov}\,(X,X) = \mathbb{Var}\,(X)$,
(ii) $\mathbb{Cov}\,(X,Y) = \mathbb{Cov}\,(Y,X)$,
(iii) $\mathbb{Cov}\,(X+Y,Z) = \mathbb{Cov}\,(X,Z) + \mathbb{Cov}\,(Y,Z)$,
(iv) $\mathbb{Cov}\,(a\,X,Y) = a\,\mathbb{Cov}\,(X,Y)$.

The proofs of the first two statements are left for the reader as exercises. Justification for statements (iii) and (iv) are given below.

*Proof.* Let $X$, $Y$, and $Z$ be random variables, then
$$\begin{aligned}
\mathbb{Cov}\,(X+Y,Z) &= \mathbb{E}\,((X+Y)Z) - \mathbb{E}\,(X+Y)\,\mathbb{E}\,(Z) \\
&= \mathbb{E}\,(XZ) + \mathbb{E}\,(YZ) - \mathbb{E}\,(X)\,\mathbb{E}\,(Z) - \mathbb{E}\,(Y)\,\mathbb{E}\,(Z) \\
&= \mathbb{E}\,(XZ) - \mathbb{E}\,(X)\,\mathbb{E}\,(Z) + \mathbb{E}\,(YZ) - \mathbb{E}\,(Y)\,\mathbb{E}\,(Z) \\
&= \mathbb{Cov}\,(X,Z) + \mathbb{Cov}\,(Y,Z).
\end{aligned}$$
Now if $a$ is a constant, then
$$\begin{aligned}
\mathbb{Cov}\,(a\,X,Y) &= \mathbb{E}\,(a\,XY) - \mathbb{E}\,(a\,X)\,\mathbb{E}\,(Y) = a\,\mathbb{E}\,(XY) - a\,\mathbb{E}\,(X)\,\mathbb{E}\,(Y) \\
&= a\,(\mathbb{E}\,(XY) - a\,\mathbb{E}\,(X)\,\mathbb{E}\,(Y)) = a\,\mathbb{Cov}\,(X,Y). \qquad \square
\end{aligned}$$

The third statement of Thm. 2.12 can be generalized as in the following corollary.

**Corollary 2.3.** *Suppose $\{X_1, X_2, \ldots, X_n\}$ and $\{Y_1, Y_2, \ldots, Y_m\}$ are sets of random variables where $n, m \geq 1$, then*
$$\mathbb{Cov}\left(\sum_{i=1}^{n} X_i, \sum_{i=1}^{m} Y_i\right) = \sum_{i=1}^{n}\sum_{j=1}^{m} \mathbb{Cov}\,(X_i, Y_j). \tag{2.16}$$

*Proof.* Consider the special case of this corollary when $m = 1$. The corollary holds trivially when $n = 1$ and follows from the third statement of Theorem 2.12 when $n = 2$. Suppose the claim holds when $n \leq k$ where $k \in \mathbb{N}$. Let $\{X_1, X_2, \ldots, X_k, X_{k+1}\}$ be random variables.

$$
\mathrm{Cov}\left(\sum_{i=1}^{k+1} X_i, Y_1\right) = \mathrm{Cov}\left(\sum_{i=1}^{k} X_i, Y_1\right) + \mathrm{Cov}\left(X_{k+1}, Y_1\right)
$$

$$
= \sum_{i=1}^{k} \mathrm{Cov}\left(X_i, Y_1\right) + \mathrm{Cov}\left(X_{k+1}, Y_1\right)
$$

$$
= \sum_{i=1}^{k+1} \mathrm{Cov}\left(X_i, Y_1\right).
$$

Therefore by induction the result is true for any finite, integer value of $n$ (at least when $m = 1$). When $m$ is an integer larger than 1,

$$
\mathrm{Cov}\left(\sum_{i=1}^{n} X_i, \sum_{j=1}^{m} Y_j\right) = \sum_{i=1}^{n} \mathrm{Cov}\left(X_i, \sum_{j=1}^{m} Y_j\right) \quad \text{(shown above)}
$$

$$
= \sum_{i=1}^{n} \mathrm{Cov}\left(\sum_{j=1}^{m} Y_j, X_i\right) \quad \text{(by Thm. 2.12 (ii))}
$$

$$
= \sum_{i=1}^{n}\sum_{j=1}^{m} \mathrm{Cov}\left(Y_j, X_i\right)
$$

$$
= \sum_{i=1}^{n}\sum_{j=1}^{m} \mathrm{Cov}\left(X_i, Y_j\right).
$$

Therefore Eq. (2.16) holds for all finite, positive integer values of $m$ and $n$.
$\square$

Yet another corollary follows from Corollary 2.3. This corollary generalizes statement (i) of Thm. 2.12.

**Corollary 2.4.** *If $\{X_1, X_2, \ldots, X_n\}$ are random variables then*

$$
\mathrm{Var}\left(\sum_{i=1}^{n} X_i\right) = \sum_{i=1}^{n} \mathrm{Var}\left(X_i\right) + \sum_{i=1}^{n}\sum_{\substack{j=1 \\ j \neq i}}^{n} \mathrm{Cov}\left(X_i, X_j\right). \tag{2.17}
$$

*Proof.* Let $Y = \sum_{i=1}^{n} X_i$, then according to the first statement of Thm. 2.12,

$$\mathbb{Var}(Y) = \mathbb{Cov}(Y, Y)$$

$$\mathbb{Var}\left(\sum_{i=1}^{n} X_i\right) = \mathbb{Cov}\left(\sum_{i=1}^{n} X_i, \sum_{j=1}^{n} X_j\right)$$

$$= \sum_{i=1}^{n}\sum_{j=1}^{n} \mathbb{Cov}(X_i, X_j) \text{ (by Corollary 2.3)}$$

$$= \sum_{i=1}^{n} \mathbb{Cov}(X_i, X_i) + \sum_{i=1}^{n}\sum_{\substack{j=1\\j\neq i}}^{n} \mathbb{Cov}(X_i, X_j)$$

$$= \sum_{i=1}^{n} \mathbb{Var}(X_i) + \sum_{i=1}^{n}\sum_{\substack{j=1\\j\neq i}}^{n} \mathbb{Cov}(X_i, X_j).$$

Once more the first statement of Thm. 2.12 was used to reintroduce the variance in the last line of the derivation. □

Often a quantity related to covariance is used as a measure of the degree to which increasing values of a random variable $X$ are associated with increasing values of another random variable $Y$. This quantity is known as **correlation** and is denoted $\mathbb{Cor}(X, Y)$ or sometimes $\rho_{X,Y}$. The correlation of two random variables is defined as

$$\mathbb{Cor}(X, Y) = \frac{\mathbb{Cov}(X, Y)}{\sqrt{\mathbb{Var}(X)\mathbb{Var}(Y)}} = \rho_{X,Y}. \tag{2.18}$$

The correlation of two random variables can be interpreted as a measure of the degree to which monotonic changes (increases or decreases) in one of the variables are reflected in similar changes (increases with increases and decreases with decreases) in the other variable. The correlation is more than just a simple re-scaling of the covariance. While the covariance of $X$ and $Y$ may numerically be positive, negative, or zero, the correlation always lies in the interval $[-1, 1]$. Once again, independent random variables have a correlation of zero and hence are described as **uncorrelated**.

**Theorem 2.13.** *Suppose $X$ and $Y$ are random variables such that $Y = aX + b$ where $a, b \in \mathbb{R}$ with $a \neq 0$. If $a > 0$ then $\mathbb{Cor}(X, Y) = 1$, while if $a < 0$ then $\mathbb{Cor}(X, Y) = -1$.*

*Proof.* Start by calculating the covariance of $X$ and $Y$.

$$\begin{aligned}
\mathbb{Cov}\,(X,Y) &= \mathbb{Cov}\,(X, aX + b) \\
&= \mathbb{E}\,(X(aX + b)) - \mathbb{E}\,(X)\,\mathbb{E}\,(aX + b) \\
&= \mathbb{E}\,(aX^2 + bX)) - \mathbb{E}\,(X)\,(a\,\mathbb{E}\,(X) + b) \\
&= a\,\mathbb{E}\,(X^2) + b\,\mathbb{E}\,(X) - a\,\mathbb{E}\,(X)\,\mathbb{E}\,(X) - b\,\mathbb{E}\,(X) \\
&= a\left(\mathbb{E}\,(X^2) - \mathbb{E}\,(X)^2\right) \\
&= a\,\mathbb{Var}\,(X).
\end{aligned}$$

The reader should make note of the use of Thm. 2.6. Therefore using the result of Exercise 2.8.4,

$$\mathbb{Cor}\,(X,Y) = \frac{a\,\mathbb{Var}\,(X)}{\sqrt{\mathbb{Var}\,(X) \cdot a^2\,\mathbb{Var}\,(X)}} = \frac{a}{|a|},$$

which is $-1$ when $a < 0$ and $1$ when $a > 0$. $\qquad\qquad\square$

The converse of Thm. 2.13 is false. A correlation close to unity does not indicate a linear relationship between the random variables. See Exercise 2.9.5.

Before bounding the correlation in the interval $[-1, 1]$ the following lemma is needed.

**Lemma 2.1 (Schwarz Inequality).** *If $X$ and $Y$ are random variables then* $(\mathbb{E}\,(XY))^2 \leq \mathbb{E}\,(X^2)\,\mathbb{E}\,(Y^2)$.

*Proof.* The cases in which $\mathbb{E}\,(X^2) = 0$, $\mathbb{E}\,(X^2) = \infty$, $\mathbb{E}\,(Y^2) = 0$, or $\mathbb{E}\,(Y^2) = \infty$ are left as exercises. Suppose for the purposes of this proof that $0 < \mathbb{E}\,(X^2) < \infty$ and $0 < \mathbb{E}\,(Y^2) < \infty$. If $a$ and $b$ are real numbers then the following two inequalities hold:

$$\begin{aligned}
0 &\leq \mathbb{E}\,((aX + bY)^2) = a^2\,\mathbb{E}\,(X^2) + 2ab\,\mathbb{E}\,(XY) + b^2\,\mathbb{E}\,(Y^2) \\
0 &\leq \mathbb{E}\,((aX - bY)^2) = a^2\,\mathbb{E}\,(X^2) - 2ab\,\mathbb{E}\,(XY) + b^2\,\mathbb{E}\,(Y^2).
\end{aligned}$$

If $a^2 = \mathbb{E}\,(Y^2)$ and $b^2 = \mathbb{E}\,(X^2)$ then the first inequality above yields

$$\begin{aligned}
0 &\leq 2\,\mathbb{E}\,(Y^2)\,\mathbb{E}\,(X^2) + 2\sqrt{\mathbb{E}\,(Y^2)\,\mathbb{E}\,(X^2)}\,\mathbb{E}\,(XY) - \mathbb{E}\,(X^2)\,\mathbb{E}\,(Y^2) \\
&\leq \sqrt{\mathbb{E}\,(Y^2)\,\mathbb{E}\,(X^2)}\,\mathbb{E}\,(XY) - \sqrt{\mathbb{E}\,(X^2)\,\mathbb{E}\,(Y^2)} \\
&\leq \mathbb{E}\,(XY).
\end{aligned}$$

By a similar set of steps the second inequality produces

$$\mathbb{E}\left(XY\right) \le \sqrt{\mathbb{E}\left(X^2\right)\mathbb{E}\left(Y^2\right)}.$$

Therefore, since

$$-\sqrt{\mathbb{E}\left(X^2\right)\mathbb{E}\left(Y^2\right)} \le \mathbb{E}\left(XY\right)$$

$$\le \sqrt{\mathbb{E}\left(X^2\right)\mathbb{E}\left(Y^2\right)}\left(\mathbb{E}\left(XY\right)\right)^2$$

$$\le \mathbb{E}\left(X^2\right)\mathbb{E}\left(Y^2\right).$$

Occasionally $\left(\mathbb{E}\left(XY\right)\right)^2$ is written as $\mathbb{E}\left(XY\right)^2$ or even as $\mathbb{E}^2\left(XY\right)$. A careful reading of the expression will avoid any possible confusion. $\square$

The background is now in place to prove the following theorem.

**Theorem 2.14.** *If $X$ and $Y$ are random variables then* $-1 \le \mathbb{C}\mathrm{or}\left(X,Y\right) \le 1$.

*Proof.* Consider the covariance of $X$ and $Y$.

$$\left(\mathbb{C}\mathrm{ov}\left(X,Y\right)\right)^2 = \left(\mathbb{E}\left(\left(X - \mathbb{E}\left(X\right)\right)\left(Y - \mathbb{E}\left(Y\right)\right)\right)\right)^2$$

$$\le \mathbb{E}\left(\left(X - \mathbb{E}\left(X\right)\right)^2\right)\mathbb{E}\left(\left(Y - \mathbb{E}\left(Y\right)\right)^2\right) \quad \text{(by Lemma 2.1)}$$

$$= \mathbb{V}\mathrm{ar}\left(X\right)\mathbb{V}\mathrm{ar}\left(Y\right).$$

Thus $\left|\mathbb{C}\mathrm{ov}\left(X,Y\right)\right| \le \sqrt{\mathbb{V}\mathrm{ar}\left(X\right)\mathbb{V}\mathrm{ar}\left(Y\right)}$, which is equivalent to the inequality,

$$-1 \le \frac{\mathbb{C}\mathrm{ov}\left(X,Y\right)}{\sqrt{\mathbb{V}\mathrm{ar}\left(X\right)\mathbb{V}\mathrm{ar}\left(Y\right)}} \le 1$$

$$-1 \le \mathbb{C}\mathrm{or}\left(X,Y\right) \le 1,$$

which follows from the definition of correlation. $\square$

**Example 2.14.** Using the data in Table 2.1 find the correlation between height and arm span.

**Solution.** Using the previously calculated covariance between height and arm span, the correlation is calculated as

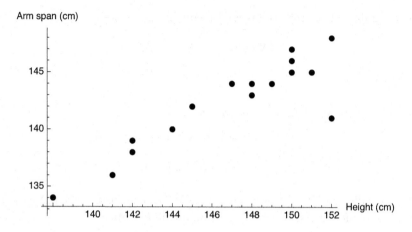

Fig. 2.6   A scatter plot of height *versus* arm span for a sample of twenty children.

$$\mathbb{C}\text{or}\,(X,Y) = \frac{\mathbb{C}\text{ov}\,(X,Y)}{\sqrt{\mathbb{V}\text{ar}\,(X)\,\mathbb{V}\text{ar}\,(Y)}} = \frac{15.445}{\sqrt{(18.6605)(16.8526)}} \approx 0.870948.$$

For this data set there seems to be an especially strong relationship between height and weight. Figure 2.6 illustrates this as well.

***Exercises***

**2.9.1** Using the definition of covariance and variance prove the first two statements of Thm. 2.12.

**2.9.2** Show that for any two random variables $X$ and $Y$

$$\mathbb{V}\text{ar}\,(X + Y) = \mathbb{V}\text{ar}\,(X) + \mathbb{V}\text{ar}\,(Y) + 2\,\mathbb{C}\text{ov}\,(X,Y).$$

If $X$ and $Y$ are independent random variables, show this formula reduces to the result mentioned in Thm. 2.10.

**2.9.3** Fill in the remaining details of the proof of Lemma 2.1 for the cases in which $\mathbb{E}\left(X^2\right) = 0$, $\mathbb{E}\left(X^2\right) = \infty$, $\mathbb{E}\left(Y^2\right) = 0$, or $\mathbb{E}\left(Y^2\right) = \infty$.

**2.9.4** The data shown in the table below was originally published in the *New York Times* (Section 3, p. 1, 31 May 1998). It lists the name of a corporate CEO, the CEO's corporation, the CEO's golf handicap, and a rating of the stock of the CEO's corporation. Determine the covariance and correlation between the CEO's golf handicap, and the corporation's stock rating.

| Name | Corp. | Handicap | Rating |
|------|-------|----------|--------|
| Terrence Murray | Fleet Financial | 10.1 | 67 |
| William T. Esrey | Sprint | 10.1 | 66 |
| Hugh L. McColl Jr. | Nationsbank | 11.0 | 64 |
| James E. Cayne | Bear Stearns | 12.6 | 64 |
| John R. Stafford | Amer. Home Prod. | 10.9 | 58 |
| John B. McCoy | Banc One | 7.6 | 58 |
| Frank C. Herringer | Transamerica | 10.6 | 55 |
| Ralph S. Larsen | Johnson&Johnson | 16.1 | 54 |
| Paul Hazen | Wells Fargo | 10.9 | 54 |
| Lawrence A. Bossidy | Allied Signal | 12.6 | 51 |
| Charles R. Shoemate | Bestfoods | 17.6 | 49 |
| James E. Perrella | Ingersoll-Rand | 12.8 | 49 |
| William P. Stiritz | Ralston Purina | 13.0 | 48 |

**2.9.5** Consider the match paired data in the table below.

| $X$ | 0 | 1 | 1 | 1 | 1 | 1 | 1 | 1 | 1 | 1 |
|-----|---|---|---|---|---|---|---|---|---|---|
| $Y$ | $y$ | 1 | 2 | 3 | 4 | 5 | 6 | 7 | 8 | 9 |

Show that while the functional relationship between $X$ and $Y$ is not linear, as $y \to -\infty$, the correlation between $X$ and $Y$ approaches $9/10$.

**2.9.6** Suppose that $X$ and $Y$ are random variables with $\mathbb{Cov}(X,Y) = -2$. Find $\mathbb{Cov}(5X+4, 2Y+3)$.

## 2.10  Odds and Wagering

Chapter 3 will introduce a fundamental financial principle underlying much of the reasoning behind the rational, fair pricing of financial instruments, namely the concept of arbitrage. Arbitrage is closely tied to the notions of the time value of money introduced in Chapter 1 and elements of probability discussed in the present chapter. Arbitrage will be introduced via a discussion of wagering. Usually in situations of wagering (sporting events, elections, and other contests) the probability of an event or outcome is not stated in the form of a real number $p \in [0,1]$ but rather as a ratio of two integers forming the **odds** either for or against the outcome. This section will introduce the concept of odds and demonstrate how to convert between odds and probabilities.

Expressing the relative likelihood of an event occurring in the form of odds is embraced by gamblers and the gambling industry to simplify the process of placing and paying off bets. Suppose a gambler wishes to place a wager on event A occurring and is told that the odds of event A are "2 : 1 against". The gambling convention is then that for each dollar wagered on event A, if event A occurs, the gambler will receive two dollars in return plus the original dollar bet. Thus the gambler has a net profit of two dollars for each dollar wagered. The probability of event A can be calculated from the odds against by interpreting the 2 : 1 odds against event A as stating that in $2 + 1 = 3$ repetitions of the experiment, event A will occur once and the complement of A will occur twice. In general if the odds against event A are $m : n$ against, then $\mathbb{P}(A) = n/(m+n)$ and $\mathbb{P}(A^c) = m/(m+n)$. Note that $\mathbb{P}(A) + \mathbb{P}(A^c) = 1$.

If the odds surrounding event A are stated as "$n : m$ in favor" then in $m + n$ repetitions of the experiment in which A is an outcome, A occurs $n$ times, thus $\mathbb{P}(A) = n/(m+n)$ and $\mathbb{P}(A^c) = m/(m+n)$.

**Example 2.15.** Chris and Pat will play against one another in tennis. The odds against Chris defeating Pat are 7 : 5. What is the probability that Chris wins the match?

**Solution.** Since the odds against Chris winning are 7 : 5,

$$\mathbb{P}(\text{Chris wins}) = \frac{5}{7+5} = \frac{5}{12}.$$

Conversion from probabilities to odds can also be accomplished. If the $\mathbb{P}(A) = p$ then the odds against event A are $(1-p) : p$ and the odds in favor of event A are $p : (1-p)$.

**Example 2.16.** Suppose the probability of Chris defeating Pat in a game of darts is $1/4$, what are the odds against Chris winning the game?

**Solution.** Since the probability of Chris winning the game is $p = 1/4$ then the odds against Chris can be expressed as

$$\frac{3}{4} : \frac{1}{4} \text{ or } 3 : 1.$$

## Exercises

**2.10.1** If a fair die is tossed what are the odds in favor of the die landing with ⚅ showing on the top face?

**2.10.2** An olympic swimmer will compete in three different individual races. The outcomes of races are all mutually independent. If the probability of the swimmer winning each race is $p = 3/5$, what are the odds against the swimmer winning all three events?

**2.10.3** At the race track the horse named "Nuance" is listed with odds of $15 : 2$ to win (these are odds against winning). If a gambler places a bet of $10 on Nuance to win, what is the gambler's net profit if Nuance wins?

**2.10.4** Show that if the odds against event A are $(1 - p) : p$, then $\mathbb{P}(A) = p$.

**2.10.5** If two cards are drawn from a well-shuffled, standard 52-card deck without replacement, what are the odds against both cards being ♡?

Chapter 3

# The Arbitrage Theorem

The concept known as **arbitrage** is subtle and can seem counter-intuitive. Formally there are two types of arbitrage (creatively named type A and type B). Type A arbitrage is an investment strategy that results in an initial positive cash flow (the investor receives a positive amount of income initially) and there is no risk of loss in the future. Type B arbitrage is a investment strategy in which the investor needs no initial cash investment, has no risk of future loss, and has a positive probability of profit in the future [Cornuejols and Tütüncü (2007)]. Often Type B arbitrage is enabled by loans to the investor, so that the investor uses none of their own money and the investment will generate enough return to repay the principal and interest of the loan and there exists a chance of excess return which the investor keeps. Informally arbitrage exists whenever two financial instruments are mis-priced relative to one another. Due to the mis-pricing, it becomes possible to make a financial gain. For example, suppose bank Alpha issues loans at a 5% interest rate and bank Beta offers a savings account which pays 6% interest. A person could take out a loan from bank Alpha and place the principal into savings with bank Beta. When it becomes time to repay bank Alpha for the loan, the person closes the savings account with bank Beta, repays the principal and interest and still has 1% of the principal as profit. To model financial markets it is assumed that prices reflect all publicly available information related to assets. This assumption is sometimes called the semi-strong form of the efficient market hypothesis. The assumption that financial markets are efficient prevents such obvious arbitrage opportunities from being commonplace. When arbitrage opportunities arise, investors wanting to make a profit flock to the mis-priced instruments and the financial market reacts by correcting the pricing of the instruments.

Financial instruments such as options, bonds, and stocks must be priced so as to be "arbitrage free". It is the absence of arbitrage which forms the basis of the derivation of the Black–Scholes equation found in Chapter 6.

Suppose there is a set of possible outcomes for some experiment and that wagers can be placed on whether particular outcomes occur. The **Arbitrage Theorem** states that either the probabilities of the outcomes are such that all bets are fair, or there is a betting scheme which produces a positive gain independent of the outcome of the experiment. The processes of wagering and determining payoffs from wagers was introduced in Sec. 2.10. A simple example will illustrate the Arbitrage Theorem.

**Example 3.1.** Suppose the odds against Nico defeating Piper in a tennis match are 3 : 1 and the odds against Piper defeating Nico are 1 : 1. Determine a betting (investment) strategy which yields a positive profit for a gambler (investor) regardless of the winner of the tennis match.

**Solution.** Converting the odds to probabilities, Nico defeats Piper with probability 1/4 while Piper defeats Nico with probability 1/2. There is obviously something wrong with these probabilities since they should add to one, but do not. The Arbitrage Theorem implies there is a betting strategy which generates a positive gain regardless of the outcome of the tennis match. Now suppose a gambler wagers $1 on Nico to win and $2 on Piper to win. If Nico wins the gambler has a net profit of $1 ($3 profit on the first bet and a loss of $2 on the second bet). If Piper wins the gambler also has a net profit of $1 (the loss of $1 on the first bet and a gain of $2 on the second). No matter which player wins the tennis match, the gambler has a positive profit.

This example is hardly transparent, so study of the Arbitrage Theorem would be beneficial in avoiding arbitrage opportunities in more complex financial situations. Before a statement and proof of the Arbitrage Theorem is presented an introduction to linear programming is needed.

## 3.1 An Introduction to Linear Programming

This section will introduce the common forms of linear programs and show their equivalencies. Linear programming is the name given to a branch of mathematics often applied in business and economics in which a linear function of some (usually large) number of variables must be optimized (either maximized or minimized) subject to a set of linear equations or inequalities.

To sample the types of linear programming problems solved by businesses, consider this simple example. A bank may invest its deposits in loans which earn 6% interest per year and in the purchase of stocks which increase in value by 13% per year. The bank wishes to maximize the total return on its investments. Assume the bank can invest a proportion $x$ in loans and proportion $y$ in stocks and must retain 10% in the form of cash (in other words 10% of the portfolio has a rate of return of 0). Any remaining proportion is simply held by the bank. The total return is therefore $0.06x + 0.13y$. Suppose that government regulations require that the bank invest no more than 60% its deposits in stocks. As a good business practice the bank wishes to devote at least 25% of its deposits to loans. These constraints impose some inequalities on the bank's investment strategy. The inequalities are $x \geq 0$, $y \geq 0$, $x + y \leq 0.90$ (non-negative proportions of the deposits are invested and the total amount invested is no greater than 90% of the total amount on deposit), $y \leq 0.6$ (government regulation), and $x \geq 0.25$ (the bank's desire to loan to businesses). An investment strategy can be represented by a vector $\langle x, y \rangle$. The bank's linear programming problem is to find the vector which satisfies the constraints and maximizes the total return. The set of vectors $\langle x, y \rangle$ (which are identified with the ordered pairs $(x, y)$ in the plane) which satisfy the constraint mentioned above make up the **feasible region**. The lines described by the equations $0.06x + 0.13y = k$ where $k$ is a constant are the **level sets** of the return on the bank's investments. Since this problem is two dimensional a plot can reveal the solution. As seen in Fig. 3.1 the optimal solution occurs when $x = 0.3$ and $y = 0.6$.

The remainder of this section introduces in a more formal way linear programming and the Duality Theorem. Other accessible introductions can be found in [Franklin (1980)], [Noble and Daniel (1988)], and [Strang (1986)]. Readers without a background in matrix algebra and vectors may wish to consult App. A. Readers wishing for a more detailed introduction not only to linear programming, but also to linear algebra, should consult [Gale (1960)].

The general linear programming problem, or **linear program**, consists of a set of linear equations or inequalities (constraints), possibly sign conditions on the solution (more constraints), and a linear expression (a weighted sum) which must be optimized (either maximized or minimized). Bold letters are used here to represent vectors and the Euclidean inner product to

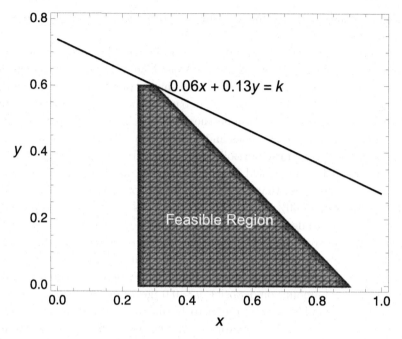

Fig. 3.1    The shaded region represents the set of possible solutions to the bank's investment decision. The total return is maximized at $\langle 0.3, 0.6 \rangle$.

express linear combinations. Vectors are column matrices. If $\mathbf{c}$ and $\mathbf{x}$ are vectors with $n$ components each, the notation

$$\mathbf{c}^T \mathbf{x} = c_1 x_1 + c_2 x_2 + \cdots + c_n x_n$$

represents a linear expression, a weighted sum of the components of vector $\mathbf{x}$ where the weights are given in vector $\mathbf{c}$. The components of vector $\mathbf{x}$ are called the **decision variables**. The linear expression $\mathbf{c}^T \mathbf{x}$ is called the **objective function**. Readers with a background in linear algebra will recognize the objective function as the Euclidean inner product of the weight vector and the decision variable vector.

For denoting the comparison of vectors a convenient extension of the equation and inequality symbols is adopted. The vector $\mathbf{u}$ is less than (less than or equal to) vector $\mathbf{v}$ if the vectors have the same number of elements and $u_i < v_i$ ($u_i \leq v_i$) for all $i$. Similarly vector $\mathbf{u}$ is greater than (greater than or equal to) vector $\mathbf{v}$ if $u_i > v_i$ ($u_i \geq v_i$) for all $i$. These inequalities will be denoted as appropriate $\mathbf{u} < \mathbf{v}$, $\mathbf{u} \leq \mathbf{v}$, $\mathbf{u} > \mathbf{v}$, or $\mathbf{u} \geq \mathbf{v}$.

The positive part of a real number $x$ will be denoted $x^+$ and is defined as

$$x^+ = \begin{cases} x \text{ if } x \geq 0 \\ 0 \text{ if } x < 0. \end{cases}$$

Similarly the negative part of $x$ is denoted $x^-$ and defined as

$$x^- = \begin{cases} -x \text{ if } x \leq 0 \\ 0 \quad \text{ if } x > 0. \end{cases}$$

The positive part, as well as the negative part, of $x$ is always a non-negative real number. Any real number $x$ can be expressed as $x = x^+ - x^-$. This notation is extended to vectors by applying it to each component of the vector. This new notation will simplify the statements of the theorems in this chapter. For instance the requirement that $\mathbf{x}$ be a vector with non-negative components can be succinctly stated as $\mathbf{x} \geq \mathbf{0}$.

The starting point of most linear programs is its statement to "maximize $\mathbf{c}^T\mathbf{x}$" subject to one or more constraints. The process of maximizing can be replaced that of minimizing by multiplying the objective function by $-1$ since

$$- \left( \max_{\mathbf{x}}(\mathbf{c}^T\mathbf{x}) \right) = \min_{\mathbf{x}}(-\mathbf{c}^T\mathbf{x}).$$

If there are several constraints placed on a solution to a linear programming problem, matrix notation is used to describe them succinctly. Suppose there are $m$ linear inequality constraints on the solution $\mathbf{x}$, a vector with $n$ components.

$$\mathbf{a}_1^T\mathbf{x} \leq b_1$$
$$\mathbf{a}_2^T\mathbf{x} \leq b_2$$
$$\vdots$$
$$\mathbf{a}_m^T\mathbf{x} \leq b_m.$$

This will be expressed as the vector inequality $A\mathbf{x} \leq \mathbf{b}$ where

$$A\mathbf{x} = \begin{bmatrix} a_{11} & a_{12} & \cdots & a_{1n} \\ a_{21} & a_{22} & \cdots & a_{2n} \\ \vdots & \vdots & & \vdots \\ a_{m1} & a_{m2} & \cdots & a_{mn} \end{bmatrix} \begin{bmatrix} x_1 \\ x_2 \\ \vdots \\ x_n \end{bmatrix} \leq \begin{bmatrix} b_1 \\ b_2 \\ \vdots \\ b_m \end{bmatrix} = \mathbf{b}.$$

The inequality $\leq$ could be replaced with $=$, $<$, $\geq$, or $>$ as needed.

Linear programs may be stated in forms that contain inequality constraints (mixing greater than or equal and less than or equal) and equality constraints. Linear programs may impose sign constraints on the decision variables. Variables whose signs are not specified for determining feasible solutions are said to be unrestricted in sign or free variables. These linear programs can be referred to as **general linear programs** and may be stated in the format of maximizing $\mathbf{c}^T\mathbf{x}$ subject to the constraints

$$A\mathbf{x} \leq \mathbf{b} \tag{3.1}$$

$$\hat{A}\mathbf{x} \geq \hat{\mathbf{b}} \tag{3.2}$$

$$\tilde{A}\mathbf{x} = \tilde{\mathbf{b}}, \tag{3.3}$$

where $A$ is a $r \times n$ matrix, $\mathbf{b}$ is a vector with $r$ components, $\hat{A}$ is a $s \times n$ matrix, $\hat{\mathbf{b}}$ is a vector with $s$ components, $\tilde{A}$ is a $t \times n$ matrix, and $\tilde{\mathbf{b}}$ is a vector with $t$ components.

Many of the results cited in this chapter are stated for linear programs with specialized requirements for the decision variables and constraints. In particular the **normal form** or **symmetric form** [Williams (1970)] possesses decision variables that are non-negative and constraints that are inequalities (all purely $\leq$ or all purely $\geq$). The symmetric form of a linear program is often convenient when proving a property or theoretical result about a linear program. In practice there is no need to worry about whether a linear program is stated in general form or symmetric form, since there are algorithmic ways to convert one format into the other. Trivially every linear program in symmetric form is a linear program in general form. It is only necessary to show that a general linear program can be recast as a linear program in symmetric form.

If $\mathbf{x}$ is unrestricted in sign then $\mathbf{x} = \mathbf{x}^+ - \mathbf{x}^-$ where $\mathbf{x}^+ \geq \mathbf{0}$ and $\mathbf{x}^- \geq \mathbf{0}$. The constraints in Eqs. (3.1)–(3.3) can now be written as

$$A(\mathbf{x}^+ - \mathbf{x}^-) \leq \mathbf{b}$$

$$-\hat{A}(\mathbf{x}^+ - \mathbf{x}^-) \leq -\hat{\mathbf{b}}$$

$$\tilde{A}(\mathbf{x}^+ - \mathbf{x}^-) \leq \tilde{\mathbf{b}}$$

$$-\tilde{A}(\mathbf{x}^+ - \mathbf{x}^-) \leq -\tilde{\mathbf{b}}.$$

In matrix/vector form this system of inequalities can be expressed as

$$\begin{bmatrix} A & -A \\ -\hat{A} & \hat{A} \\ \tilde{A} & -\tilde{A} \\ -\tilde{A} & \tilde{A} \end{bmatrix} \begin{bmatrix} \mathbf{x}^+ \\ \mathbf{x}^- \end{bmatrix} \leq \begin{bmatrix} \mathbf{b} \\ -\hat{\mathbf{b}} \\ \tilde{\mathbf{b}} \\ -\tilde{\mathbf{b}} \end{bmatrix}. \tag{3.4}$$

The matrix on the left-hand side of Eq. (3.4) is expressed in block form and the vectors are the concatenation of the appropriate decision variables and constants found in the general linear programs. The linear expression for the objective function of the general linear program can be written as

$$(\mathbf{c}, -\mathbf{c})^T (\mathbf{x}^+, \mathbf{x}^-) = \mathbf{c}^T \mathbf{x}^+ - \mathbf{c}^T \mathbf{x}^- = \mathbf{c}^T (\mathbf{x}^+ - \mathbf{x}^-) = \mathbf{c}^T \mathbf{x}.$$

Thus, the general linear program has been recast in the form of a symmetric linear program. Vector $(\mathbf{x}^+, \mathbf{x}^-)$ is a solution to the symmetric linear program if and only if $\mathbf{x} = \mathbf{x}^+ - \mathbf{x}^-$ is a solution to the general linear program. Thus the two linear programming problems, the one with the non-negativity constraint and the one without it, are equivalent [González-Díaz *et al.* (2010)].

The vector $\mathbf{x}$ is a **feasible vector** or **feasible solution** to the symmetric linear program if $\mathbf{x} \geq \mathbf{0}$ and $\mathbf{x}$ satisfies the set of constraint inequalities $A\mathbf{x} \leq \mathbf{b}$. The feasible solutions to a symmetric linear program form a **convex set**. A set is convex if for every pair of points $P$ and $Q$ contained in the set, the line segment connecting them also lies completely in the set. Algebraically the set $C$ is convex if and only if for all $\mathbf{x}, \mathbf{y} \in C$ it is the case that $\lambda \mathbf{x} + (1 - \lambda)\mathbf{y} \in C$ for all $\lambda \in [0, 1]$. A subset $D \in \mathbb{R}^n$ is **bounded** if there exists $M > 0$ such that

$$D \subset \{\mathbf{x} \in \mathbb{R}^n \mid \|\mathbf{x}\| < M\}.$$

The symmetric linear program is called **feasible** if there exists a feasible solution to it. If the vector $\mathbf{x}$ is feasible and maximizes the objective function, then $\mathbf{x}$ is also called an **optimal solution**.

Another common type of linear program is known as the **standard form** and involves finding a vector $\mathbf{x}$ with non-negative components such that for a given vector $\mathbf{c}$ the linear expression $\mathbf{c}^T \mathbf{x}$ is optimized (maximized or minimized) subject to equality constraints of the form $\mathbf{a}^T \mathbf{x} = b$ where $\mathbf{a}$ is a given vector and $b$ is a constant [Winston (1994)]. The standard form of a linear program is often preferred when solving the problem numerically (such as by the simplex method).

Every linear program in symmetric form can be put into the format of a linear program in standard form. Starting with a linear program in symmetric form,

$$\text{maximize } \mathbf{c}^T \mathbf{x} \text{ subject to } A\mathbf{x} \leq \mathbf{b} \text{ and } \mathbf{x} \geq \mathbf{0},$$

introduce **slack variables**. To illustrate, the following inequality constraint of a linear program

$$x_1 + x_2 + x_3 \leq 1 \text{ becomes } x_1 + x_2 + x_3 + \hat{x}_4 = 1,$$

where slack variable, $\hat{x}_4 \geq 0$ "takes up the slack" to produce the equality constraint of a standard linear program. No modification of the objective function is necessary. The weight assigned to a slack variable in the objective function is zero. If there are multiple linear inequality constraints in the symmetric linear program, several slack variables may be introduced. Suppose $A$ is an $m \times n$ matrix, $\mathbf{x}$ is a vector of $n$ components, and $\mathbf{b}$ is a vector of $m$ components, then by augmenting $\mathbf{x}$ with $m$ slack variables and $A$ with the $m \times m$ identity matrix the inequality constraint $A\mathbf{x} \leq \mathbf{b}$ is equivalent to

$$
\left[\begin{array}{cccc|cccc}
a_{11} & a_{12} & \cdots & a_{1n} & 1 & 0 & \cdots & 0 \\
a_{21} & a_{22} & \cdots & a_{2n} & 0 & 1 & \cdots & 0 \\
\vdots & \vdots & & \vdots & \vdots & \vdots & \ddots & \vdots \\
a_{m1} & a_{m2} & \cdots & a_{mn} & 0 & 0 & \cdots & 1
\end{array}\right]
\left[\begin{array}{c}
x_1 \\ x_2 \\ \vdots \\ x_n \\ \hat{x}_{n+1} \\ \hat{x}_{n+2} \\ \vdots \\ \hat{x}_{n+m}
\end{array}\right]
=
\left[\begin{array}{c}
b_1 \\ b_2 \\ \vdots \\ b_m
\end{array}\right]
$$

$$
\left[\, A \,|\, I_m \,\right] \left[\begin{array}{c} \mathbf{x} \\ \hat{\mathbf{x}} \end{array}\right] = \overline{A}\,\overline{\mathbf{x}} = \mathbf{b}.
$$

The symmetric form of a linear program has been shown to be equivalent to the standard linear program "maximize $\mathbf{c}^T\mathbf{x}$ subject to $\overline{A}\,\overline{\mathbf{x}} = \mathbf{b}$ and $\overline{\mathbf{x}} \geq \mathbf{0}$", where it is understood that $\overline{A}$ is an $m \times (n+m)$ matrix consisting of the original $m \times n$ constraint matrix from the standard linear program augmented with the $m \times m$ identity matrix and $\overline{\mathbf{x}}$ is the solution vector to the standard linear program augmented with the slack variables.

Now start with a linear program in standard form,

$$\text{maximize } \mathbf{c}^T\mathbf{x} \text{ subject to } A\mathbf{x} = \mathbf{b} \text{ and } \mathbf{x} \geq \mathbf{0}.$$

The equality constraint $A\mathbf{x} = \mathbf{b}$ is equivalent to the pair of inequalities $A\mathbf{x} \leq \mathbf{b}$ and $A\mathbf{x} \geq \mathbf{b}$. The latter inequality involving the greater than or equal comparison is equivalent to the inequality $-A\mathbf{x} \leq -\mathbf{b}$ which involves the less than or equal comparison. Therefore the original linear program is equivalent to the linear program in symmetric form,

$$\text{maximize } \mathbf{c}^T\mathbf{x} \text{ subject to } \left[\begin{array}{c} A \\ -A \end{array}\right] \mathbf{x} \leq \left[\begin{array}{c} \mathbf{b} \\ -\mathbf{b} \end{array}\right] \text{ and } \mathbf{x} \geq \mathbf{0}.$$

Any vector $\mathbf{x}$ which is a feasible solution to a standard linear program program is also a feasible solution to the equivalent symmetric program.

Likewise if **z** is a feasible vector for a symmetric program it is also a feasible vector for the related standard linear program.

**Example 3.2.** Maximize $5x_1 + 4x_2 + 8x_3$ subject to the constraints $x_1 + x_2 + x_3 \leq 1$ and $\mathbf{x} \geq \mathbf{0}$.

**Solution.** The constraints require the solution of the optimization problem to lie in a subset of the positive orthant of $\mathbb{R}^3$. This subset is a tetrahedron. See Fig. 3.2(a). If the maximum of the objective function is the, as yet unknown, value $k$, then $5x_1 + 4x_2 + 8x_3 = k$ defines a plane. The largest value of $k$ for which the level set of the objective function intersects the constrained set of points will be the maximum of the objective function. Thus the maximum of the objective function is 8 as can be seen in Fig. 3.2(b). The objective function is maximized at the point $(x_1, x_2, x_3) = (0, 0, 1)$.

If the dimensions of a linear program are not too large the optimization task is easily accomplished. Graphical techniques often suffice for linear programs with three or fewer unknowns. When the dimensions of a linear program are large this is impractical and thus more sophisticated methods must be used. Some of these methods will be explored in the next section.

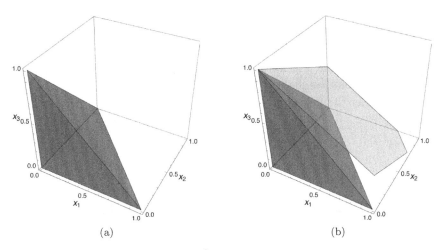

(a)                                          (b)

Fig. 3.2 The feasible set of points in $\mathbb{R}^3$ where the objective function's maximum may occur is illustrated in (a). The maximum of the objective function is the last level set of the objective function (shown in (b)) to intersect the tetrahedron of constrained, feasible solution points.

## Exercises

**3.1.1** Suppose the odds against the three possible outcomes of an experiment are as given in the table below.

| Outcome | Odds |
|---------|------|
| A | 2:1 |
| B | 3:1 |
| C | 1:1 |

Find a betting strategy which produces a positive net profit regardless of the outcome of the experiment.

**3.1.2** Sketch the region in the plane which satisfies the following inequalities:

$$x_1 + 3x_2 \geq 6$$
$$3x_1 + x_2 \geq 6$$
$$x_1 \geq 0$$
$$x_2 \geq 0.$$

Is the region convex? What are the coordinates of the points at the corners of the region?

**3.1.3** Minimize the objective function $x_1 + x_2$ on the region described in Exercise 3.1.2.

**3.1.4** Find inequality constraints on $x_1$ and $x_2$ which will describe the rectangular region with corners at $(0,0)$, $(2,0)$, $(2,3)$, and $(0,3)$.

**3.1.5** Introduce slack variables in the inequality constraints found in Exercise 3.1.4 to produce equality constraints $A\mathbf{x} = \mathbf{b}$. What are $A$ and $\mathbf{b}$?

**3.1.6** Consider the following region in the plane.

$$x_1 + 2x_2 \geq 6$$
$$2x_1 + x_2 \geq 6$$
$$x_1 \geq 0$$
$$x_2 \geq 0.$$

On this feasible set, minimize the following cost functions (if the minimum exists).

(a) $x_1 + x_2$
(b) $3x_1 + x_2$
(c) $x_1 - x_2$

**3.1.7** Maximize the cost function $2x_1 + 7x_2 + 5x_3$ over the set where $x_1 + 2x_2 + 3x_3 = 6$ and $\mathbf{x} \geq \mathbf{0}$.

**3.1.8** Find all solutions $\mathbf{x}$ which minimize the cost function $x_1 + x_2 + 3x_3$ subject to the constraint $x_1 + x_2 + x_3 = 1$ and $\mathbf{x} \geq \mathbf{0}$.

**3.1.9** Consider the linear program stated as "minimize $7x_1 + 9x_2 + 16x_3$ subject to $\mathbf{x} \geq \mathbf{0}$ and $x_1 + 2x_2 + 9x_3 \geq 2$". Re-state this linear program in symmetric form.

**3.1.10** Consider the linear program "maximize $x_1 + x_2 + x_3$ subject to $\mathbf{x} \geq \mathbf{0}$ and

$$\begin{bmatrix} 1 & 2 & 3 \\ 6 & 5 & 4 \end{bmatrix} \begin{bmatrix} x_1 \\ x_2 \\ x_3 \end{bmatrix} = \begin{bmatrix} 1 \\ 2 \end{bmatrix}".$$

Re-state this linear program in symmetric form.

**3.1.11** Consider the linear program stated as "minimize $7y_1 + 2y_2 + 11y_3$ subject to

$$21y_1 + 13y_2 + 4y_3 \geq 11$$
$$17y_1 + 4y_2 + 8y_3 \geq 23$$
$$14y_1 + 2y_2 + 27y_3 \geq 15$$
$$11y_1 + 8y_2 + 18y_3 \geq 29,$$

and $\langle y_1, y_2, y_3 \rangle \geq \langle 0, 0, 0 \rangle$". Re-state this linear program in standard form. Express it in matrix/vector form.

## 3.2 Primal and Dual Problems

This section derives and discusses a result from which the Arbitrage Theorem follows. The **Duality Theorem** is a familiar result to readers having studied linear programming and operations research. Mathematicians frequently benefit from the ability to solve a problem by means of finding the solution to a related, but simpler, problem. This is certainly true of linear programming problems. For every linear programming problem there is an associated linear programming problem known as its **dual**. Henceforth the original problem will be known as the **primal**. For linear programs stated in symmetric form, these paired optimization problems are related in the following ways.

**Primal:** maximize $\mathbf{c}^T \mathbf{x}$ subject to $A\mathbf{x} \leq \mathbf{b}$ and $\mathbf{x} \geq \mathbf{0}$.
**Dual:** minimize $\mathbf{b}^T \mathbf{y}$ subject to $A^T \mathbf{y} \geq \mathbf{c}$ and $\mathbf{y} \geq \mathbf{0}$.

Table 3.1    The symmetric relationship between the primal and dual linear programs.

$$\max_{\mathbf{x}} \mathbf{c}^T \mathbf{x}$$

|              |              |       | $(x_1 \geq 0)$ | $(x_2 \geq 0)$ | $\cdots$ | $(x_n \geq 0)$ |            |
|--------------|--------------|-------|----------------|----------------|----------|----------------|------------|
|              |              |       | $x_1$          | $x_2$          | $\cdots$ | $x_n$          |            |
|              | $(y_1 \geq 0)$ | $y_1$ | $a_{11}$       | $a_{12}$       | $\cdots$ | $a_{1n}$       | $\leq b_1$ |
| $\min_{\mathbf{y}} \mathbf{b}^T \mathbf{y}$ | $(y_2 \geq 0)$ | $y_2$ | $a_{21}$       | $a_{22}$       | $\cdots$ | $a_{2n}$       | $\leq b_2$ |
|              | $\vdots$     | $\vdots$ | $\vdots$     | $\vdots$       | $\vdots$ | $\vdots$       | $\vdots$   |
|              | $(y_m \geq 0)$ | $y_m$ | $a_{m1}$       | $a_{m2}$       | $\cdots$ | $a_{mn}$       | $\leq b_m$ |
|              |              |       | $\geq c_1$     | $\geq c_2$     | $\cdots$ | $\geq c_n$     |            |

Note that:

(1) The objective of maximization in the primal is replaced with the objective of minimization in the dual.

(2) The unknown of the dual is a vector $\mathbf{y}$ with $m$ components where $m$ is the number of inequality constraints of the primal.

(3) The vector $\mathbf{b}$ moves from the constraint of the primal to the objective function of the dual.

(4) The vector $\mathbf{c}$ moves from the objective of the primal to the constraint of the dual.

(5) The constraints of the dual are greater than or equal inequalities and there are $n$ of them where $n$ is the dimension of the decision vector $\mathbf{x}$.

The relationship between the primal and the dual for symmetric linear programs is summarized in Table 3.1 (adapted from [Winston (1994)]).

**Example 3.3.** Express the dual of the following primal problem.

Maximize $x_1 - 3x_2$ subject to $\mathbf{x} \geq \mathbf{0}$ and

$$x_1 + x_2 \leq 3$$
$$x_1 - x_2 \leq 3$$
$$x_1 + 3x_2 \leq 5.$$

**Solution.** The dual problem is to minimize $3y_1 + 3y_2 + 5y_3$ subject to $\mathbf{y} \geq \mathbf{0}$ and

$$y_1 + y_2 + y_3 \geq 1$$
$$y_1 - y_2 + 3y_3 \geq -3.$$

When the primal problem is stated in general form, the dual problem can also be formulated (after the primal is first converted to its equivalent symmetric form). Consider the general linear program of maximizing $\mathbf{c}^T\mathbf{x}$ subject to the constraints in Eqs. (3.1)–(3.3) and for which the decision variables are unrestricted in sign. The general linear problem can be recast as a problem in symmetric form with the constraints in Eq. (3.4). The dual of this linear program is

$$\text{minimize } (\mathbf{b}, -\hat{\mathbf{b}}, \tilde{\mathbf{b}}, -\tilde{\mathbf{b}})^T(\mathbf{y}, \hat{\mathbf{y}}, \tilde{\mathbf{y}}^+, \tilde{\mathbf{y}}^-),$$

subject to the inequality constraints,

$$\begin{bmatrix} A^T & -\hat{A}^T & \tilde{A}^T & -\tilde{A}^T \\ -A^T & \hat{A}^T & -\tilde{A}^T & \tilde{A}^T \end{bmatrix} \begin{bmatrix} \mathbf{y} \\ \hat{\mathbf{y}} \\ \tilde{\mathbf{y}}^+ \\ \tilde{\mathbf{y}}^- \end{bmatrix} \geq \begin{bmatrix} \mathbf{c} \\ -\mathbf{c} \end{bmatrix},$$

where $\mathbf{y} \geq \mathbf{0}$ (with $r$ components), $\hat{\mathbf{y}} \geq \mathbf{0}$ (with $s$ components), $\tilde{\mathbf{y}}^+ \geq \mathbf{0}$ (with $t$ components), and $\tilde{\mathbf{y}}^- \geq \mathbf{0}$ (with $t$ components also). If $\tilde{\mathbf{y}}$ is defined as $\tilde{\mathbf{y}} = \tilde{\mathbf{y}}^+ - \tilde{\mathbf{y}}^-$ then $\tilde{\mathbf{y}}$ is unrestricted in sign and the dual constraints can be expressed as

$$\begin{bmatrix} A^T & -\hat{A}^T & \tilde{A}^T \\ -A^T & \hat{A}^T & -\tilde{A}^T \end{bmatrix} \begin{bmatrix} \mathbf{y} \\ \hat{\mathbf{y}} \\ \tilde{\mathbf{y}} \end{bmatrix} \geq \begin{bmatrix} \mathbf{c} \\ -\mathbf{c} \end{bmatrix}.$$

Note that this set of inequalities implies

$$\begin{bmatrix} A^T & -\hat{A}^T & \tilde{A}^T \end{bmatrix} \begin{bmatrix} \mathbf{y} \\ \hat{\mathbf{y}} \\ \tilde{\mathbf{y}} \end{bmatrix} = \mathbf{c}.$$

Two observations from this exercise will be important later when proving the Arbitrage Theorem. First, decision variables unrestricted in sign in the primal (dual) problem induce equality constraints in the dual (primal) problem. Second, equality constraints in the primal (dual) problem induce decision variables which are unrestricted in sign in the dual (primal) problem.

Given a primal or a dual problem it is a routine matter to construct its partner. The primal and the dual form a set of "fraternal twin" problems. The following theorem will shed some light on their relationship.

**Theorem 3.1.** *The dual of the dual is the primal.*

*Proof.* Starting with the dual problem in symmetric form,

$$\text{minimize } \mathbf{b}^T\mathbf{y} \text{ subject to } A^T\mathbf{y} \geq \mathbf{c} \text{ and } \mathbf{y} \geq \mathbf{0},$$

re-write the dual as a maximization problem with less than or equal constraints:

$$\text{maximize } (-\mathbf{b})^T \mathbf{y} \text{ subject to } (-A)^T \mathbf{y} \leq -\mathbf{c} \text{ and } \mathbf{y} \geq \mathbf{0}.$$

Now the dual of this problem (*i.e.*, the dual of the dual) is

$$\text{minimize } (-\mathbf{c})^T \mathbf{x} \text{ subject to } ((-A)^T)^T \mathbf{x} \geq -\mathbf{b} \text{ and } \mathbf{x} \geq \mathbf{0}.$$

This problem is logically equivalent to the problem,

$$\text{maximize } \mathbf{c}^T \mathbf{x} \text{ subject to } A\mathbf{x} \leq \mathbf{b} \text{ and } \mathbf{x} \geq \mathbf{0},$$

which is the primal problem in symmetric form.        □

It should be no surprise that the solutions to the primal and dual are related.

**Theorem 3.2 (Weak Duality Theorem).** *If* $\mathbf{x}$ *and* $\mathbf{y}$ *are the feasible solutions of the primal and dual problems respectively, then* $\mathbf{c}^T \mathbf{x} \leq \mathbf{b}^T \mathbf{y}$. *If* $\mathbf{c}^T \mathbf{x} = \mathbf{b}^T \mathbf{y}$ *then these solutions are optimal for their respective problems.*

*Proof.* Feasible solutions to the primal and the dual problems must satisfy the constraints $A\mathbf{x} \leq \mathbf{b}$ with $\mathbf{x} \geq \mathbf{0}$ (for the primal problem) and $A^T \mathbf{y} \geq \mathbf{c}$ with $\mathbf{y} \geq \mathbf{0}$ (for the dual). Multiply the constraint in the dual by $\mathbf{x}^T$ to obtain

$$\mathbf{x}^T A^T \mathbf{y} \geq \mathbf{x}^T \mathbf{c} \iff \mathbf{c}^T \mathbf{x} \leq \mathbf{y}^T A \mathbf{x}.$$

Multiply the constraint in the primal by $\mathbf{y}^T$ to find

$$\mathbf{y}^T A \mathbf{x} \leq \mathbf{y}^T \mathbf{b} = \mathbf{b}^T \mathbf{y}.$$

Note that the directions of the inequalities are preserved because each component of $\mathbf{x}$ and $\mathbf{y}$ is non-negative. Combining these last two inequalities produces

$$\mathbf{c}^T \mathbf{x} \leq \mathbf{y}^T A \mathbf{x} \leq \mathbf{b}^T \mathbf{y}. \tag{3.5}$$

Therefore $\mathbf{c}^T \mathbf{x} \leq \mathbf{b}^T \mathbf{y}$.

Now suppose there exist solutions $\mathbf{x}^*$ and $\mathbf{y}^*$ to the primal and dual problems respectively for which $\mathbf{c}^T \mathbf{x}^* = \mathbf{b}^T \mathbf{y}^*$. Let $\mathbf{x}$ be any feasible solution to the primal problem, then

$$\mathbf{c}^T \mathbf{x} \leq \mathbf{b}^T \mathbf{y}^* = \mathbf{c}^T \mathbf{x}^*,$$

by Eq. (3.5) and implying that the maximum of the objective function for the primal problem is achieved at $\mathbf{x}^*$. Likewise if $\mathbf{y}$ is any feasible solution to the dual problem, then by Eq. (3.5)

$$\mathbf{b}^T \mathbf{y} \geq \mathbf{c}^T \mathbf{x}^* = \mathbf{b}^T \mathbf{y}^*,$$

meaning that the minimum of the objective function for the dual problem is achieved at $\mathbf{y}^*$.        □

One consequence of the Weak Duality Theorem is that if $\mathbf{x}$ is a feasible solution for the primal and $\mathbf{y}$ is a feasible solution for the dual, then the linear expression for the primal $\mathbf{c}^T\mathbf{x}$ is bounded above and the linear expression for the dual $\mathbf{b}^T\mathbf{y}$ is bounded below. When trying to maximize (minimize) a linear expression for the primal (dual) problem, it is helpful to know that there exists an upper (a lower) bound on this expression. The converse of Thm. 3.2 is not true. The Weak Duality Theorem does not guarantee there exist feasible solutions $\mathbf{x}$ and $\mathbf{y}$ to the primal and dual linear programs for which $\mathbf{c}^T\mathbf{x} = \mathbf{b}^T\mathbf{y}$. However the Strong Duality Theorem or simply, the Duality Theorem presented next does imply this condition. This theorem was originally proved in [Gale *et al.* (1951)].

**Theorem 3.3 (Duality Theorem).** *One and only one of the following four cases can be true.*

(i) *There exist optimal solutions for both the primal and dual problems with the maximum of $\mathbf{c}^T\mathbf{x}$ equal to the minimum of $\mathbf{b}^T\mathbf{y}$.*

(ii) *There exists no feasible solution to the primal problem and the dual problem has feasible solutions for which the minimum of $\mathbf{b}^T\mathbf{y}$ approaches $-\infty$.*

(iii) *There exists no feasible solution to the dual problem and the primal problem has feasible solutions for which the maximum of $\mathbf{c}^T\mathbf{x}$ approaches $\infty$.*

(iv) *Neither the primal nor the dual problem has a feasible solution.*

The reader should observe that the four conditions are mutually exclusive, and thus only one can be true for a given pair of primal and dual problems. The proof of the Duality Theorem relies on **Farkas' Lemma** [Franklin (1980)]. Background theory supporting Farkas' Lemma, a statement of the lemma (Lemma A.1), and a useful corollary (Cor. A.1) are found in App. A.

*Proof.* First consider the primal and dual problems stated in symmetric form.

**Primal:** maximize $\mathbf{c}^T\mathbf{x}$ subject to $A\mathbf{x} \leq \mathbf{b}$ and $\mathbf{x} \geq \mathbf{0}$.
**Dual:** minimize $\mathbf{b}^T\mathbf{y}$ subject to $A^T\mathbf{y} \geq \mathbf{c}$ and $\mathbf{y} \geq \mathbf{0}$.

Suppose there exist feasible solutions $\mathbf{x}$ and $\mathbf{y}$ to the primal and dual linear programs respectively. By Thm. 3.2 $\mathbf{c}^T\mathbf{x} \leq \mathbf{b}^T\mathbf{y}$ and thus the objective functions of the primal and dual problems are bounded. Let the minimum of the dual linear program be attained at $\mathbf{d}$ and suppose that for all feasible solutions $\mathbf{x}$ to the primal it is the case that $\mathbf{c}^T\mathbf{x} < \mathbf{b}^T\mathbf{d}$. Consequently, there exists no feasible solution $\mathbf{x}$ to the primal linear program for which $\mathbf{c}^T\mathbf{x} \geq \mathbf{b}^T\mathbf{d}$. This is equivalent to the system of inequalities,

$$\begin{bmatrix} A \\ -\mathbf{c}^T \end{bmatrix} \mathbf{x} \leq \begin{bmatrix} \mathbf{b} \\ -\mathbf{b}^T\mathbf{d} \end{bmatrix}. \tag{3.6}$$

where the real matrix on the left-hand side of Eq. (3.6) is $(m+1) \times n$ and the vector on the right-hand side has $m+1$ real components. Using the assumption that there exists no $\mathbf{x}$ such that $\mathbf{c}^T\mathbf{x} \geq \mathbf{b}^T\mathbf{d}$ and applying Cor. A.1 to Eq. (3.6) demonstrates that statement (i) of Cor. A.1 does not hold and therefore statement (ii) of Cor. A.1 is true. As a result there exists $\mathbf{y} \in \mathbb{R}^{m+1}$ such that $\mathbf{y} \geq \mathbf{0}$ and

$$\begin{bmatrix} A^T \big| -\mathbf{c} \end{bmatrix} \begin{bmatrix} y_1 \\ y_2 \\ \vdots \\ y_m \\ y_{m+1} \end{bmatrix} = \mathbf{0} \text{ and } \begin{bmatrix} \mathbf{b}^T \big| -\mathbf{b}^T\mathbf{d} \end{bmatrix} \begin{bmatrix} y_1 \\ y_2 \\ \vdots \\ y_m \\ y_{m+1} \end{bmatrix} = -1.$$

If $y_{m+1} = 0$ then the two equations above can be re-written as

$$A^T \begin{bmatrix} y_1 \\ y_2 \\ \vdots \\ y_m \end{bmatrix} = \mathbf{0} \text{ and } \mathbf{b}^T \begin{bmatrix} y_1 \\ y_2 \\ \vdots \\ y_m \end{bmatrix} = -1.$$

Thus if $y_{m+1} = 0$ then statement (ii) of Cor. A.1 holds (*i.e.*, there exists $\mathbf{y} \in \mathbb{R}^m$ with $\mathbf{y} \geq \mathbf{0}$ such that $\mathbf{b}^T\mathbf{y} = -1$ and $A^T\mathbf{y} = \mathbf{0}$) which implies statement (i) of Cor. A.1 does not hold, in other words, there exists no $\mathbf{x} \in \mathbb{R}^n$ for which $A\mathbf{x} \leq \mathbf{b}$. This contradicts the assumption that a feasible solution to the primal linear program exists, therefore $y_{m+1} > 0$. Define the vector $\mathbf{z} \in \mathbb{R}^m$ as

$$\mathbf{z} = \frac{1}{y_{m+1}} \begin{bmatrix} y_1 \\ y_2 \\ \vdots \\ y_m \end{bmatrix}.$$

Note $\mathbf{z} \geq \mathbf{0}$ and

$$\begin{bmatrix} A^T \big| -\mathbf{c} \end{bmatrix} \begin{bmatrix} \mathbf{z} \\ 1 \end{bmatrix} = A^T\mathbf{z} - \mathbf{c} = \mathbf{0} \implies A^T\mathbf{z} \geq \mathbf{c},$$

and consequently $\mathbf{z}$ is a feasible solution for the dual linear program. However,

$$\left[\,\mathbf{b}^T\,\middle|-\mathbf{b}^T\mathbf{d}\,\right]\begin{bmatrix}\mathbf{z}\\1\end{bmatrix}=\frac{1}{y_{m+1}}\left[\,\mathbf{b}^T\,\middle|-\mathbf{b}^T\mathbf{d}\,\right]\begin{bmatrix}y_1\\y_2\\\vdots\\y_m\\y_{m+1}\end{bmatrix}=\frac{-1}{y_{m+1}}$$

$$\mathbf{b}^T\mathbf{z}=\mathbf{b}^T\mathbf{d}-\frac{1}{y_{m+1}}<\mathbf{b}^T\mathbf{d},$$

which contradicts the assumption that the minimum of the dual program is attained at $\mathbf{d}$. Therefore the optimal values of the primal and dual linear programs must be equal.

Now suppose the primal linear program has no feasible solutions (*i.e.*, for all $\mathbf{x}\in\mathbb{R}^n$ with $\mathbf{x}\geq\mathbf{0}$ that $A\mathbf{x}>\mathbf{b}$) and $\mathbf{y}$ is a feasible solution to the dual. In this case statement (A.1) of Cor. A.1 does not hold and consequently statement (A.1) of Cor. A.1 does hold. Thus there exists $\mathbf{z}\in\mathbb{R}^m$ such that $\mathbf{z}\geq\mathbf{0}$, $A^T\mathbf{z}=\mathbf{0}$, and $\mathbf{b}^T\mathbf{z}=-1$. The vectors $\mathbf{w}=\mathbf{y}+\lambda\mathbf{z}$ are feasible for the dual program when $\lambda\geq0$ since $\mathbf{w}\geq\mathbf{0}$ and

$$A^T\mathbf{w}=A^T\mathbf{y}+\lambda A^T\mathbf{z}=A^T\mathbf{y}+\lambda(\mathbf{0})=A^T\mathbf{y}\geq\mathbf{c}.$$

The objective function of the dual linear program is unbounded since

$$\lim_{\lambda\to\infty}\mathbf{b}^T\mathbf{w}=\lim_{\lambda\to\infty}\left(\mathbf{b}^T\mathbf{y}+\lambda\mathbf{b}^T\mathbf{z}\right)=\lim_{\lambda\to\infty}\left(\mathbf{b}^T\mathbf{y}-\lambda\right)=-\infty.$$

The cases of feasible primal linear program/infeasible dual linear program and infeasible primal and dual linear programs are left as exercises. $\square$

**Example 3.4.** Consider the linear program of maximizing $4x_1+3x_2$ subject to $x_1+x_2\leq2$ and $x_1,x_2\geq0$. Use the dual of this linear program to solve the primal program.

**Solution.** The dual linear program can be expressed as minimize $2y_1$ subject to $y_1\geq3$ and $y_1\geq4$ and $y_1\geq0$. While both the primal and dual problems may appear simple, the dual is trivial. The minimum value of $y_1$ subject to the constraints must be $y_1=4$. According to the Weak Duality Theorem, the minimum of the objective function of the primal must be at least $2(4)=8$. Comparison of the feasible set of solutions to the primal with the level set $4x_1+3x_2=8$ yields the solution $\mathbf{x}=(2,0)^T$. See Fig. 3.3.

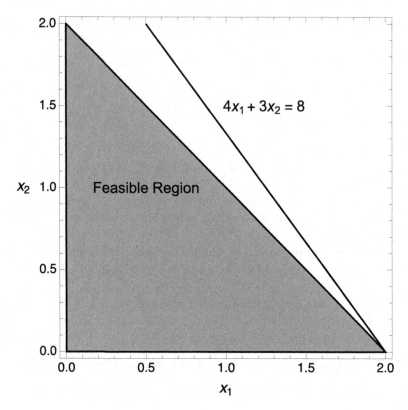

Fig. 3.3 The maximum of the objective function $4x_1 + 3x_2$ subject to the constraints $x_1 + x_2 \leq 2$ and $x_1, x_2 \geq 0$, occurs at the point with coordinates $(x_1, x_2) = (2, 0)$.

Another consequence of Thm. 3.3 is the strengthening of the inequality derived in Eq. (3.5). When $\mathbf{x}$ and $\mathbf{y}$ are optimal solutions to the primal and dual problems respectively, then

$$\mathbf{c}^T \mathbf{x} = \mathbf{y}^T A \mathbf{x} = \mathbf{b}^T \mathbf{y}. \tag{3.7}$$

This implies the following relationship between the optimal solutions.

$$(A^T \mathbf{y} - \mathbf{c})^T \mathbf{x} = 0. \tag{3.8}$$

Since $\mathbf{x}$ is feasible then each of its components is non-negative. Likewise the assumption in the dual, $A^T \mathbf{y} \geq \mathbf{c}$, implies that each component of the vector $A^T \mathbf{y} - \mathbf{c}$ is non-negative. As a consequence of Eq. (3.8) vector $\mathbf{x}$ must be zero in every component for which vector $A^T \mathbf{y} - \mathbf{c}$ is positive and vice versa. This can be stated in the form of a theorem.

**Theorem 3.4 (Complementary Slackness).** *Optimality in the primal and dual problems requires either* $x_j = 0$ *or* $(A^T\mathbf{y} - \mathbf{c})_j = 0$ *for each* $j = 1, 2, \ldots, n$.

Due to the symmetry between the primal and dual problems it can also be stated that either $y_i = 0$ or $(A\mathbf{x} - \mathbf{b})_i = 0$ for each $i = 1, 2, \ldots, m$. The variable $x_j$ is the complementary variable to the variable $(A^T\mathbf{y} - \mathbf{c})_j$. Likewise the variable $y_i$ is the complementary variable to the variable $(A\mathbf{x} - \mathbf{b})_i$.

**Example 3.5.** Find the optimal value of the linear programming problem, maximize $\mathbf{c}^T\mathbf{x} = -3x_1 + 2x_2 - x_3 + 3x_4$ subject to $\mathbf{x} \geq \mathbf{0}$ and

$$\begin{bmatrix} 1 & 1 & -1 & 0 \\ -2 & 0 & 1 & 1 \end{bmatrix} \begin{bmatrix} x_1 \\ x_2 \\ x_3 \\ x_4 \end{bmatrix} \leq \begin{bmatrix} 5 \\ 3 \end{bmatrix}. \tag{3.9}$$

**Solution.** Since the set of points described by the constraints exists in four-dimensional space, it will be more difficult to use geometrical and graphical thinking to analyze this problem. The dual of this linear program may used to solve this primal problem. The dual problem is to minimize $\mathbf{b}^T\mathbf{y} = 5y_1 + 3y_2$ subject to

$$\begin{bmatrix} 1 & -2 \\ 1 & 0 \\ -1 & 1 \\ 0 & 1 \end{bmatrix} \begin{bmatrix} y_1 \\ y_2 \end{bmatrix} \geq \begin{bmatrix} -3 \\ 2 \\ -1 \\ 3 \end{bmatrix}. \tag{3.10}$$

The space of points on which the dual is to be optimized exists in two-dimensional space and thus is easily pictured. The constraints of the dual can be thought of as a system of inequalities.

$$y_1 - 2y_2 \geq -3$$
$$y_1 \geq 2$$
$$-y_1 + y_2 \geq -1$$
$$y_2 \geq 3.$$

The solution to this set of inequalities can be pictured as the shaded region shown in Fig. 3.4. The minimum of the objective function for the dual occurs at the point with coordinates $(y_1, y_2) = (3, 3)$. Thus the minimum value is 24 which will also be the maximum value of the primal problem's objective

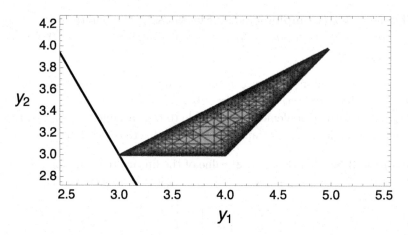

Fig. 3.4   The shaded region in the plot denotes the constrained set of points from Eq. (3.10) for the dual problem.

function. At the optimal point for the dual problem, strict inequality is present in the second and third constraints since

$$y_1 = 3 > 2$$
$$-y_1 + y_2 = 0 > -1.$$

By Thm. 3.4 the second and third components of $\mathbf{x}$ in the primal problem must be zero. Therefore the primal can be recast as maximizing $-3x_1 + 3x_4$ subject to $x_1 \geq 0$, $x_4 \geq 0$ and

$$\begin{bmatrix} 1 & 1 & -1 & 0 \\ -2 & 0 & 1 & 1 \end{bmatrix} \begin{bmatrix} x_1 \\ 0 \\ 0 \\ x_4 \end{bmatrix} = \begin{bmatrix} x_1 \\ -2x_1 + x_4 \end{bmatrix} \leq \begin{bmatrix} 5 \\ 3 \end{bmatrix}.$$

By inspection $x_1 = 5$ and $x_4 = 13$. Consequently the maximum of the objective function for the primal is seen to be 24 and it occurs at the point with coordinates $(x_1, x_2, x_3, x_4) = (5, 0, 0, 13)$.

In this chapter a great deal of background knowledge in linear programming has been developed. Linear programming is a vast field of study in its own right. The previous material is merely an introduction intended to support the proof of the Fundamental Theorem of Finance in the next section.

## Exercises

**3.2.1** Show that if feasible solutions exist to the primal linear program and the dual program is infeasible then the objective function of the primal is unbounded.

**3.2.2** Given an example of a primal/dual pair of linear programs in symmetric form, both of which are infeasible.

**3.2.3** Consider the standard linear program of maximizing $2x_1 - 3x_2$ subject to the constraints:

$$3x_1 + 5x_2 = 4$$
$$-2x_1 + 4x_2 = -3.$$

Show that this linear program is not feasible.

**3.2.4** Consider the linear program of maximizing $2x_1 - 3x_2$ subject to the constraints:

$$-2x_1 + x_2 \leq -3$$
$$x_1 - 2x_2 \leq 3.$$

Show that this linear program is feasible, but has no optimal solution.

**3.2.5** State the dual of the linear program in Exercise 3.2.4. Since the primal problem has no optimal solution, what does the Duality Theorem (Thm. 3.3) imply about the dual problem? Verify this condition directly from the dual problem.

**3.2.6** Consider the linear program of maximizing $x_1 + 3x_2$ subject to $\langle x_1, x_2 \rangle \geq \mathbf{0}$ and

$$-3x_1 + 6x_2 \leq -1$$
$$x_1 - 3x_2 \leq 2.$$

State the dual to this primal program.

**3.2.7** Consider the linear program of maximizing $2x_1 - x_2$ subject to $\langle x_1, x_2 \rangle \geq \mathbf{0}$ and

$$-4x_1 + x_2 \leq 2$$
$$x_1 - x_2 \leq 1$$
$$2x_1 + x_2 \leq 5.$$

State the dual to this primal program.

**3.2.8** Consider the linear program of maximizing $x_1 + x_2$ subject to $\langle x_1, x_2 \rangle \geq \mathbf{0}$ and

$$-2x_1 - x_2 \leq 2$$
$$-x_1 - x_2 \leq 1.$$

Does the linear program have an optimal solution? *Hint*: Use the dual linear program.

**3.2.9** Write down the dual problem of the following linear programming problem: minimize $x_1 + x_2 + x_3$ subject to $2x_1 + x_2 = 4$ and $x_3 \leq 6$ with $x_i \geq 0$ for $i = 1, 2, 3$.

**3.2.10** Find the solutions to the primal and dual problems of Exercise 3.2.9.

**3.2.11** Write down the dual problem to the following linear programming problem: maximize $2y_1 + 4y_3$ subject to $y_1 + y_2 \leq 1$, $y_2 \geq 0$, and $y_2 + 2y_3 \leq 1$.

**3.2.12** Find the solutions to the primal and dual problems of Exercise 3.2.11.

**3.2.13** Minimize the objective function $x_1 + x_2 + 2x_3$ on the set where $x_1 + 2x_2 + 3x_3 \geq 15$ and $x_i \geq 0$ for $i = 1, 2, 3$.

**3.2.14** Minimize the objective function $7x_2 + 9x_3$ subject to the constraints $x_1 + x_2 + x_3 \geq 5$ and $x_2 + x_3 + 2x_4 \geq 1$ with $x_i \geq 0$ for $i = 1, 2, 3$.

**3.2.15** Consider the linear program of maximizing $x_1 + 2x_2 + x_3 + x_4$ subject to $\langle x_1, x_2, x_3, x_4 \rangle \geq \mathbf{0}$ and

$$x_1 + x_2 + 2x_3 + x_4 \leq 3$$
$$3x_1 + 2x_2 + 2x_3 + 3x_4 \leq 7.$$

Find the optimal solution. *Hint*: Use the dual linear program.

**3.2.16** A metal worker can make knives or scissors. A knife takes 15 minutes to make and a pair of scissors takes 25 minutes to make. Each knife requires 6 ounces of steel to make, while a pair of scissors requires 8 ounces of steel. The metal worker can sell each knife for a profit of $2.25 and each pair of scissors for a profit of $3.50. If the worker has 40 hours to make knives and scissors and 52 pounds of steel, how many knives and scissors should be made to maximize the worker's profit?

**3.2.17** A farmer has 440 acres of land on which two crops may be planted, corn and saw grass. Each acre of corn costs $75 to plant and each acre of saw grass costs $50 to plant. Each acre of corn planted requires 110 bushels of storage while each acres of saw grass planted requires 30 bushels of storage. The farmer has access to 30,000 bushels of storage space. Each acre of corn will generate a profit of $100 and each acre of saw grass will generate a profit of $80. The farmer has $50,000 with which to plant. How much of each crop should be planted to maximize the profit of the farmer?

**3.2.18** A company sells two sizes of tablet personal computer, a 7-inch model and a 10-inch model. The tablets can be manufactured in either Mausia or Dianfa (fictional countries). The manufacturing facility in Mausia has an operating budget of $50,000 per day and can produce a total of 250 tablet PCs per day and the facility in Dianfa has an operating budget of $60,000 per day and can make 290 tablet PCs per day. In Mausia it costs

$175 to make a 7-inch tablet and $250 to make a 10-inch tablet. In Dianfa it costs $205 to make a 7-inch tablet and $255 to make a 10-inch tablet. Each 7-inch tablet sold generates a profit of $65 and each 10-inch tablet sold brings in a profit of $85. The company needs no more than 250 of the 7-inch tablet and 290 of the 10-inch tablet each day. Describe the linear program of finding the maximum profit in the form of a primal problem.

**3.2.19** Let $\mathbf{x}$ be a non-negative vector with $n$ components, $\mathbf{y}$ be a non-negative vector of $m$ components, and $A$ be a fixed $m \times n$ matrix. Define the function

$$\phi(\mathbf{x}, \mathbf{y}) = \mathbf{c}^T \mathbf{x} + \mathbf{b}^T \mathbf{y} - \mathbf{x}^T A^T \mathbf{y},$$

where $\mathbf{c}$ is a fixed vector of $n$ components and $\mathbf{b}$ is a fixed vector of $m$ components. The point $(\hat{\mathbf{x}}, \hat{\mathbf{y}})$ is a **saddle point** of $\phi(\mathbf{x}, \mathbf{y})$ if

$$\phi(\mathbf{x}, \hat{\mathbf{y}}) \leq \phi(\hat{\mathbf{x}}, \hat{\mathbf{y}}) \leq \phi(\hat{\mathbf{x}}, \mathbf{y}),$$

for all non-negative vectors $\mathbf{x}$ and $\mathbf{y}$.

Consider the primal/dual pair of linear programs,

**Primal:** maximize $\mathbf{c}^T \mathbf{x}$ subject to $A\mathbf{x} \leq \mathbf{b}$ and $\mathbf{x} \geq \mathbf{0}$.
**Dual:** minimize $\mathbf{b}^T \mathbf{y}$ subject to $A^T \mathbf{y} \geq \mathbf{c}$ and $\mathbf{y} \geq \mathbf{0}$.

Show that if $\hat{\mathbf{x}}$ and $\hat{\mathbf{y}}$ are optimal solutions to the primal and dual problems respectively, then $(\hat{\mathbf{x}}, \hat{\mathbf{y}})$ is a saddle point of $\phi(\mathbf{x}, \mathbf{y})$.

**3.2.20** Show that if $(\hat{\mathbf{x}}, \hat{\mathbf{y}})$ is a saddle point of $\phi(\mathbf{x}, \mathbf{y})$ defined in Exercise 3.2.19 then $\hat{\mathbf{x}}$ and $\hat{\mathbf{y}}$ are optimal solutions of their corresponding primal and dual problems.

## 3.3 The Fundamental Theorem of Finance

The linear programming developed earlier in this chapter now permits the statement and justification of the main result, the Arbitrage Theorem, often thought of as the Fundamental Theorem of Finance.

Consider an experiment with $m$ possible outcomes numbered 1 through $m$. Suppose $n$ wagers (numbered 1 through $n$) can be placed on the outcomes of the experiment. Let $r_{ji}$ be the return for a unit bet on wager $i$ when the outcome of the experiment is $j$. For the sake of conciseness let the $m \times n$ matrix $R = [r_{ji}]$ for $j = 1, 2, \ldots, m$ and $i = 1, 2, \ldots, n$. The vector $\mathbf{x} = \langle x_1, x_2, \ldots, x_n \rangle$ is called a **betting strategy**. Component $x_i$ is the amount placed on wager $i$. In the context of finance, the components of vector $\mathbf{x}$ may be thought of as the purchases and sales of potential investments. The aggregate return from this betting strategy when outcome $j$

occurs is then $(R\mathbf{x})_j$. The Arbitrage Theorem to be proved in this section states that the probabilities of the $m$ outcomes of the experiment are such that for each bet the expected value of the payoff is zero, or there exists a betting strategy for which the payoff is positive regardless of the outcome of the experiment.

**Theorem 3.5.** *If one of the following statements is true, the other must be false.*

(i) *There is a $m$-dimensional vector of probabilities* $\mathbf{p} = \langle p_1, p_2, \ldots, p_m \rangle$ *for which* $\mathbf{p}^T R = \mathbf{0}^T$.

(ii) *There is a betting strategy* $\mathbf{x} = \langle x_1, x_2, \ldots, x_n \rangle$ *for which* $R\mathbf{x} > \mathbf{0}$.

*Proof.* Suppose (i) is true and let $\mathbf{x} = \langle x_1, x_2, \ldots, x_n \rangle$ be any betting strategy. Then $\mathbf{p}^T R \mathbf{x} = 0$ and since $\mathbf{p} \geq \mathbf{0}$ and $\mathbf{p} \neq \mathbf{0}$ this implies $R\mathbf{x} \leq \mathbf{0}$ and (ii) is not true.

Now suppose (ii) is true. If statement (i) is also true then (ii) is false, a contradiction. Thus at most one of the two statements is true. □

The situation in which statement (i) is true may be interpreted as stating that all betting strategies have an expected return of zero since

$$\mathbf{p}^T R \mathbf{x} = \sum_{j=1}^{m} p_j \left( \sum_{i=1}^{n} x_i r_{ji} \right) = \mathbb{E}\left( \sum_{i=1}^{n} x_i r_{ji} \right) = \mathbb{E}\,(R\mathbf{x}) = 0.$$

Consider a collection of $n+1$ investments whose values at time $t = 0$ are denoted as $s_i(0)$ for $i = 0, 1, 2, \ldots, n$. The investment $s_0$ is assumed to have a riskless return. Its accumulated future value will be $s_0(\tau) = s_0(0)e^{r\tau}$ for all $\tau > 0$ where $r$ is the risk-free interest rate. Without loss of generality let $s_0(0) = 1$. Let $y_i$ be the number of shares of investment $i$ bought or sold at time $t = 0$ for $i = 0, 1, 2, \ldots, n$. If $y_i < 0$ then the investor has sold the $i$th asset, while if $y_i > 0$ the investor has purchased the asset. Since the return on investment $s_0$ is deterministic (non-random) it will be treated separately from the remaining $n$ investments. Let vector $\mathbf{s}(0) = \langle s_1(0), s_2(0), \ldots, s_n(0) \rangle$ be the vector of investment or asset prices while vector $\mathbf{y} = \langle y_1, y_2, \ldots, y_n \rangle$ (a vector of unknowns) is called the investor's **portfolio**. The value of the investor's portfolio without the risk-free investment will be $(\mathbf{s}(0))^T \mathbf{y}$.

At time $t = \tau$ the marketplace of investment values will be in one of $m$ mutually exclusive states described by the sample space $\Omega = \{\omega_1, \omega_2, \ldots, \omega_m\}$. The value of investment $i$ in state $j$ of the market at time

$t = \tau > 0$ can be described by the random variables $S_i^j(\tau)$ for $i = 1, 2, \ldots, n$. Keep in mind that $S_0^j(\tau) = (1)e^{r\tau}$ with probability 1 for all $j = 1, 2, \ldots, m$. The value of the investor's portfolio at time $t = \tau$ when the market is in state $j$ can be expressed as $(\mathbf{s}^j(\tau))^T\mathbf{y}$. An **arbitrage portfolio** is a $\mathbf{y}$ such that $(\mathbf{s}(0))^T\mathbf{y} \le 0$ and $(\mathbf{s}^j(\tau))^T\mathbf{y} > 0$ for all $j = 1, 2, \ldots, m$. An arbitrage portfolio can be interpreted as a portfolio with a non-positive value at $t = 0$ which will have a positive value at $t = \tau > 0$ regardless of the state of the market. An arbitrage portfolio creates wealth out of nothing.

Assume the probability the market will be in state $\omega_j$ is $p_j > 0$ for $j = 1, 2, \ldots, m$. The probabilities are assumed to be strictly positive since there is no need to consider future market states that have zero probability of occurrence. Naturally it is still assumed that $\sum_{j=1}^m p_j = 1$. A probability distribution $p_j^* = \mathbb{P}(\omega_j)$ is called a **risk-neutral probability distribution** if the price of investment $s_i(0)$ (the price at time $t = 0$) is the present value of the expected value of the asset at time $t = \tau > 0$. This condition must hold for $i = 0, 1, 2, \ldots, n$. The present value is calculated under the risk-free interest rate compounded continuously. The expected value is found using the risk-neutral probability distribution. This can be expressed as

$$s_i(0) = e^{-r\tau} \sum_{j=1}^m p_j^* S_i^j(\tau) = e^{-r\tau}\, \mathbb{E}^*\,(S_i(\tau)). \qquad (3.11)$$

Suppose $\mathbf{p} = \langle p_1, p_2, \ldots, p_m \rangle$ is a risk-neutral probability distribution, then by Eq. (3.11)

$$s_i(0)y_i = e^{-r\tau} \sum_{j=1}^m p_j S_i^j(\tau)y_i \text{ for } i = 0, 1, 2, \ldots, n$$

$$(\mathbf{s}(0))^T\mathbf{y} = e^{-r\tau} \sum_{j=1}^m p_j (S^j(\tau))^T\mathbf{y}.$$

If $(S^j(\tau))^T\mathbf{y} > 0$ for all $j = 1, 2, \ldots, m$ then $(\mathbf{s}(0))^T\mathbf{y} > 0$ as well, or in other words, arbitrage cannot exist. The Arbitrage Theorem stated below implies the converse is true as well.

**Theorem 3.6 (Arbitrage Theorem).** *A risk-neutral probability measure exists if and only if there is no arbitrage.*

*Proof.* Half of this biconditional was established in the paragraph preceding the statement of the theorem. Assume now that no arbitrage is possible for any portfolio.

Define the $n \times m$ matrix $S(\tau)$ as

$$S(\tau) = \begin{bmatrix} S_1^1(\tau) & S_1^2(\tau) & \cdots & S_1^m(\tau) \\ S_2^1(\tau) & S_2^2(\tau) & \cdots & S_2^m(\tau) \\ \vdots & \vdots & & \vdots \\ S_n^1(\tau) & S_n^2(\tau) & \cdots & S_n^m(\tau) \end{bmatrix}.$$

The $i, j$th entry of the matrix is a random variable representing the value of investment $i$ in state $j$ of the market at time $t = \tau > 0$. Define set $U \subset \mathbb{R}^n$ as the set of all vectors of the form

$$\mathbf{u} = e^{-r\tau} S(\tau) \mathbf{p},$$

where $\mathbf{p} = \langle p_1, p_2, \ldots, p_m \rangle$ is some probability distribution on $\Omega$. Vector $\mathbf{u}$ is an $n$-vector of present-valued, expected values of the $n$ risky investments. Set $U$ is bounded (see Sec. A.2). Set $U$ is also convex, since if vectors $\mathbf{u}, \mathbf{v} \in U$ then there exist probability distributions $\mathbf{p}$ and $\mathbf{q}$ such that $\mathbf{u} = e^{-r\tau} S(\tau) \mathbf{p}$ and $\mathbf{v} = e^{-r\tau} S(\tau) \mathbf{q}$ respectively. If $\theta \in (0, 1)$ then $\pi = (1 - \theta)\mathbf{p} + \theta\mathbf{q}$ is a probability distribution on $\Omega$ since $0 \leq \pi_j \leq 1$ for $j = 1, 2, \ldots, m$ and

$$\sum_{j=1}^m \pi_j = \sum_{j=1}^m [(1 - \theta)p_j + \theta q_j] = (1 - \theta) \sum_{j=1}^m p_j + \theta \sum_{j=1}^m q_j = 1 - \theta + \theta = 1.$$

Consequently, $(1 - \theta)\mathbf{u} + \theta\mathbf{v} \in U$. Finally the set $U$ is closed. To confirm closure let $\{\mathbf{u}^{(k)}\}_{k=1}^\infty$ be convergent sequence of vectors in $U$ with limit $\mathbf{u}$. For each vector $\mathbf{u}^{(k)}$ in the sequence there is a corresponding vector of probabilities on $\Omega$ denoted $\mathbf{p}^{(k)}$ such that $\mathbf{u}^{(k)} = e^{-r\tau} S(\tau) \mathbf{p}^{(k)}$ and

$$\lim_{k \to \infty} \mathbf{u}^{(k)} = \lim_{k \to \infty} e^{-r\tau} S(\tau) \mathbf{p}^{(k)}$$
$$= e^{-r\tau} S(\tau) \lim_{k \to \infty} \mathbf{p}^{(k)} = e^{-r\tau} S(\tau) \mathbf{p} \in U.$$

To finish the proof one must show that when no arbitrage portfolio exists then every vector of the $n$ risky investments $\mathbf{s}(0) \in U$. By way of contradiction, suppose no arbitrage portfolio exists and $\mathbf{s}(0) \notin U$. By the Separating Hyperplane theorem (Thm. A.8) there exists a non-zero vector $\mathbf{y} \in \mathbb{R}^n$ and a real number $b$ such that $(\mathbf{s}(0))^T \mathbf{y} < b < \mathbf{u}^T \mathbf{y}$ for all $\mathbf{u} \in U$. This inequality can be expressed in terms of the definition of $\mathbf{u}$ as

$$\sum_{i=1}^n s_i(0)y_i < b < e^{-r\tau} \sum_{i=1}^n (S(\tau)\mathbf{p}_j)_i y_i.$$

Since set $U$ is closed then the inequality holds for probability distributions on $\Omega$ of the form $\mathbf{p} = \mathbf{e}_j$, the $j$th basis vector in the standard basis for $\mathbb{R}^m$. Therefore

$$\sum_{i=1}^{n} s_i(0)y_i < b < e^{-r\tau} \sum_{i=1}^{n} S_i^j(\tau)y_i.$$

Choose quantity $y_0 = -b$ and the following inequalities hold.

$$y_0 + \sum_{i=1}^{n} s_i(0)y_i < 0 < e^{-r\tau}y_0 e^{r\tau} + e^{-r\tau} \sum_{i=1}^{n} S_i^j(\tau)y_i$$

$$\sum_{i=0}^{n} s_i(0)y_i < 0 < e^{-r\tau} \sum_{i=0}^{n} S_i^j(\tau)y_i.$$

This implies $\mathbf{y} = \langle y_0, y_1, y_2, \ldots, y_n \rangle$ is an arbitrage portfolio and hence contradicts the starting assumption. $\quad\square$

## Exercises

**3.3.1** Assume that $S_i^j(\tau) > 0$ for all $j = 1, 2, \ldots, m$ and use Eq. (3.11) to show that $s_i(0) > 0$ for $i = 1, 2, \ldots, n$.

**3.3.2** Suppose that $\mathbf{p}$ is a probability distribution for which the value of the $i$th investment satisfies the following equation.

$$s_i(0) = e^{-r\tau} \sum_{j=1}^{m} S_i^j(\tau)p_j.$$

Show that the rate of return of the $i$th investment is

$$r = -\frac{1}{\tau} \ln \left( \frac{s_i(0)}{\sum_{j=1}^{m} p_j S_i^j(\tau)} \right).$$

**3.3.3** Suppose the $i$th investment has value 1 in all possible future states of the market. Show that the rate of return of the $i$th investment is

$$r = -\frac{1}{\tau} \ln(s_i(0)).$$

**3.3.4** Suppose the value of an asset 1 month from now is given by the random variable $X$ where the probability distribution of $X$ is given in the table below. If the current value of the asset is 53 and assuming no arbitrage, what is the implied risk-free annual interest rate compounded continuously?

| $x$ | 40 | 50 | 60 | 70 |
|---|---|---|---|---|
| $\mathbb{P}(X = x)$ | 0.20 | 0.40 | 0.25 | 0.15 |

**3.3.5** Suppose the value of an asset 2 months from now is given by the random variable $X$ where the probability distribution of $X$ is given in the table below. The risk-free annual interest rate compounded continuously is $r = 3.5\%$. What is the no arbitrage value of the asset?

| $x$ | 90 | 100 | 105 | 110 |
|---|---|---|---|---|
| $\mathbb{P}\,(X = x)$ | 0.25 | 0.35 | 0.20 | 0.20 |

## 3.4   Remarks

The approach taken to establishing the Arbitrage Theorem in this chapter is based on the article [Dybvig and Ross (2008)]. While Thm. 3.6 is called the "Arbitrage Theorem" in this text, some authors apply this name to results equivalent to Thm. 3.5 (for example [Ross (1999)]. Readers should not be concerned over which of the two theorems should rightly be called the Arbitrage Theorem, but instead should keep in mind the important implications. If an asset is not priced correctly, then arbitrage exists. For example if an investment is priced too low, it may be possible to create an arbitrage opportunity by borrowing funds at the risk-free rate in order to purchase the investment with plans to sell it at a later time at a profit. If an asset is priced too high, perhaps it should be sold (in a process called "shorting") and with some of the proceeds of the sale, re-purchased when the price has been set correctly.

Readers often wonder if any true arbitrage opportunities exist. The recent heightened volatility in the financial markets has provided observers with interesting examples of mis-priced financial products. One such example can be drawn from the behavior of the stock of Accenture, PLC (a business consulting corporation, ticker symbol ACN) on May 6, 2010. During the two-minute-long period from 2:47PM to 2:49PM, Accenture stock traded on the New York Stock Exchange (NYSE) fell from nearly $40.00 per share to $0.01 per share and then returned to approximately $40.00. It is unlikely that a fundamental change in the value of the corporation was responsible for the 99.975% drop in the value of its stock in one minute, thus almost certainly for nearly a minute the stock was mis-priced. The reason for the temporary mis-pricing has been attributed to a combination of human error and the un-intended effects of computerized trading. This put downward pressure on the price of Proctor and Gamble stock (ticker symbol PG), one of the 30 stocks making up the Dow Jones Industrial (DJI) average, and thus this average fell as well. Computers monitoring the fall of the DJI average, executed stop-loss sell orders to prevent further losses in case the DJI fell further. Some firms using computer-driven trading suspected that something was amiss in the prices they were monitoring and slowed or stopped electronic trading. This had the effect of removing buyers and sellers from the stock market which led to further disruptions [Mehta *et al.* (2010)].

Many algorithmic trading firms serve as market makers, keeping track of the order book of bid and ask prices for stocks. A common practice of market makers is to enter "stub quotes" as place holders in the order book for the stock for which they create a market. Stub quotes can serve as bid prices, usually set very low, for example at $0.01, so that during the normal course of price fluctuations, there is always an order to buy the stock no matter how low the price may fall. This prevents the occurrence of a situation where no investor is willing to buy the stock at any price. Likewise stub quotes can serve as ask prices, set as high as $100,000 per share, so that there are always parties willing to sell in the market. Stocks should never be traded at the stub quotes; however, when algorithmic trading exhausted the legitimate buyers of Accenture stock on May 6, 2010 the highest remaining bids in the market were the stub quotes at $0.01 and in the absence of human intervention trades were executed at this price. Accenture and its stockholders were not ruined on that day since the exchange overseeing trading canceled these orders. In the end an arbitrage opportunity did not exist for investors in Accenture stock, since buy orders at the temporary low price were not allowed to be executed.

Chapter 4

# Optimal Portfolio Choice

The process and inputs involved in deciding which investments should be part of a portfolio is very complex. This chapter will explore some of the measures of stock and portfolio performance and investor behaviors and preferences which influence these decisions. The notions of interest theory, probability, statistics, and arbitrage introduced in earlier chapters will come together to provide a rational approach to portfolio selection.

There are several factors involved in the selection of a portfolio of securities, options, bonds, cash, *etc.* Investors will likely be interested in the rate of return on the portfolio and the possible variations in this rate of return. A preferred portfolio could be one with a maximal rate of return or it could mean a portfolio for which the probability of a large fluctuation (usually in the downward direction) in the value of the portfolio is minimized. Alternatively a portfolio for some investors could combine the two notions allowing the investor to specify, for example an acceptable level of return and then designing the portfolio with the minimum deviation from that return. The probability of deviating from a desired rate of return is part of our definition of risk associated with a portfolio. This chapter will approach portfolio selection as a task of selecting the investor's average rate of return while minimizing the variance in the rate of return, and hence minimizing the risk to the investor. An important component of this theory is the Capital Assets Pricing Model which relates the rate of return of a specific investment to the rate of return for the entire market of investments. This model states the difference in the expected rates of return for a specific security and the risk-free interest rate is proportional to the difference in the expected rates of return for the market and the risk-free interest rate.

## 4.1 Return and Risk

Suppose the price of a security such as a stock has been observed regularly for an extended period of time. Let the raw data be the sequence $\{S_0, S_1, \ldots, S_n\}$. Consider the following list of closing prices for Apple, Inc. (ticker symbol AAPL) shown in Table 4.1. The stock price was observed roughly every 30 days. The **return** between two consecutive observation times of the stock price is the sum of the price of the stock and any dividend paid at the later observation time divided by the price of the stock at the earlier observation time minus one. The return over the interval from October 8, 2020 to November 6, 2020 is

$$\frac{118.69 + 0.205}{114.97} - 1 = 0.034139 = 3.4139\%.$$

The monthly return is generally not constant and thus data such as that in Table 4.1 provide a sample of a random variable. The mean of the sample is a point estimate of the mean of the random variable and the standard deviation of the sample is a point estimate of the standard deviation of the random variable. Using the stock price observations in Table 4.1 the month over month returns (expressed as percentages) for AAPL stock are

$$R = \{4.0717, -2.0031, 3.4139, 4.2632, 5.7939, 4.6173, -14.9166,$$
$$9.9175, 1.9781, -3.3177, 12.8128\}.$$

Thus the average monthly return for the period of data collection is

$$\mathbb{E}\,(R) = \frac{1}{11}\sum_R R_i = \frac{26.6310}{11} = 2.421\%.$$

Table 4.1   Closing price of Apple, Inc. (AAPL) stock and dividends (if any paid) during late 2020 and the first half of 2021.

| Date | Close ($) | Dividend ($) |
| --- | --- | --- |
| August 10, 2020 | 112.73 | |
| September 09, 2020 | 117.32 | |
| October 08, 2020 | 114.97 | |
| November 06, 2020 | 118.69 | 0.205 |
| December 07, 2020 | 123.75 | |
| January 07, 2021 | 130.92 | |
| February 05, 2021 | 136.76 | 0.205 |
| March 08, 2021 | 116.36 | |
| April 07, 2021 | 127.90 | |
| May 07, 2021 | 130.21 | 0.220 |
| June 04, 2021 | 125.89 | |
| July 06, 2021 | 142.02 | |

A point estimate of the annual return for the stock is $r = (12)(2.421) = 29.052\%$. The variance of the monthly return is

$$\mathbb{Var}\,(R) = \frac{1}{11-1}\sum_{R}(R_i - \mathbb{E}\,(R))^2 = \frac{540.787}{10} = 54.0787\%^2.$$

Thus the standard deviation in the monthly return is $\sqrt{54.0787} = 7.3538\%$. Assuming returns in non-overlapping intervals of time are independent, this quantity can be adjusted to a yearly standard deviation by multiplying by $\sqrt{12}$. Hence the annual standard deviation in the return on the stock is $s = (7.3538)\sqrt{12} = 25.4744\%$. This standard deviation in the return (typically expressed on an annual basis) is known as the **volatility** of the stock. Figure 4.1 contains a scatter plot of the volatility and return for the stocks making up the Standard and Poor's 500 index (S&P500).

Volatility is related to the "riskiness" of the investment. The risk is due to at least two sources. One source consists of factors related to the company itself, for example, its current market share, competition from similar companies, new products or services it may be introducing, the people involved in the operation of the company, *etc.* These types of factors make up the **idiosyncratic risk**. The other major source of risk factors is the

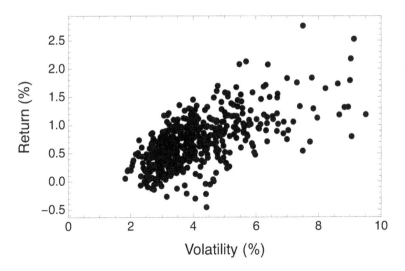

Fig. 4.1 Each dot in the graphic represents one of the firms in Standard and Poor's 500 stock index (S&P 500). The coordinates of the dots were determined by collecting daily closing price data on the stocks for the trading days between July 1, 2020 and June 30, 2021.

larger economic and political climate. This includes whether the national or global economy is strong or weak, whether the national or global political leadership is stable or unstable, whether natural disasters (earthquakes, climate change, pandemics, and so on) are affecting business operations, *etc.* Since these factors would tend to effect every company in the market regardless of the company's particular products or services, these factors make up **systematic risk**.

One of the often cited goals of investors is the desire to accurately predict future returns on securities or other investments. This type of prediction is sometimes done through the development of confidence intervals for a quantity. The sample mean return and sample volatility are point estimates of the true mean and volatility of the stock. Since there are only 11 sample observations of the return on AAPL stock the 95%-confidence interval [Sullivan (2018), Table VII] estimate for the true return is

$$(r - 2.228s, \ r + 2.228s) = (-27.7049\%, \ 85.8089\%).$$

This result should be interpreted as stating that if multiple samples of 11 returns of AAPL stock could be obtained from 2020–2021 and the 95%-confidence intervals constructed from each sample, the true return would be found in 95% of these 95%-confidence interval estimates. It should not be interpreted as stating that there is a 95% chance that the true return is inside this particular interval, though it may be. Even if investors could know the probability that the true return lies in this interval, the interval is so wide as to provide little predictive value. One way to narrow the predicted interval would be to collect more observations on the stock returns. There are at least two problems with this approach. First, the stock return is unlikely to remain constant over time. Second, Apple Inc. became a public company in late 1980 so there is a limited amount of data available from which to make predictions. The situation for other companies may be even worse. Google (or Alphabet Inc.) had its initial public offering in 2004.

As can be seen in Fig. 4.1 many stocks have a large volatility relative to their return. One method investors have for reducing volatility is to diversify their stock holdings. For example, the S&P500 index had an average return during the period of July 1, 2020 to June 30, 2021 of 0.5310% and a volatility of 1.9116%. Compared to the individual stocks in the index, this is a higher return and lower volatility, making the index an attractive investor. The effect of diversification can be explained mathematically.

Consider an idealized portfolio of $n$ risky securities and assume

- the random return on each security is $R_i$ for $i = 1, 2, \ldots, n$,
- the variance in the return on each security is $\mathbb{V}\text{ar}\,(R_i) = \sigma_i^2$ for $i = 1, 2, \ldots, n$,
- the pairwise covariances on the returns are $\mathbb{C}\text{ov}\,(R_i, R_j) = \sigma_{ij}$ for $i \neq j$,
- the portfolio weights are uniform, in other words $x_i = 1/n$ for $i = 1, 2, \ldots, n$.

The return on the portfolio is $R_P = \frac{1}{n} \sum_{i=1}^{n} R_i$. Under these assumptions the variance of the return of the uniformly weighted portfolio is

$$\mathbb{V}\text{ar}\,(R_P) = \frac{1}{n}\left(\frac{1}{n}\sum_{i=1}^{n}\sigma_i^2\right) + \left(1 - \frac{1}{n}\right)\frac{1}{n^2 - n}\sum_{i=1}^{n}\sum_{\substack{j=1 \\ j \neq i}}^{n}\sigma_{ij}. \qquad (4.1)$$

Note that the expression $\frac{1}{n}\sum_{i=1}^{n}\sigma_i^2$ is the average of the variances of the securities and

$$\frac{1}{n^2 - n}\sum_{i=1}^{n}\sum_{\substack{j=1 \\ j \neq i}}^{n}\sigma_{ij},$$

is the average of the covariances of the securities. As $n$ increases, $\mathbb{V}\text{ar}\,(R_P)$ decreases and asymptotically reaches the average covariance. Thus the volatility of the portfolio can be minimized by diversification. In fact if the average covariance is zero, then the portfolio volatility can be eliminated.

Suppose a portfolio consists of $n \geq 2$ securities with random returns $R_1, R_2, \ldots, R_n$. Suppose the weight of the $i$th security in the portfolio is $x_i$ with $x_1 + x_2 + \cdots + x_n = 1$. The mean return of the portfolio will be

$$r_P = x_1 r_1 + x_2 r_2 + \cdots + x_n r_n. \qquad (4.2)$$

Note that $r_i = \mathbb{E}\,(R_i)$ is the mean return of the $i$th security. The variance of the portfolio can be expressed as

$$\mathbb{V}\text{ar}\,(R_P) = \sum_{i=1}^{n}\sum_{j=1}^{n} x_i x_j \,\mathbb{C}\text{ov}\,(R_i, R_j)$$

$$= \mathbf{x}^T \begin{bmatrix} \mathbb{C}\text{ov}\,(R_1, R_1) & \mathbb{C}\text{ov}\,(R_1, R_2) & \cdots & \mathbb{C}\text{ov}\,(R_1, R_n) \\ \mathbb{C}\text{ov}\,(R_2, R_1) & \mathbb{C}\text{ov}\,(R_2, R_2) & \cdots & \mathbb{C}\text{ov}\,(R_2, R_n) \\ \vdots & \vdots & \ddots & \vdots \\ \mathbb{C}\text{ov}\,(R_n, R_1) & \mathbb{C}\text{ov}\,(R_n, R_2) & \cdots & \mathbb{C}\text{ov}\,(R_n, R_n) \end{bmatrix} \mathbf{x}. \qquad (4.3)$$

The expression $\mathbf{x}$ is the $n$-vector of portfolio weights. Recall that $\mathbb{Cov}(R_i, R_j) = \mathbb{Cor}(R_i, R_j)\sigma_{R_i}\sigma_{R_j}$ and since $-1 \leq \mathbb{Cor}(R_i, R_j) \leq 1$ then $\mathbb{Cov}(R_i, R_j) \leq \sigma_{R_1}\sigma_{R_j}$. This inequality is strict under the mild assumption that no pair of stocks has returns that are perfectly correlated. This implies the volatility of the portfolio,

$$\mathbb{Var}(R_P) < \sum_{i=1}^{n}\sum_{j=1}^{n} x_i x_j \sigma_{R_i}\sigma_{R_j} = (x_1\sigma_{R_1} + x_2\sigma_{R_2} + \cdots + x_n\sigma_{R_n})^2$$

$$\sigma_{R_P} < x_1\sigma_{R_1} + x_2\sigma_{R_2} + \cdots + x_n\sigma_{R_n},$$

the weighted average of the volatilities of the individual securities. Thus as a portfolio is made more diverse the volatility of the portfolio is decreased.

When a portfolio of investments has minimal risk, the risk that remains is the systematic risk. All of the firm-specific or idiosyncratic risk has been diversified away. Such a portfolio is called an **efficient portfolio**. This raises the question, do efficient portfolios exist? The answer depends upon an investor's estimates of the return for the securities available in the market, but an often used stand-in for an efficient portfolio is the market portfolio. One market portfolio has already been introduced, the S&P500 (ticker symbol ^SPX). Another common market portfolio surrogate is the Wilshire 5000 index (ticker symbol ^W5000) of the market value of all American stocks traded in the United States.

In general a risky investment has a return which is greater than the risk-free interest rate. The difference between the return on a risky investment and the risk-free interest rate is called the **risk premium**. The next section will explore the relationship between the risk premium for an efficient portfolio and the risk premium for an individual stock.

### Exercises

**4.1.1** The table below contains additional data on the closing price and dividends paid on Apple, Inc. (AAPL) stock. Find the annual rate of return and the volatility of the stock. Are these values the same as found in 2020–2021?

**4.1.2** Suppose there are two banks. The first bank has 1,000 outstanding loans of $100,000 each. There is a 2% chance that a borrower will default, in which case the first bank collects $0 on the loan. The second bank has

| Date | Close ($) | Dividend ($) |
|------|-----------|--------------|
| January 10, 2018 | 43.57 | |
| February 09, 2018 | 39.10 | 0.1575 |
| March 09, 2018 | 44.99 | |
| April 06, 2018 | 42.10 | |
| May 11, 2018 | 47.15 | 0.1825 |
| June 08, 2018 | 47.92 | |
| July 09, 2018 | 47.65 | |
| August 10, 2018 | 51.88 | 0.1825 |
| September 07, 2018 | 55.33 | |
| October 08, 2018 | 55.94 | |
| November 08, 2018 | 52.12 | 0.1825 |
| December 10, 2018 | 42.40 | |

100 outstanding loans of $1,000,000 each. For it, there is also a 2% chance that a borrower will default and the second bank will collect nothing. Assume all borrowers are independent of each other.

(a) Find the expected amount each bank will collect on its loans.
(b) Find the standard deviation in the amount each bank will collect on its loans.
(c) Which bank faces the most risk and why?

**4.1.3** Suppose the annual effective, risk-free rate of interest is 7.5% and suppose the S&P500 index will return either 37% or $-15\%$ per year with equal probabilities and different year's returns being independent. Suppose an investor has $1,000 to invest.

(a) Find the expected accumulated value if investment is made at the risk-free rate for 2 years.
(b) Find the expected accumulated value if investment is made at the risk-free rate for the first year and in the stock index for the second year.
(c) Find the expected accumulated value if investment is made at the in the stock index for the 2 years.

**4.1.4** Justify Eq. (4.1).

**4.1.5** Consider the following information about the rates of return of stocks from three different companies $A$, $B$, and $C$.

| State of Economy | Prob. | Rate of Return on Stock $A$ | Rate of Return on Stock $B$ | Rate of Return on Stock $C$ |
|---|---|---|---|---|
| Good | 0.55 | 0.06 | 0.17 | 0.35 |
| Bad | 0.45 | 0.14 | −0.01 | −0.09 |

(a) Find the mean return and variance for each stock.
(b) Find the three covariances of the returns.
(c) Find the volatility of a portfolio consisting of 25% stock $A$, 25% stock $B$, and 50% stock $C$.

## 4.2 The Efficient Frontier

This section will build on the ideas of the previous section by exploring the risk and return profiles of portfolios consisting of, to start, small numbers of risky assets, then larger numbers of risky assets, and then portfolios diverse enough to be considered efficient plus risk-free investments. Suppose an investor creates a portfolio of two risky assets $A$ and $B$ with expected returns $r_A$ and $r_B$ and volatilities $\sigma_A$ and $\sigma_B$. Suppose the correlation between the returns on the assets is $\rho = \mathbb{C}\text{or}\,(R_A, R_B)$. The portfolio assigns weight $x$ to the first asset and $1 - x$ to the second asset, where naturally $x + (1 - x) = 1$. According to Eqs. (4.2) and (4.3) the expected return and variance of the portfolio can be expressed as

$$\mathbb{E}\,(R_P) = r_P = xr_A + (1 - x)r_B \tag{4.4}$$

$$\mathbb{V}\text{ar}\,(R_P) = \sigma_P^2 = x^2\sigma_A^2 + (1 - x)^2\sigma_B^2 + 2x(1 - x)\rho\sigma_A\sigma_B. \tag{4.5}$$

The points in the $(\sigma, r)$-plane (the two-dimensional space for which the first coordinate of an ordered pair specifies the volatility of the portfolio and the second specified the expected return) form a curve whose shape will depend on $x$ and $\rho$. As a simple case, suppose $\rho = 1$, then

$$\sigma_P = \sqrt{(x\sigma_A + (1 - x)\sigma_B)^2} = x\sigma_A + (1 - x)\sigma_B,$$

assuming $0 \leq x \leq 1$. For $\rho = -1$ the curve consists of two straight line segments (see Exercise 4.2.1) and for $-1 < \rho < 1$ the curve is a segment of a hyperbola.

$$\sigma_P^2 - \alpha(r_P - \beta)^2 = \gamma, \tag{4.6}$$

where $\alpha$, $\beta$, and $\gamma$ are constants with $\alpha > 0$. Several exercises at the end of this section are devoted to re-writing Eq. (4.5) in the form of Eq. (4.6). Typical behavior is shown in Fig. 4.2. On each of the curves in the

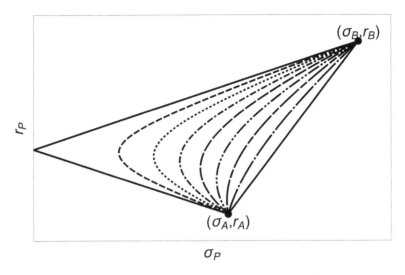

Fig. 4.2   This family of curves in the $(\sigma, r)$-plane represents the risks and returns for portfolios an investor can create from two correlated risky assets. The straight line on the right connecting points $(\sigma_A, r_A)$ and $(\sigma_B, r_B)$ corresponds to the correlation of returns being $\rho = 1$. As the correlation decreases the curves bend to the left.

$(\sigma, r)$-plane there is a portfolio with the minimum volatility or risk. Suppose that portfolio is represented by the ordered pair $(\sigma_{\min}, r_{\min})$. Thus investors may create portfolios with volatilities that are bounded below. For a given level of risk above $\sigma_{\min}$, there will be two portfolios with the same volatility but with two distinct returns. A rational investor will select the portfolio with the higher of the two returns. Thus the portfolios that investors will actually hold lie along just one arm of the hyperbola. This curve segment is known as the **efficient frontier**. It is illustrated in Fig. 4.3. Portfolios represented by risk-return ordered pairs $(\sigma, r)$ on the efficient frontier are known as **efficient portfolios**. These ideas of minimum volatility, maximum return analysis for portfolios are credited to Harry Markowitz [Markowitz (1952)].

The investment choices are expanded if, in addition to the two risky assets, portfolios may also include a risk-free asset. Furthermore assume that borrowing/lending at the risk-free interest rate and short sales of risky assets are permitted. An investor may place of portion of their wealth in a portfolio of the two risky assets, and the remainder in the risk-free asset. For a given level of risk (volatility in return), the investor will desire the highest return. Hence the portfolio of risky assets will be selected from the efficient frontier. Let $r_f$ be the risk-free interest rate. If $1 - x$ is invested

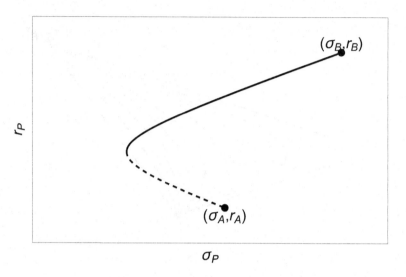

Fig. 4.3   Investors will only hold portfolios of assets $A$ and $B$ which lie along the solid portion of the curve, the efficient frontier. Portfolios on the dashed portion of the curve have the same risk as portfolios on the efficient frontier, but lower returns and thus would not be held by rational investors.

at the risk-free rate and $x$ is invested in an efficient portfolio $P$, then the expected return on the investment is

$$\mathbb{E}\left((1-x)r_f + xR_P\right) = (1-x)r_f + xr_P,$$

where $r_p$ is the expected return on the two-asset risky portfolio. The variance in the return on the investment is

$$\mathbb{Var}\left((1-x)r_f + xR_P\right) = x^2\,\mathbb{Var}\left(R_P\right) = (x\,\sigma_P)^2.$$

This holds since the volatility in the return of the risk-free asset is zero. Hence the volatility of this investment is proportional to the volatility of the risky portfolio with proportionality constant $|x|$. The locus of points in the $(\sigma, r)$-plane representing this investment will be a straight line through the points $(0, r_f)$ and $(\sigma_P, r_P)$. Several such lines are illustrated in Fig. 4.4 corresponding to different investor choices of efficient portfolios. Again, rational investors prefer higher returns for a given level of risk, thus among the efficient portfolios there exists an optimal portfolio for which the line through $(0, r_f)$ has maximum slope or equivalently, the line through $(0, r_f)$ is tangent to the efficient frontier formed by the two risky assets. The slope of the straight line through $(0, r_f)$ and $(\sigma_P, r_P)$ is

$$\frac{r_P - r_f}{\sigma_P},$$

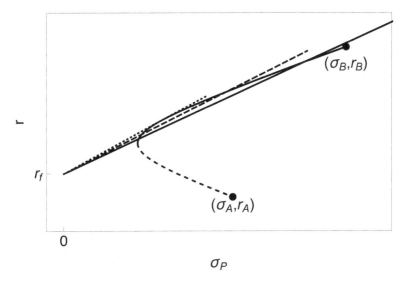

Fig. 4.4   If an investor can invest $x$ in an efficient portfolio and $1 - x$ in a risk-free asset (such as 1-month US Treasury bonds), the risk-return profile resembles a straight line through the point $(0, r_f)$.

which is known as the **Sharpe Ratio** [Sharpe (1966)]. It can be thought of as the ratio of the excess return on the portfolio (difference between the portfolio's rate of return and the risk-free rate of return) and the risk associated with the portfolio. The optimal portfolio will have the maximum Sharpe Ratio and will be called *the* efficient portfolio. A procedure for finding the tangent portfolio will be presented in a moment after a further generalization of the investment portfolio is conducted. The idea that all investors seek the same risky, tangent portfolio and adjust their appetite for risk by holding the efficient portfolio and risk-free assets is due to James Tobin [Tobin (1958)].

Suppose investors can create risky portfolios from $n > 2$ risky assets rather than just two. The locus of points of the form $(\sigma_P, r_P)$ representing the risky portfolios occupy a two-dimensional region instead of a curve. The upper boundary of this region is still referred to as the efficient frontier. To make the discussion more concrete consider the stocks of the five corporations: Generac Holdings Inc. (GNRC), Oracle (ORCL), Accenture (ACN), Ford (F), and Fiserv (FISV). The volatilities of returns, returns themselves, and covariances of the returns for the five corporations based on monthly closing prices collected over the period of January 2010 to December 2020

Table 4.2    The average annual returns, volatilities, and correlations between the stocks of five randomly selected stocks. The data was collected on the last trading day pf each month during the interval January 1, 2010 and December 31, 2020.

| Symbol | $\sigma$ | $r$ | Correlation | | | | |
|---|---|---|---|---|---|---|---|
| GNRC | 0.354596 | 0.322670 | 1.000000 | −0.170766 | −0.044920 | −0.075889 | −0.120644 |
| ORCL | 0.215907 | 0.117724 | −0.170766 | 1.000000 | 0.616496 | 0.398347 | 0.472117 |
| ACN | 0.204072 | 0.191479 | −0.044920 | 0.616496 | 1.000000 | 0.463382 | 0.607387 |
| F | 0.292920 | 0.023958 | −0.075889 | 0.398347 | 0.463382 | 1.000000 | 0.445994 |
| FISV | 0.177580 | 0.229110 | −0.120644 | 0.472117 | 0.607387 | 0.445994 | 1.000000 |

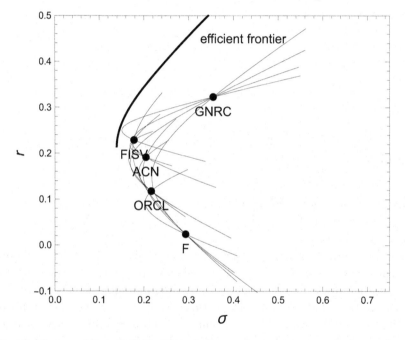

Fig. 4.5    As a portfolio is diversified by including stocks that are not perfectly correlated, the efficient frontier moves to the left in the $(\sigma, r)$-plane.

are listed in Table 4.2. The efficient frontier for portfolios which can be created from combinations of the five stocks is shown in Fig. 4.5. The hyperbolic curves which represent the portfolios constructed from just two of the five stocks lie to the right of the efficient frontier. The same relationship would exist for portfolios created from three and four of the five stocks. As the portfolio is diversified, the volatility is decreased and hence the efficient frontier moves to the left.

In general the optimal risky portfolio of $n$ assets can be determined by maximizing the Sharpe ratio of the portfolio. This is a constrained optimization problem and is solved as follows. Suppose there are $n$ risky assets with expected returns and return volatilities $r_i$ and $\sigma_i$ for $i = 1, 2, \ldots, n$. Let the weight assigned to the $i$th asset be $x_i$ for $i = 1, 2, \ldots, n$. Let the objective function be

$$f(x_1, x_2, \ldots, x_n) = \frac{r_P - r_f}{\sigma_P},$$

where $r_f$ is the risk-free rate and $r_P$ and $\sigma_P$ are given by Eqs. (4.2) and (4.3) respectively. The constraint equation is that the sum of the portfolio weights must be 1. Thus the objective function can be re-written as

$$f(x_1, x_2, \ldots, x_n) = \frac{\sum_{i=1}^{n} x_i(r_i - r_f)}{\left(\sum_{i=1}^{n} \sum_{j=1}^{n} x_i x_j \, \mathbb{C}\text{ov}\,(R_i, R_j)\right)^{1/2}}. \tag{4.7}$$

At a maximum of function $f$, $\partial f/\partial x_i = 0$ for $i = 1, 2, \ldots, n$. Let $k \in \{1, 2, \ldots, n\}$ and consider the partial derivative of the objective function with respect to $x_k$.

$$\frac{\partial f}{\partial x_k} = \frac{(r_k - r_f \sigma_P - (\sum_{i=1}^{n} x_i(r_i - r_f))\frac{1}{2}\sigma_P^{-1} 2 \sum_{i=1}^{n} x_i \, \mathbb{C}\text{ov}\,(R_i, R_k)}{\sigma_P^2}. \tag{4.8}$$

If the subscript $M$ denotes the portfolio with the optimal choice of weights then for $k = 1, 2, \ldots, n$

$$0 = \frac{(r_k - r_f \sigma_M - (r_M - r_f)\sigma_M^{-1} \sum_{i=1}^{n} x_i \, \mathbb{C}\text{ov}\,(R_i, R_k)}{\sigma_M^2} r_k - r_f$$

$$= \frac{r_M - r_f}{\sigma_M^2} \sum_{i=1}^{n} x_i \, \mathbb{C}\text{ov}\,(R_i, R_k). \tag{4.9}$$

This forms a system of $n$ linear equations in the $n$ unknowns $x_1$, $x_2$, $\ldots$, $x_n$. Every equation contains the expression $(r_M - r_f)/\sigma_M^2$ which is known as the **market price of risk**. For the sake of convenience replace this expression with the symbol $\eta$ and define $w_i = \eta x_i$ for $i = 1, 2, \ldots, n$. The linear system of equations can be expressed in matrix-vector form as

$$\begin{bmatrix} \mathbb{C}\text{ov}\,(R_1, R_1) & \mathbb{C}\text{ov}\,(R_1, R_2) & \cdots & \mathbb{C}\text{ov}\,(R_1, R_n) \\ \mathbb{C}\text{ov}\,(R_2, R_1) & \mathbb{C}\text{ov}\,(R_2, R_2) & \cdots & \mathbb{C}\text{ov}\,(R_2, R_n) \\ \vdots & \vdots & \ddots & \vdots \\ \mathbb{C}\text{ov}\,(R_n, R_1) & \mathbb{C}\text{ov}\,(R_n, R_2) & \cdots & \mathbb{C}\text{ov}\,(R_n, R_n) \end{bmatrix} \begin{bmatrix} w_1 \\ w_2 \\ \vdots \\ w_n \end{bmatrix} = \begin{bmatrix} r_1 - r_f \\ r_2 - r_f \\ \vdots \\ r_n - r_f \end{bmatrix}.$$

Once this system of equations is solved, note that

$$w_1 + w_2 + \cdots + w_n = \eta(x_1 + x_2 + \cdots + x_n) = \eta.$$

Hence the market price of risk is determined and the optimal portfolio weights are found as $x_i = w_i/\eta$ for $i = 1, 2, \ldots, n$.

**Example 4.1.** Suppose a stock market consists of only three stocks $A$, $B$, and $C$. The expected returns are respectively $r_A = 0.30$, $r_B = 0.34$, and $r_C = 0.34$. The risk-free interest rate is $r_f = 0.12$. The covariances between the returns on the stocks are given in the following table.

| | Covariance | | |
| --- | --- | --- | --- |
| | **Stock $A$** | **Stock $B$** | **Stock $C$** |
| **Stock $A$** | 0.004 | 0.002 | 0.000 |
| **Stock $B$** | 0.002 | 0.004 | 0.002 |
| **Stock $C$** | 0.000 | 0.002 | 0.004 |

Find the weights of the stocks in the minimum variance portfolio.

**Solution.** The system of linear equations for the stocks can be written as

$$0.004w_1 + 0.002w_2 + 0.000w_3 = 0.30 - 0.12$$
$$0.002w_1 + 0.004w_2 + 0.002w_3 = 0.34 - 0.12$$
$$0.000w_1 + 0.002w_2 + 0.004w_3 = 0.34 - 0.12.$$

The solution is $(w_1, w_2, w_3) = (40, 10, 50)$ (with sum $\eta = 100$). After rescaling the solution becomes $(x_1, x_2, x_3) = (2/5, 1/10, 1/2)$. Thus, 40% of an investor's wealth should be placed in stock $A$, 10% in stock $B$, and 50% in stock $C$, if the investor wishes to hold only risky assets.

Once the optimal risky portfolio has been identified, investors may divide their investments between risk-free investing, such as US Treasury bonds, and the optimal portfolio. This enables an investor to adjust the risk or the return in their investment strategy. The line through points $(0, r_f)$ and $(\sigma_M, r_M)$ is known as the **capital market line (CML)**. All investors will hold the same mixture of risky stocks. The difference will be in the proportion of the investor's total portfolio invested in risky stocks (while the remainder is invested risk-free). Every portfolio represented on the CML is an efficient portfolio. Figure 4.6 pictures the capital market line and the efficient frontier for the five stocks used in an earlier example. The risk-free rate of interest was assumed to be 5%. The volatility of the optimal risky portfolio is 17.5583% per year while the expected return is 31.0866% per year. Investors may then choose their preferred level of risk (or volatility in return) and locate the corresponding point on the capital market line. Risk adverse investors will choose points on the capital market line closer to $(0, r_f)$ while less risk adverse investors will choose points on the capital market line further to the right.

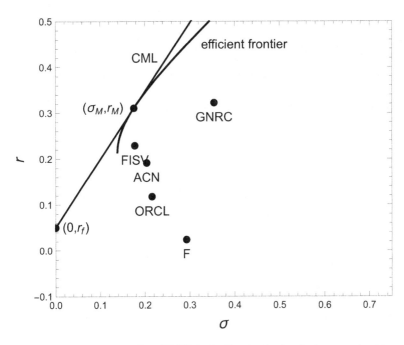

Fig. 4.6 The capital market line (CML) is the line in the $(\sigma, r)$-plane passing through $(0, r_f)$ and $(\sigma_M, r_M)$ where the latter coordinates represent the portfolio of risky investments with the maximum Sharpe ratio. All investors will pick their preferred balance of risky and risk-free investments from the CML.

Readers may wonder how difficult it is to identify the optimal portfolio in practice. Most finance professionals use the total supply of securities available in a market (sometimes referred to as the market portfolio) as the optimal portfolio. The weight of the $i$th security is just the market value of the $i$ security divided by the total market value of all securities in the market. The Wilshire 5000 (symbol ^W5000) is a security index representing the value-weighted portfolio of all stocks traded in the major US exchanges.

*Exercises*

**4.2.1** When the correlation between the two risky asset returns is $-1$ show the curve in the $(\sigma, r)$-plane can be expressed as

$$\left( \begin{cases} (1-x)\sigma_B - x\sigma_A \text{ if } 0 \le x < \frac{\sigma_B}{\sigma_A + \sigma_B} \\ x\sigma_A - (1-x)\sigma_B \text{ if } \frac{\sigma_B}{\sigma_A + \sigma_B} \le x \le 1 \end{cases} , xr_A + (1-x)r_B \right).$$

**4.2.2** Suppose stock $A$ has an expected return of 20% and a volatility of 45% while stock $B$ has an expected return of 16% and a volatility of 50%. The correlation between the returns on the two stocks is $-1$. What portfolio weights on the two stocks produce a risk-free portfolio? What is the risk-free interest rate in this situation?

**4.2.3** Derive the following equation from Eq. (4.5).

$$
\begin{aligned}
(r_A - r_B)^2 \sigma_P^2 \\
= \sigma_A^2 \left( x^2 r_A^2 + 2x(1-x)r_A r_B + (1-x)^2 r_B^2 + r_B^2 - 2r_B(xr_A + (1-x)r_B) \right) \\
+ \sigma_B^2 \left( x^2 r_A^2 + 2x(1-x)r_A r_B + (1-x)^2 r_B^2 + r_A^2 - 2r_A(xr_A + (1-x)r_B) \right) \\
- 2\operatorname{Cov}(R_A, R_B) \left( x^2 r_A^2 + 2x(1-x)r_A r_B + (1-x)^2 r_B^2 \right. \\
\left. + r_A r_B - r_A(xr_A + (1-x)r_B) - r_B(xr_A + (1-x)r_B) \right).
\end{aligned}
$$

**4.2.4** Use the result of Exercise 4.2.3 to show that

$$
\begin{aligned}
(r_A - r_B)^2 \sigma_P^2 = \operatorname{Var}(R_A - R_B)r_P^2 + \operatorname{Var}(r_B R_A - r_A R_B) \\
- 2\operatorname{Cov}(R_A - R_B, r_B R_A - r_A R_B)r_P.
\end{aligned}
$$

**4.2.5** Use the result of Exercise 4.2.4 to show that

$$
\begin{aligned}
(r_A - r_B)^2 \sigma_P^2 = \operatorname{Var}(R_A - R_B)r_P^2 + r_A^2 r_B^2 \operatorname{Var}\left( \frac{1}{r_A}R_A - \frac{1}{r_B}R_B \right) \\
- 2r_A r_B \operatorname{Cov}\left( R_A - R_B, \frac{1}{r_A}R_A - \frac{1}{r_B}R_B \right)r_P.
\end{aligned}
$$

**4.2.6** Consider a portfolio consisting to two risky assets. The correlation of the returns on the assets is $-1 < \rho < 1$. Find the expected return and volatility of the portfolio with the minimum risk.

**4.2.7** Suppose an investor has sold \$3,000,000 of stock $A$ and used the proceeds along with an additional \$10,000,000 to purchase \$13,000,000 of stock $B$. The expected return and volatility of the two stocks are in the following table.

| Stock | Expected Return | Volatility |
|-------|-----------------|------------|
| $A$   | 14.50%          | 42.00%     |
| $B$   | 16.75%          | 38.00%     |

The correlation between the returns on the stocks is 0.63. What are the expected return and volatility of the portfolio?

**4.2.8** Suppose the correlation between the returns on the two stocks described in Exercise 4.2.7 increases. Does the portfolio become more or less riskier? Why?

**4.2.9** Suppose a stock market consists of only three risky stocks $A$, $B$, and $C$. The expected returns are respectively $r_A = 0.15$, $r_B = 0.10$, and $r_C = 0.08$. The risk-free interest rate is $r_f = 0.06$. The covariances between the returns on the stocks are given in the following table.

Covariance

|  | Stock $A$ | Stock $B$ | Stock $C$ |
|---|---|---|---|
| Stock $A$ | 0.019 | 0.007 | 0.001 |
| Stock $B$ | 0.007 | 0.038 | −0.002 |
| Stock $C$ | 0.001 | −0.002 | 0.022 |

Find the weights of the stocks in the minimum variance portfolio.

**4.2.10** Kelly currently has \$500,000 invested in a portfolio whose volatility and expected return are respectively 25% and 8% per year. The optimal (market) portfolio has a volatility of 17% per year and an expected return of 12% per year. The risk-free interest rate is 5%.

(a) If Kelly wishes to maintain their current rate of return while minimizing risk, in what portfolio should they invest?

(b) If Kelly wishes to keep their current risk level while maximizing return, in what portfolio should they invest?

## 4.3 The Capital Asset Pricing Model

In the previous section the notions of the efficient frontier and the efficient portfolio (or tangent portfolio) were introduced. The present section will address the issue of determining the return an asset is required to produce in order for investors to purchase shares in the asset. As the reader may expect this will depend on factors such as the correlation of the asset's return with the return on the efficient portfolio. The main purpose of this section is the introduction of a model of risk and return first proposed by William Sharpe [Sharpe (1964)]. The model rests on the following three assumptions.

(1) Investors can buy or sell assets without taxes or transaction fees at competitive market prices. Borrowing or lending takes place at the risk-free interest rate.

(2) Investors will hold efficient portfolios yielding the maximum expected return for their preferred level of risk.

(3) Investors have the same estimates of rates of return and volatilities of returns on traded assets based on public information.

The last assumption sometimes describes investors has having homogeneous expectations.

According to the second assumption all investors will want to hold the portfolio with the maximum Sharpe ratio, the efficient portfolio illustrated in Fig. 4.6. By the third assumption (investors having homogeneous expectations) all investors will want to hold the same efficient portfolio. The aggregate of all investors' efficient portfolios will be the portfolio of all assets available in the market and thus the market portfolio is the efficient, tangent portfolio. Investors are assumed to hold a mixture of risk-free investments and the market portfolio somewhere on the capital market line depending on their tolerance for risk. Evaluating Eq. (4.9) for the efficient market portfolio $M$ yields the equation,

$$r_k - r_f = \frac{r_M - r_f}{\sigma_M^2} \, \text{Cov}\,(R_M, R_k).$$

Re-arranging the terms results in the equation,

$$\mathbb{E}\,(R_k) = r_k = r_f + \frac{\sigma_k \, \text{Cor}\,(R_k, R_M)}{\sigma_M}(\mathbb{E}\,(R_M) - r_f). \tag{4.10}$$

The expression $\sigma_k \, \text{Cor}\,(R_k, R_M)/\sigma_M$ is often denoted $\beta_k$. It measures the sensitivity of the expected return on the $k$th asset to the return on the market portfolio. For every unit change in the return on the market portfolio, the return on the $k$th asset will change by $\beta_k$. Equation (4.10) is known as the **Capital Asset Pricing Model**. The linear equation for $r_k$ in Eq. (4.10) is called the equation of the **security market line (SML)**. It has many implications. The quantity $r_k$ is known as the required return of asset $k$. If the return on the $k$th asset is smaller than $r_k$ then including more of the $k$th asset in the market portfolio decreases the Sharpe ratio of the portfolio. The opposite is also true. If the return on the $k$th asset is greater than $r_k$ then including more of the $k$th asset in the market portfolio increases the Sharpe ratio of portfolio.

The concepts of the capital market line and the security market line are easy to confuse with one another. Keep in mind that the capital market line is the straight line through the points $(0, r_f)$ and $(\sigma_M, r_M)$ and is tangent to the efficient frontier in the $(\sigma, r)$-plane. The security market line is the straight line through the points $(0, r_f)$ and $(1, r_M)$ in the $(\beta, r)$-plane. Thus all stocks are represented by points along the SML by ordered pairs $(\beta_k, r_k)$ where the coordinate $\beta_k$ measures the sensitivity of the excess return on the $k$th stock to the excess return on the market portfolio. Figure 4.7 illustrates the security market line for the hypothetical market of just five stocks used in examples in this chapter. The CML relates for a given level of risk the

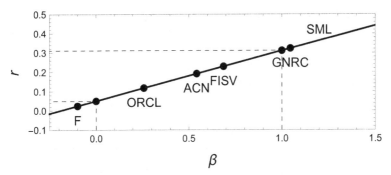

Fig. 4.7   The security market line (SML) illustrates the linear equation of the capital asset pricing model (CAPM). The expected return of an asset is a linear function of its $\beta$, the sensitivity of the return to the return on the efficient market portfolio. Risk-free assets such as US Treasury bills have a $\beta = 0$ and naturally return the risk-free rate. The market portfolio has $\beta = 1$ and thus returns $r_M$. All stocks in the market have returns and betas which lie somewhere on the SML.

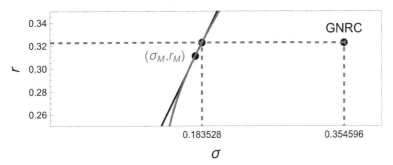

Fig. 4.8   The point on the capital market line having the same expected return as that of a single security splits the volatility or risk of that security into the systematic portion and firm-specific portion. In this example the risk-free rate is assumed to be 5%.

maximum return an investor can expect to receive, as measured by the total volatility $\sigma$ of a portfolio. The SML relates for a given level of risk, as measured by $\beta$, the return an investor can expect. The CML and SML employ two different measures of risk.

Comparison of the capital market line and the security market line also yield insight into the total amount of risk and the amount of systematic risk associated with a security. Consider the single security Generac Holdings Inc. (GNRC) in Fig. 4.8. GNRC has a estimated volatility of $\sigma_{GNRC} = 35.4596\%$ and an estimated return of $r_{GNRC} = 32.267\%$. The security does not lie on the CML, but there is a combination of the tangent portfolio and the risk-free asset which has the same return. The volatility of this investment would be $\sigma = 18.3528\%$. The difference between the volatility of

GNRC and the portfolio on the CML is the idiosyncratic or GNRC-specific risk for which investors in the GNRC security alone are not compensated. Specifically an investor in GNRC is only compensated for risk at the level of 18.3528%. Recall that the slope of the CML is the Sharpe ratio of the tangent portfolio. This implies,

$$\frac{r_M - r_f}{\sigma_M} = \frac{r_{GNRC} - r_f}{\sigma}.$$

Solving this equation for $\sigma$ and using the CAPM from Eq. (4.10) produce

$$\sigma = \frac{(r_{GNRC} - r_f)\sigma_M}{r_M - r_f} = \frac{\beta_{GNRC}(r_M - r_f)\sigma_M}{r_M - r_f} = \beta_{GNRC}\sigma_M. \qquad (4.11)$$

Therefore the systematic risk in the Generac Holdings Inc. security is the firm's beta multiplied by the volatility of the market portfolio. A similar analysis can be performed on any security in the market.

A portfolio of securities has a beta as well. Suppose a portfolio consists of $n$ assets with weights $x_1, x_2, \ldots, x_n$. In the comment following Eq. (4.10) the beta of a single asset is defined as

$$\beta_k = \frac{\sigma_k \, \text{Cor}\,(R_k, R_M)}{\sigma_M} = \frac{\text{Cov}\,(R_k, R_M)}{\text{Var}\,(R_M)}.$$

Since the return on a portfolio is just the weighted returns of its individual assets, then

$$\begin{aligned}
\beta_P &= \frac{\text{Cov}\,(R_P, R_M)}{\text{Var}\,(R_M)} = \frac{\text{Cov}\,(x_1 R_1 + x_2 R_2 + \cdots + x_n R_n, R_M)}{\text{Var}\,(R_M)} \\
&= \frac{x_1 \, \text{Cov}\,(R_1, R_M) + x_2 \, \text{Cov}\,(R_2, R_M) + \cdots + x_n \, \text{Cov}\,(R_n, R_M)}{\text{Var}\,(R_M)} \\
&= x_1 \beta_1 + x_2 \beta_2 + \cdots + x_n \beta_n.
\end{aligned}$$

Thus the beta of a portfolio of investments is the weighted average of the betas of the individual assets in the portfolio.

### Exercises

**4.3.1** When the economy is good the market portfolio usually increases by 35% and when the economy is bad the market portfolio decreases by 25%. A specific asset increases in value by 40% when the economy is good and decreases in value by 20% when the economy is bad. Estimate the beta of this asset.

**4.3.2** Suppose the risk-free interest rate is 5%. The market portfolio has an expected return of $r_M = 12\%$ and a volatility of $\sigma_M = 15\%$. A specific asset has returns which have a correlation of $\rho = 0.34$ with the return on the market portfolio. The variance in the returns for the asset is $\sigma^2 = 36\%$. Find the expected return for the asset.

**4.3.3** The asset and market described in Exercise 4.3.2 what are the expected return and volatility of an efficient portfolio having the same systematic risk as the asset?

**4.3.4** Suppose an investor has invested \$10,000 in just one asset. The asset has an expected return of 14% and a volatility of 35%. The risk-free interest rate is 4%. The market portfolio has an expected return of 11% and a volatility of 17%.

(a) Find an alternative investment with the minimum volatility that has the same rate of return as the original investment.

(b) Find an alternative investment with the maximum rate of return and the same volatility as the original investment.

**4.3.5** Consider the description of the returns on the market portfolio and securities $A$ and $B$ summarized in the following table.

### Returns

| Market | Stock $A$ | Stock $B$ |
|--------|-----------|-----------|
| −12% | −7% | −18% |
| −6% | −4% | −10% |
| 6% | 8% | 4% |
| 13% | 17% | 8% |

Each of the four scenarios described in the table is equally likely to occur.

(a) Find the volatilities of the returns of the market portfolio and the stocks $A$ and $B$.

(b) Find the covariances between return on the market portfolio and the returns on the two stocks.

(c) Find the betas of the two stocks.

**4.3.6** Consider the following three securities.

| Stock $i$ | Volatility | $\text{Cor}\,(R_i, R_M)$ |
|-----------|------------|--------------------------|
| $A$ | 33% | 0.30 |
| $B$ | 20% | 0.55 |
| $C$ | 45% | 0.65 |

The volatility and expected return of the market portfolio are 15% and 10% respectively. The risk-free interest rate is 3%. Suppose an investor creates a portfolio consisting of 20% each of stocks $A$ and $B$ and 60% of stock $C$.

(a) Find the expected returns of the individual stocks.
(b) Find the beta of the portfolio.
(c) Find the expected return of the portfolio.
(d) Find the volatility of a stock that lies on the efficient frontier and has an equal amount of systematic risk as the portfolio.

**4.3.7** Suppose that stock $A$ has a beta of 1.45 and stock $B$ has a beta of 0.90. The risk-free interest rate is 5% and the expected return of the market portfolio is 17%. The expected return of a portfolio of stocks $A$ and $B$ is 22%. Find the weights of the two stocks in the portfolio.

**4.3.8** Suppose that a stock has an expected return of 18%. The market portfolio has an expected return of 12% and a volatility of 35%. The risk-free interest rate is 5%. What is the minimum value for the volatility of the stock?

**4.3.9** Suppose that a stock has an expected return of 15%, a volatility of 35% and a beta of 0.95. The market portfolio has an expected return of 10%. The risk-free interest rate is 3%. Is this stock correctly priced? Why or why not?

**4.3.10** Suppose Brooke has an equally weighted portfolio of stocks $A$ and $B$ with a portfolio beta of 1.1. Cameron has an equally weighted portfolio of stocks $A$, $B$, and $C$ where the beta of $C$ is 1.8. The expected return of the market portfolio is 15% and the risk-free interest rate is 5%. Find the expected return of Cameron's portfolio.

**4.3.11** Suppose the risk-free interest rate is 4.75% per year, the expected rate of return on the stock market is 7.65% per year, and the standard deviation in the return on the market is 22% per year. If the covariance in the expected returns on a particular stock and the market is 15%, determine the expected rate of return on the stock.

## 4.4   Uncorrelated Returns

Consider the special case of Eq. (4.5) when the rates of return of the two investments are uncorrelated. Under this assumption $\rho = 0$ and the critical value of $x$ at which the variance in the portfolio return is minimized reduces to

$$x^* = \frac{\sigma_B^2}{\sigma_A^2 + \sigma_B^2} = \frac{\frac{1}{\sigma_A^2}}{\frac{1}{\sigma_A^2} + \frac{1}{\sigma_B^2}}.$$

There is an appealing simplicity and symmetry to this critical value. This is not a coincidence, it is in fact a pattern seen in the more general case of allocating a unit of investment among $n$ different potential investments.

**Theorem 4.1.** *Suppose that $0 \leq x_i \leq 1$ will be invested in security $i$ for $i = 1, 2, \ldots, n$ subject to the constraint that $x_1 + x_2 + \cdots + x_n = 1$. Suppose the rate of return of security $i$ is a random variable $R_i$ and that all the rates of return are mutually uncorrelated. The optimal, minimum variance portfolio described by the allocation vector $\langle x_1^*, x_2^*, \ldots, x_n^* \rangle$, is the one for which*

$$x_i^* = \frac{\frac{1}{\sigma_i^2}}{\sum_{j=1}^n \frac{1}{\sigma_j^2}} \; for \; i = 1, 2, \ldots, n,$$

*where $\sigma_i^2 = \mathsf{Var}\,(R_i)$.*

*Proof.* Let $W = x_1 R_1 + x_2 R_2 + \cdots + x_n R_n$ be the return of the portfolio. Since the rates of return $R_i$ are uncorrelated then the variance in the rate of return $W$ is

$$\mathsf{Var}\,(W) = \sum_{i=1}^n x_i^2 \sigma_i^2,$$

and is subject to the constraint that $1 = \sum_{i=1}^n x_i$. Minimizing the variance subject to the constraint is accomplished by use of Lagrange Multipliers [Stewart (1999)]. This technique states that $\mathsf{Var}\,(W)$ will be optimized at one of the solutions $(x_1, x_2, \ldots, x_n, \lambda)$ of the set of simultaneous equations:

$$\nabla \left( \sum_{i=1}^n x_i^2 \sigma_i^2 \right) = \lambda \nabla \left( \sum_{i=1}^n x_i \right)$$

$$\sum_{i=1}^n x_i = 1.$$

The symbol $\nabla$ denotes the gradient operator. These equations are equivalent to respectively:

$$2 x_i \sigma_i^2 = \lambda \; \text{for} \; i = 1, 2, \ldots, n, \; \text{and} \; \sum_{i=1}^n x_i = 1.$$

Solving for $x_i$ in the first equation and substituting into the second equation determines that

$$\lambda = \frac{2}{\sum_{j=1}^{n} \frac{1}{\sigma_j^2}}.$$

Substituting this expression for $\lambda$ into the first equation yields

$$x_i = \frac{\frac{1}{\sigma_i^2}}{\sum_{j=1}^{n} \frac{1}{\sigma_j^2}} \text{ for } i = 1, 2, \ldots, n.$$

The reader can confirm that for each $i$, $x_i \in [0, 1]$. □

The previous discussion can be generalized yet again to the situation in which the portfolio of securities is financed with borrowed capital. Let $\mathbf{w} = \langle w_1, w_2, \ldots, w_n \rangle$ represent the portfolio of investments where each $w_i$ represents the fraction of the portfolio invested in asset $i$. As before, $R_i$ will represent the rate of return of the $i$th security. For the sake of simplicity we will assume this is a one-period model in which investments are purchased by borrowing money which must be paid back at simple interest rate $r$. The net wealth generated by the portfolio financed with borrowed capital after one time period is

$$R(\mathbf{w}) = \sum_{i=1}^{n} w_i(1 + R_i) - (1 + r) \sum_{i=1}^{n} w_i = \sum_{i=1}^{n} w_i(R_i - r). \tag{4.12}$$

The expected value of the net wealth generated by the portfolio and the variance in the net wealth are functions of the vector $\mathbf{w}$ and are defined to be respectively,

$$r(\mathbf{w}) = \mathbb{E}\left(R(\mathbf{w})\right) \tag{4.13}$$

$$\sigma^2(\mathbf{w}) = \mathbb{V}\text{ar}\left(R(\mathbf{w})\right). \tag{4.14}$$

In this situation, the borrowed amounts $w_1, w_2, \ldots, w_n$ do not have to sum to unity, or to any other prescribed value.

**Lemma 4.1.** *Assuming the rates of return on the securities are uncorrelated, the optimal portfolio generating an expected unit amount of net wealth with the minimum variance in the net wealth is*

$$\mathbf{w}^* = \left\langle \frac{\frac{r_1-r}{\sigma_1^2}}{\sum_{j=1}^{n} \frac{(r_j-r)^2}{\sigma_j^2}}, \frac{\frac{r_2-r}{\sigma_2^2}}{\sum_{j=1}^{n} \frac{(r_j-r)^2}{\sigma_j^2}}, \ldots, \frac{\frac{r_n-r}{\sigma_n^2}}{\sum_{j=1}^{n} \frac{(r_j-r)^2}{\sigma_j^2}} \right\rangle, \tag{4.15}$$

*where $r_i = \mathbb{E}(R_i)$ and $\sigma_i^2 = \mathbb{V}\text{ar}(R_i)$ for $i = 1, 2, \ldots, n$.*

The proof of this lemma is similar to the proof of Thm. 4.1 and is left as an exercise.

The existence of the optimal portfolio $\mathbf{w}^*$ provided by Lemma 4.1 will be used in the following result known as the **Portfolio Separation Theorem**. The name is suggestive of the result which indicates to an investor how portions of a portfolio should be invested so as to minimize the variance in the wealth generated.

**Theorem 4.2. (*Portfolio Separation Theorem*)** *If $b$ is any positive scalar, the variance of all portfolios with expected wealth generated equal to $b$ is minimized by portfolio $b\,\mathbf{w}^*$ where $\mathbf{w}^*$ is described in Eq. (4.15).*

*Proof.* Suppose $\mathbf{x}$ is a portfolio for which $r(\mathbf{x}) = b$, then

$$\frac{1}{b}r(\mathbf{x}) = r\left(\frac{1}{b}\mathbf{x}\right) \quad \text{(by Exercise 4.4.5)}$$
$$= 1.$$

Thus we see that $\frac{1}{b}\mathbf{x}$ is a portfolio with unit expected rate of return. For the portfolio $\mathbf{w}^*$,

$$\sigma^2(b\,\mathbf{w}^*) = b^2\sigma^2(\mathbf{w}^*)$$
$$\leq b^2\sigma^2\left(\frac{1}{b}\mathbf{x}\right) \quad \text{(by Lemma 4.1)}$$
$$= \sigma^2(\mathbf{x}),$$

by Exercise 4.4.5. $\qquad\qquad\qquad\qquad\qquad\qquad\qquad\qquad\qquad\qquad\quad \square$

As a consequence of the Portfolio Separation Theorem all expected wealths generated by portfolios can be normalized to unity when determining the optimal portfolio.

### *Exercises*

**4.4.1** Suppose that an investor will split a unit of wealth between the following securities possessing variances in their rates of return as listed in the table below. Assuming that the rates of return are uncorrelated, determine the proportion of the portfolio which will be allocated to each security so that the variance in the returned wealth is minimized.

| Security $i$ | $\text{Var}\,(R_i)$ |
|:---:|:---:|
| A | 0.24 |
| B | 0.41 |
| C | 0.27 |
| D | 0.16 |
| E | 0.33 |

**4.4.2** Prove Lemma 4.1.

**4.4.3** This exercise extends Exercise 4.4.1 by determining the optimal portfolio which minimizes the variance in the return under the condition that the portfolio is financed with money borrowed at the interest rate of 11%.

| Security $i$ | $\mathbb{E}\,(R_i)$ | $\text{Var}\,(R_i)$ |
|:---:|:---:|:---:|
| A | 0.13 | 0.24 |
| B | 0.12 | 0.41 |
| C | 0.15 | 0.27 |
| D | 0.11 | 0.16 |
| E | 0.17 | 0.33 |

**4.4.4** Determine the optimal portfolio allocation which minimizes the variance in the return under the condition that the portfolio is financed with money borrowed or lent at the interest rate of 9%.

| Security $i$ | $\mathbb{E}\,(R_i)$ | $\text{Var}\,(R_i)$ |
|:---:|:---:|:---:|
| A | 0.10 | 0.42 |
| B | 0.18 | 0.45 |
| C | 0.17 | 0.25 |
| D | 0.05 | 0.20 |
| E | 0.21 | 0.35 |

**4.4.5** For any real scalar $c$ show that for $r(\mathbf{w})$ and $\sigma^2(\mathbf{w})$ as defined in Eqs. (4.13) and (4.14) the following results hold.

$$r(c\,\mathbf{w}) = c\,r(\mathbf{w})$$
$$\sigma^2(c\,\mathbf{w}) = c^2\sigma^2(\mathbf{w}).$$

## 4.5   Utility Functions

This section will focus on defining and understanding a class of functions which can be used as the basis of rational decision making. Suppose that an investor is faced with the choice of two different investment products. Suppose further that the set of outcomes resulting from investing in either of the products is $\{C_1, C_2, \ldots, C_n\}$. The reader should think of this set of outcomes as the union of the possible outcomes of investing in the first product and the possible outcomes resulting from investing in the second product. The probability of outcome $C_i$ coming about as the result of investing in the first product will be denoted $p_i$ for $i = 1, 2, \ldots, n$. If outcome $C_i$ cannot result from investing in the first product then, of course, $p_i = 0$. For the second product the values of the probabilities will be denoted $q_i$.

The investor can rank the outcomes in order of desirability. Without loss of generality suppose the outcomes have been ranked from least to most desirable as

$$C_1 \leq C_2 \leq \cdots \leq C_n.$$

To each of the possible outcomes a **utility** can be assigned. The utility function $u(C_i)$ is defined as follows. To start, $u(C_1) = C_1$ and $u(C_n) = C_n$. The values of $u(C_i)$ for $i = 2, 3, \ldots, n - 1$ will be defined by referring to $C_1$ and $C_n$. Suppose that for each such $i$ the investor is given the following choice: participate in another random experiment in which they receive outcome $C_i$ with certainty, or participate in a random experiment where they will receive $C_1$ with probability $\phi_i$ or receive $C_n$ with probability $1 - \phi_i$. The expected value of the outcome of the first experiment is $C_i$ and the expected value of the outcome of the second experiment is $E_i = \phi_i C_1 + (1 - \phi_i) C_n$. If $\phi_i = 1$ then the investor will surely participate in the random experiment with the certain outcome of $C_i$, since they rank this as a more desirable outcome than $C_1$. If $\phi_i = 0$ then the investor will participate in the second random experiment since its expected outcome is $C_n$. They would not better their result by taking the sure outcome $C_i$. At some value of $\phi_i \in [0, 1]$ the investor will be indifferent to the choice. The utility of $C_i$, in other words $u(C_i)$ is defined to be the $E_i$ for which the investor is indifferent to the choice of experiment.

Utility functions are specific to individual investors, like personality traits. In general a utility function is an increasing function. Most rational investors assign greater utility to more preferable outcomes. It is assumed that the value of $\phi_i$ at which the investor is indifferent to the choice of

experiment is unique. The rational investor, once having decided to participate in the second random experiment, will not at a higher expected value
decide to once again accept the certain result. Thus the utility function is
well-defined.

The utility function provides a means by which two outcomes of unequal
desirability may be compared. The utility of outcome $C_i$ is equivalent to
receiving $C_1$ with probability $\phi_i$ or receiving $C_n$ with probability $1 - \phi_i$. The
reader should note that there are multiple notions of probability at work
here. The utility function was defined in terms of a person's preference for
receiving outcome $C_i$ with certainty or willingness to participate in a specific
type of random experiment. However, the probability of the occurrence of
outcome $C_i$ is not $u(C_i)$. Recall we have assumed that $\mathbb{P}(C_i) = p_i$ or $q_i$
depending on which of two investment products the hypothetical investor
chooses. Returning to the original task of deciding between two different
investment products, suppose the investor decides to invest in the first
product. For the first investment product the expected value of the utility
function is then

$$\mathbb{E}\left(u(\mathbf{p})\right) = \sum_{i=1}^{n} p_i u(C_i) = C_1 \sum_{i=1}^{n} p_i \phi_i + C_n \sum_{i=1}^{n} p_i (1 - \phi_i).$$

Thus the expected utility of the first investment is equivalent to the
expected value of a simple random experiment in which the investor will
receive the least desirable outcome $C_1$ with probability $\sum_{i=1}^{n} p_i \phi_i$ and the
most desirable outcome $C_n$ with probability $\sum_{i=1}^{n} p_i(1 - \phi_i)$. Similarly the
expected value of the utility function for the second investment product is
found to be

$$\mathbb{E}\left(u(\mathbf{q})\right) = \sum_{i=1}^{n} q_i u(C_i).$$

Therefore the investor will choose the first product whenever

$$\sum_{i=1}^{n} p_i u(C_i) > \sum_{i=1}^{n} q_i u(C_i), \tag{4.16}$$

and otherwise will choose the second product.

This section treats outcomes or consequences $C_i$ of making an investment decision. More concretely these outcomes can be thought of as receiving differing amounts of money for an investment (some of which may
be negative). Thus in general the **utility function** denoted $u(x)$ is the
investor's utility of receiving an amount $x$. The remainder of this section
will explore categories and properties of utility functions.

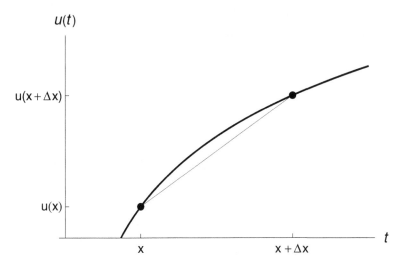

Fig. 4.9 The extra utility received, $u(x + \Delta x) - u(x)$, is a decreasing function of $x$. Many utility functions are concave functions for which the secant line will lie below the corresponding points on the graph of the function.

A general property of utility functions is that the amount of extra utility that an investor experiences when $x$ is increased to $x + \Delta x$ is non-increasing. In other words $u(x + \Delta x) - u(x)$ is a non-increasing function of $x$. This property is illustrated in Fig. 4.9. A function $f(t)$ is **concave** on an open interval $(a, b)$ if for every $x, y \in (a, b)$ and every $\lambda \in [0, 1]$ it is the case that

$$\lambda f(x) + (1 - \lambda)f(y) \leq f(\lambda x + (1 - \lambda)y). \qquad (4.17)$$

Graphically this may be interpreted as meaning that all the secant lines lie below the graph of $f(t)$. See Fig. 4.9. A utility function $u(t)$ obeying inequality (4.17) is called concave also. An often repeated statement about investing is that an investor may receive greater rewards by taking greater risks. However, for most rational investors there is a level of reward beyond which, in order to reap greater reward, they are unwilling to accept the higher level of risk. While a casino gambler may be willing to roll the dice for a chance to double an investment of \$20, it would be the rare gambler who would be willing to roll the dice in order to double \$20,000,000. The utility of that extra reward is not as high as the initial reward. These are gamblers and investors who will avoid games or investments for which the risk is (in their belief) too high even if the potential rewards are commensurately high. An investor whose utility function is concave is said to be **risk-averse**. An investor with a linear utility function of the form $u(x) = ax + b$ with $a > 0$ is said to be **risk-neutral**. An investor whose utility function increases more rapidly as the reward increases is said to be **risk-seeking**.

The following result holds for concave functions.

**Theorem 4.3.** *If $f \in C^2(a, b)$ then $f$ is concave on $(a, b)$ if and only if $f''(t) \leq 0$ for $a < t < b$.*

*Proof.* If $f$ is concave on $(a, b)$ then by definition $f$ satisfies inequality (4.17). Let $x, y \in (a, b)$. Without loss of generality we may assume $x < y$. If $w = \lambda x + (1 - \lambda)y$ and if $0 < \lambda < 1$ then $a < x < w < y < b$. Inequality (4.17) is then equivalent to

$$(1 - \lambda)\left[f(y) - f(w)\right] \leq \lambda \left[f(w) - f(x)\right]. \tag{4.18}$$

By the definition of $w$,

$$1 - \lambda = \frac{w - x}{y - x} \text{ and} \tag{4.19}$$

$$\lambda = \frac{y - w}{y - x}. \tag{4.20}$$

Substituting Eqs. (4.19) and (4.20) into inequality (4.18) and rearranging terms produces another inequality which is equivalent to the definition of a concave function stated in inequality (4.17).

$$\frac{f(y) - f(w)}{y - w} - \frac{f(w) - f(x)}{w - x} \leq 0. \tag{4.21}$$

Applying the Mean Value Theorem to each of the difference quotients of inequality (4.21) implies that for some $\alpha$ and $\beta$ satisfying with $x < \alpha < w < \beta < y$,

$$f'(\beta) - f'(\alpha) \leq 0.$$

Using the Mean Value Theorem once more proves that for some $t$ with $\alpha < t < \beta$

$$f''(t)(\beta - \alpha) \leq 0.$$

which implies $f''(t) \leq 0$. The reader should keep in mind that this last inequality is still equivalent to the inequality defining a concave function. Thus inequality (4.17) holds for a twice continuously differentiable function if and only if $f''(t) \leq 0$ for $a < t < b$. $\square$

Concave functions possess many useful properties such as Jensen's Inequality.

**Theorem 4.4. (Discrete Jensen's Inequality)** *Let $f$ be a concave function on the interval $(a, b)$, suppose $x_i \in (a, b)$ for $i = 1, 2, \ldots, n$, and suppose $\lambda_i \in [0, 1]$ for $i = 1, 2, \ldots, n$ with $\sum_{i=1}^{n} \lambda_i = 1$, then*

$$\sum_{i=1}^{n} \lambda_i f(x_i) \le f\left(\sum_{i=1}^{n} \lambda_i x_i\right). \tag{4.22}$$

*Proof.* Let $\mu = \sum_{i=1}^{n} \lambda_i x_i$ and note that since $\lambda_i \in [0, 1]$ for $i = 1, 2, \ldots, n$ and $\sum_{i=1}^{n} \lambda_i = 1$, then $a < \mu < b$. The equation of the line tangent to the graph of $f$ at the point $(\mu, f(\mu))$ is $y = f'(\mu)(x - \mu) + f(\mu)$. Since $f$ is concave on $(a, b)$ then

$$f(x_i) \le f'(\mu)(x_i - \mu) + f(\mu) \text{ for } i = 1, 2, \ldots, n.$$

Therefore

$$\sum_{i=1}^{n} \lambda_i f(x_i) \le \sum_{i=1}^{n} \left(\lambda_i \left[f'(\mu)(x_i - \mu) + f(\mu)\right]\right)$$

$$= f'(\mu) \sum_{i=1}^{n} (\lambda_i x_i - \lambda_i \mu) + f(\mu) \sum_{i=1}^{n} \lambda_i$$

$$= f(\mu) = f\left(\sum_{i=1}^{n} \lambda_i x_i\right).$$

$\square$

Jensen's inequality may be interpreted as saying that for a concave function, the mean of the values of the function applied to a set of values of a random variable is no greater than the function applied to the mean of the values of the random variable. If $u$ is a concave utility function, then $\mathbb{E}(u(X)) \le u(\mathbb{E}(X))$. Thus a risk-averse investor prefers the certain return of $C = \mathbb{E}(X)$ over receiving a random return with this expected value.

**Example 4.2.** Suppose $f(x) = \tanh x$ and let

$$\{\lambda_1, \lambda_2, \lambda_3, \lambda_4, \lambda_5\} = \{0.30, 0.07, 0.23, 0.20, 0.20\}.$$

Show that Thm. 4.4 holds.

**Solution.** The reader can verify that $f(x)$ is concave on $(0, \infty)$. If $X_i = i^2$ for $i = 1, 2, \ldots, 5$ then

$$\sum_{i=1}^{5} \lambda_i f(X_i) = 0.928431 \le 1 = f\left(\sum_{i=1}^{5} \lambda_i X_i\right).$$

There is also a continuous version of Jensen's Inequality.

**Theorem 4.5.** *(Continuous Jensen's Inequality) Let $g(t)$ be an integrable function on $[0,1]$ and let $f$ be a concave function, then*

$$\int_0^1 f(g(t))\, dt \le f\left(\int_0^1 g(t)\, dt\right). \tag{4.23}$$

*Proof.* For the sake of conciseness of notation let $\alpha = \int_0^1 g(t)\, dt$, and let $y = f'(\alpha)(x - \alpha) + f(\alpha)$, the equation of the tangent line passing through the point with coordinates $(\alpha, f(\alpha))$. Since $f$ is concave then

$$f(g(t)) \le f'(\alpha)(g(t) - \alpha) + f(\alpha),$$

which implies that

$$\int_0^1 f(g(t))\, dt \le \int_0^1 \left[f'(\alpha)(g(t) - \alpha) + f(\alpha)\right] dt$$

$$= f(\alpha) + f'(\alpha) \int_0^1 (g(t) - \alpha)\, dt$$

$$= f(\alpha) = f\left(\int_0^1 g(t)\, dt\right).$$

$\square$

As the name suggests, **expected utility** is the expected value of a utility function. Returning to the previous discussion of choosing between two possible investment instruments, we see that the investment with the greater expected utility is preferable. The inequality in (4.16) indicated that if the first investment choice results in outcomes $\{C_1, C_2, \ldots, C_n\}$ with respective probabilities $p_i$ for $i = 1, 2, \ldots n$ and the second investment choice produces the same outcomes but with probabilities $\{q_1, q_2, \ldots, q_n\}$ then a rational investor would select the first investment whenever

$$\sum_{i=1}^n p_i u(C_i) > \sum_{i=1}^n q_i u(C_i).$$

Suppose the random variable $X$ is thought of as the set of outcomes with probabilities $\{p_1, p_2, \ldots, p_n\}$ while the random variable $Y$ represents the outcomes with probabilities $\{q_1, q_2, \ldots, q_n\}$. Then the first investment is preferable whenever

$$\mathbb{E}\left(u(X)\right) > \mathbb{E}\left(u(Y)\right).$$

**Example 4.3.** An investor must choose between the following two "investments":

**A:** Flip a fair coin, if the coin lands heads up the investor receives $10, otherwise they receive nothing.

**B:** Receive an amount $M$ with certainty.

The investor is risk-averse with a utility function $u(x) = x - x^2/25$. Find the interval of $M$ for which the investor prefers investment $A$.

**Solution.** The rational investor will select the investment with the greater expected utility. The expected utility for investment A is

$$\frac{1}{2}u(10) + \frac{1}{2}u(0) = \frac{1}{2}\left(10 - \frac{10^2}{25}\right) = 3.$$

The expected utility for B is $u(M) = M - M^2/25$. Thus the investor will choose the coin flip whenever

$$3 > M - \frac{M^2}{25}$$

$$M^2 - 25M + 75 > 0$$

$$\frac{25 - 5\sqrt{13}}{2} > M.$$

Thus investment A is preferable to B whenever $M < \$3.49$.

The astute reader will note that the quadratic inequality solved above is satisfied for $M$ in the sets defined by the relationships

$$M < \frac{25 - 5\sqrt{13}}{2} \approx 3.49 \text{ or } M > \frac{25 + 5\sqrt{13}}{2} \approx 21.51.$$

Mathematically this implies the investor would prefer investment A even if the certain amount of investment B is greater than $21.51. A wise investor would never accept a chance at a maximum payoff of $10 if they can receive a guaranteed amount of $21.51 or more. This type of situation leads us to adopt the following logical convention.

If $M < 3.49$ the investor of the previous example will choose investment A, while for $M \geq \$3.49$ the investor will choose the investment B. This example illustrates the concept known as the **certainty equivalent** which is defined as the minimum value $C$ of a random variable $X$ at which $u(C) = \mathbb{E}\left(u(X)\right)$.

**Example 4.4.** Consider the following investment choices.

**A:** Flip a fair coin, if the coin lands heads up the investor receives $0 < X \leq 10$, otherwise they receive $0 < Y < X$.
**B:** Receive an amount $C$ with certainty.

Assume the investor's utility function is the same as in the previous example, and find the certainty equivalent $C$.

**Solution.** The certainty equivalent and payoffs of investment A must satisfy the following equation.

$$C - \frac{C^2}{25} = \frac{1}{2}\left(X - \frac{X^2}{25} + Y - \frac{Y^2}{25}\right)$$

$$\left(C - \frac{25}{2}\right)^2 = \frac{1}{2}\left[\left(X - \frac{25}{2}\right)^2 + \left(Y - \frac{25}{2}\right)^2\right]$$

$$C = \frac{25}{2} - \frac{1}{\sqrt{2}}\sqrt{\left(X - \frac{25}{2}\right)^2 + \left(Y - \frac{25}{2}\right)^2}$$

Again note that the certainty equivalent is the smallest value of $C$ satisfying the equation. The design of investment A specifies that $0 < Y < X < 10$. Thus the certainty equivalent can be thought of as a surface plotted over the triangular region bounded by $0 < X < 10$ with $0 < Y < X$. Figure 4.10 illustrates the certainty equivalent as a function of $X$ and $Y$.

### Exercises

**4.5.1** Which of the following utility functions are concave on their domain?

(a) $u(x) = \ln x$ for $x > 0$,
(b) $u(x) = (\ln x)^2$ for $x > 0$,
(c) $u(x) = \tan^{-1} x$ for $-\infty < x < \infty$.

**4.5.2** The mean of a discrete random variable as defined in Eq. (2.11) is sometimes called the **arithmetic mean**. The **harmonic mean** is defined as

$$\mathcal{H} = \frac{1}{\sum_X \frac{\mathbb{P}(X)}{X}}. \tag{4.24}$$

The **geometric mean** is defined as

$$\mathcal{G} = \prod_X X^{\mathbb{P}(X)}. \tag{4.25}$$

Use Jensen's inequality (4.22) to show that $\mathcal{H} \leq \mathcal{G} \leq \mathbb{E}(X)$ and that equality holds only when $X_1 = X_2 = \cdots = X_n$. Assume that all the random variables are positive.

**4.5.3** Verify inequality (4.23) for $f(x) = \tanh x$ and $g(t) = t$.

**4.5.4** Find the certainty equivalent for the choice between (a) flipping a fair coin and either winning $10 or losing $2, or (b) receiving an amount $C$ with certainty. Assume a utility function of $f(x) = x - x^2/50$.

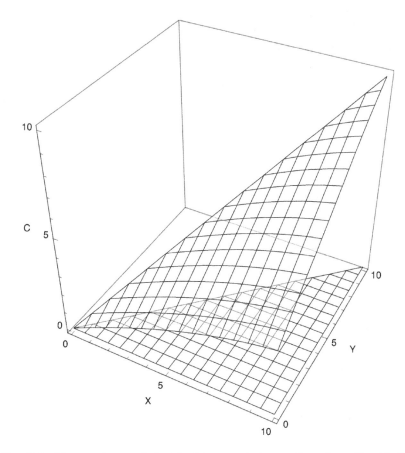

Fig. 4.10 The certainty equivalent in this example is a section of an ellipsoid. The domain of interest in the $XY$-plane is drawn as a shadow below the surface.

**4.5.5** Suppose Drew must choose between receiving an amount $C$ with certainty or playing a game in which a fair die is rolled. If a prime number results Drew wins $15, but if a composite number results Drew loses that amount of money. Assume that Drew's utility function is $f(x) = x - x^2/2$. Find the certainty equivalent $C$.

**4.5.6** Suppose a risk-averse investor with utility function $u(x) = \ln x$ will invest a proportion $\alpha$ of their total capital $x$ in an investment which will pay them either $2\alpha x$ with probability $p$ or nothing with probability $1 - p$. The amount of capital not invested will earn interest for one time period at the simple interest rate $r = 11\%$. What proportion of their capital should the investor allocate if they wish to achieve the maximum expected utility?

**4.5.7** Consider the function $u(x) = 1 - e^{-bx}$ where $b > 0$.

(a) Show that $u(x)$ is concave.

(b) Show that $0 \leq x_1 < x_2$ implies $u(x_1) < u(x_2)$.

**4.5.8** Suppose an investor whose utility function is $u(x) = 1 - e^{-x/100}$ has a total of \$1000 to invest in two securities. The expected rate of return for the first investment is $r_1 = 0.08$ with a standard deviation in the rate of return of $\sigma_1 = 0.03$. The expected rate of return for the second investment is $r_2 = 0.13$ with a standard deviation in the rate of return of $\sigma_2 = 0.09$. The correlation in the rates of return is $\rho = -0.26$. What is the optimal amount of money to place into each investment?

Chapter 5

# Forwards and Futures

This chapter introduces some of the concepts and terminology associated with the buying and selling of assets and securities such as precious metals, energy resources, and stocks. The main issue discussed will be the pricing of forward contracts and futures. The present chapter will give the reader the opportunity to apply the theory of interest, the arbitrage principle, and some elementary probability to the problem of pricing two commonly traded financial derivatives. The term "derivative" is used because the values of these financial instruments is "derived" from underlying securities or commodities. The material of this chapter can also be treated as a "warm-up" exercise for the later chapters on options and the development of the Black–Scholes option pricing formula.

## 5.1 Definition of a Forward Contract

The explanations of many of the concepts in this chapter will refer to buying or selling a financial instrument. In the financial markets it is often possible for a party to sell something which they do not yet own by borrowing the asset from a true owner. For example, investor A may borrow 100 shares of stock from investor B and sell them with the understanding that by some time in the future investor A will purchase 100 shares of the same stock and return them to investor B. Normally one buys an asset first and sells it later, but in the financial arena one can often sell first and then buy later. Borrowing and selling an asset with the agreement to re-purchase and return it later is called adopting a **short position** in the asset, or sometimes **shorting** the asset. Adopting a short position can be a profitable transaction if the investor believes the price of the asset is going to decrease in the future. Purchasing an asset first and then selling it in the future is called adopting a **long position**.

At its essence, a **forward** is an agreement between two agents (usually called the "party" and the "counter-party") to buy or sell a specified quantity of an asset or commodity at a specified price on a specified date in the future. The time at which the forward contract is created is called the agreement date and is usually treated as $t = 0$ for mathematical convenience. For a forward contract no money changes hands on the agreement date, that is, there is no "up front" cost to a forward contract. The later date on which assets or commodities must be delivered and on which payment can be made is called the settlement date or maturity date. The forward is an obligation to buy or sell at the agreed upon quantity, price, and time. The party agreeing to purchase the asset is said to hold a long forward contract. The counter-party (who must sell the asset) is in the short position. If the party or counter-party breaks the agreement, they may face legal and financial consequences. Due to the structure of forward contracts the party and counter-party assume a non-trivial risk of default. Futures contracts covered later in this chapter address the risk of default.

Consider the situation of a manufacturer producing portable digital music players. Their product depends upon an adequate supply of solid state memory to match the manufacturing output of music players. If the manufacturer is concerned that the readily available supply of memory may fall short of their needs 3 months from now, they may enter into a forward contract with a memory supplier to sell them, say, 100,000 units of memory for one million dollars in 3 months. Once the forward contract is established, the memory supplier must deliver the 100,000 units of memory in 3 months for the price of one million dollars even if another buyer would be willing to pay more for it. The music player manufacturer must buy the 100,000 units of memory in 3 months for one million dollars even if another supplier is willing to sell it for less. While there are many reasons an institution may buy or sell a forward contract, the chief reason is that the forward reduces the risk of market prices moving against the party or counter-party. The music player manufacturer may fear the business disruption which a shortage of solid state memory at affordable prices may cause, while the memory supplier may worry that increased supply or improved technology may decrease the price of memory 3 months hence. The party or counter-party of a forward contract can effectively "cancel" a forward contract by entering into an opposing forward contract. For example, suppose the music player manufacturer, after setting up the original forward contract, decides they will not need the solid state memory. They can enter into a forward contract with another memory buyer. The music player manufacturer becomes a seller of its unneeded memory. The memory supplier (perhaps concerned they will be unable to provide and adequate supply of memory) can enter into a forward contract with other solid state memory suppliers to provide the memory which will ultimately be delivered to the music player manufacturer as part of the original forward contract.

A forward contract can be set up between two agreeing parties and can be customized to the precise needs of the buyer and seller. Since highly customized or specialized forward contracts may be of limited interest or value to other parties, forwards are usually described as being less "liquid" than futures contracts. The property of liquidity is a measure of how easily an asset or investment is to sell, should the need to sell it arise. In contrast securities such as stocks (and futures as well) are often traded in a **market**. A stock market is created to increase the efficiency of the process of buying and selling stocks. Creating and maintaining this efficient environment for trading stocks is the business of **market makers**. Like any (for profit) business the market makers expect to earn money from the operation of the market. There are a number of ways this may happen. The market makers may charge buyers and sellers fees or commissions. Money can also be earned from the difference in the buying and selling price of a stock. In a stock market there may be several owners of a particular corporation's stock who are willing to sell and a number of buyers wishing to purchase that stock. Investors, both buyers and sellers may have different prices for the stock. A price a buyer is willing to pay is called a **bid** price. A price a seller of a stock is willing to accept is known as an **ask** price. The ask price is known also as the **offer** price. Typically the ask prices exceed the bid prices. At a given moment in the market the difference between the lowest ask price and the highest bid price is called the **bid/ask spread** or **bid/offer spread**. The market maker earns the bid/ask spread on stock trades as compensation for operating the market. Commissions, fees, and the bid/ask spread are examples of **transaction costs** associated with stock trading.

**Example 5.1.** Suppose the lowest ask price of a share of stock is $50.10 and the highest bid price for the stock is $50.00.

(i) What is the bid/ask spread per share of stock?
(ii) If an investor wishes to purchase 1,000 shares, how much will the investor pay in total?

(iii) If a stockholder wishes to sell 1,000 shares, how much will the seller receive in total?

(iv) If 1,000 shares are bought/sold, how much will the market maker earn?

**Solution.**

(i) The bid/ask spread is $0.10 per share.

(ii) A stock buyer who issues a buy order for 1,000 shares will pay $50,100.

(iii) The seller will receive $50,000.

(iv) The market maker will earn $100 on the trade (plus any other fees or commissions charged).

Without market makers and the markets for trading they create and operate, buyers and sellers of stocks and commodities would be responsible for locating each other. Anyone who has sold a used car knows that there are costs associated with finding potential buyers. When trading in a market, buyers and sellers also benefit from knowing the last price at which a stock traded and from the price competition among traders.

*Exercises*

**5.1.1** A commercial bakery enters into a forward contract with a farmer to purchase 15,000 bushels of wheat at a price of $3.75 per bushel in 3 months time. Assuming no default by either party, how much payment will the farmer receive at settlement?

**5.1.2** Alex has just drilled a well and has started to collect crude oil. To reduce uncertainty over the market price of oil in 6 months $S(6)$, Alex sells a forward contract for one million barrels of crude oil with delivery in 6 months and a forward price of $F$ dollars per barrel.

(a) How much is a long position in one million barrels of crude oil worth in 6 months?

(b) How much will Alex receive at settlement for the crude oil?

**5.1.3** Suppose the bid/ask spread for a share of a particular stock is $0.25. An investor buys 1,000 shares and then immediately sells them (before the bid and ask prices can change). What is the total transaction cost (the so-called **round trip cost**) to the investor?

**5.1.4** A farmer's hens have just produced 1,000 new chicks. The chicks will be ready to market in 2 months time. To reduce uncertainty over the market price of chicks in 2 months, the farmer sells the chicks in a forward contract for $2 each for settlement in 2 months. In 2 months the price of chicks is $2.20. What is the farmer's short forward contract worth at settlement?

**5.1.5** Devon wishes to buy Drew's house in 1 year. Devon enters into a long forward for delivery of Drew's house in 1 year for $100,000. At settlement the market value of Drew's house is $105,000. What is Devon's profit in this situation?

## 5.2 Pricing a Forward Contract

Consumers are accustomed to an instantaneous transfer of ownership when purchasing most items. A smartphone, for instance, becomes the buyer's property the instant the buyer transfers the appropriate amount of money to the seller. The process of purchasing assets such as financial instruments is more general and events can take place at different times separated by finite intervals. To make the initial scenario simple assume that events may take place at time $t = 0$ and also later at time $t = T > 0$. There are three components or actions involved in the simplest description of an asset purchase: (1) fixing or agreeing on the price for the asset; (2) making payment for the asset; and (3) transferring ownership of the asset from the seller to the buyer. Logically the price is fixed before payment is made so it is assumed that the price of the asset is fixed at $t = 0$. However, the remaining two actions may occur at either $t = 0$ or $t = T$ depending on the arrangement made between buyer and seller.

The traditional buyer/seller mode of purchase in which all three events occur simultaneously is called an **outright purchase** or **spot contract**. The situation in which the buyer receives ownership of the asset at time $t = 0$ but pays for the purchase at the later time $t = T$ is called a **fully leveraged purchase**. A fully leveraged purchase is equivalent to buying the asset on credit or with borrowed money. A buyer who pays for the asset at time $t = 0$ but does not receive ownership until time $t = T$ is said to have purchased the security using a **prepaid forward contract**. The prepaid forward contract on a stock allows the seller to retain certain rights associated with ownership of the stock until $t = T$. For example, the seller would retain voting rights inherent in ownership of the stock until $t = T$. Another important reserved right for the seller in the prepaid forward contract situation is the right to receive any dividend payments associated with the stock. Lastly the delivery price of the asset is determined at $t = 0$ and payment for the asset and transfer of ownership both take place at $t = T$, it is said that the stock is purchased using a **forward contract**. The timing and coordination of the payment for the purchase and the transfer of ownership have implications for the delivery price.

The simplest case to analyze is the case of outright purchase of the asset. If the asset is worth $S(0)$ at time $t = 0$ and payment and transfer of ownership will take place at time $t = 0$ then the amount paid should be $S(0)$. To pay any other amount would create an arbitrage opportunity for either the buyer or the seller (see Exercise 5.2.1). Since the fully leveraged purchase is equivalent to the buyer purchasing the asset with borrowed money, the cost of the fully leveraged purchase is $S(0)e^{r_b T}$, where the continuously compounded interest rate for borrowing is $r_b$.

Determining the price of a prepaid forward or forward contract requires a more detailed (though still elementary) justification. A rigorous argument requires the introduction of new terminology and assumptions. Assume the asset pays no **dividends**. Dividends are periodic payments paid by a corporation to owners of the corporation's stock. Some credit unions pay their members an annual dividend, since the members are essentially owners of the credit union, though the members may not think of themselves as stockholders. The funds for the dividends are paid out of a portion of the profit made by the corporation.

The price of a prepaid forward contract is justified in the proof of the following theorem.

**Theorem 5.1.** *The arbitrage-free price $F_{0,T}^P$ of a prepaid forward contract on a non-dividend paying asset worth $S(0)$ at time $t = 0$ for which ownership of the asset will be transferred to the buyer at time $t = T > 0$ is $F_{0,T}^P = S(0)$.*

*Proof.* Suppose $F_{0,T}^P < S(0)$. The investor can purchase the forward and sell the asset (the buyer has entered into a long position on the forward and a short[1] position on the asset). Since $S(0) - F_{0,T}^P > 0$ the investor has a positive cash flow at $t = 0$ which can be invested at the risk-free rate $r$. At $t = T$, the investor receives ownership of the asset and immediately unwinds their short position in the asset. The cash flow at $t = T$ is therefore zero. Thus the investor has a guaranteed profit of $(S(0) - F_{0,T}^P)e^{rT} > 0$. Hence if $F_{0,T}^P < S(0)$, Type A arbitrage is present.

Suppose $F_{0,T}^P > S(0)$. The investor can sell a prepaid forward and purchase the asset at time $t = 0$ and (the buyer has entered into a long position on the security and a short position on the forward). Since $F_{0,T}^P - S(0) > 0$ the buyer has a positive cash flow at $t = 0$ which can be invested at the

---

[1] A more detailed description of the practice of shorting stocks is contained in McDonald (2013).

risk-free rate $r$. At $t = T$, the investor must transfer ownership of the asset to the counter-party who purchased the forward. The cash flow at $t = T$ is therefore zero. The investor has made a profit of $(S(0) - F_{0,T}^P)e^{rT} > 0$. Hence if $F_{0,T}^P > S(0)$, Type A arbitrage is present. Consequently $F_{0,T}^P = S(0)$. $\square$

An alternative proof of the pricing formula for the prepaid forward contract makes use of a present value argument and the Arbitrage Theorem (Thm. 3.6). Since the stock is worth $S(0)$ at time $t = 0$ and has a risk-neutral expected value of $\mathbb{E}^*(S(T)) = S(0)e^{rT}$ at time $t = T$, then the present value of the prepaid forward is

$$F_{0,T}^P = \mathbb{E}^*(S(T))e^{-\mu T} = S(0).$$

Summarizing the prepaid forward contract, for a non-dividend paying asset the value of the prepaid forward contract is the same as the present value of the asset. The reader may question the fairness of paying "full price" for an asset for which transfer of ownership will not be made until a future time. However, this is the correct no arbitrage price. To settle any misgivings a reader might have about this, conduct the following thought experiment. As long as the stock pays no dividends, under the assumption of risk-neutral pricing, the rate of return on the stock is the same as the rate of return on risk-free savings. Think of buying the stock through the mail. Even though the mail delivery will occur a few days after the purchase, the only behavior of the stock of interest is the market fluctuation of its value, which occurs whether or not the stock is physically in the buyer's possession. From the risk-neutral viewpoint the change in the value of the stock will on average be at the risk-free savings rate.

It is more common that an investor will purchase a forward contract rather than a prepaid forward. Recall that a forward contract is similar to a prepaid forward except that the payment for the forward and the transfer of the ownership of the security take place simultaneously at $t = T > 0$ while the price $F_{0,T}$ of the forward contract is determined at time $t = 0$. In this section assume that the risk-free interest rate (which may be for example the interest rate paid on US Treasury Bonds) is denoted by $r$ and this interest is compounded continuously. The reader can easily modify the arguments given below to other interest compounding schedules. Assume that parties can borrow and lend money at rate $r$. To determine the price of a forward contract, a quick intuitive approach is that since only the timing of the payment is different between the forward contract and the prepaid forward, the no arbitrage price of a forward contract should be the future value of the prepaid forward, all other conditions being the same.

**Theorem 5.2.** *Suppose a share of a non-dividend paying security is worth $S(0)$ at time $t = 0$ and that the continuously compounded risk-free interest rate is $r$, then the price of the forward contract is*

$$F_{0,T} = S(0)e^{rT}. \tag{5.1}$$

*Proof.* Suppose $F_{0,T} < S(0)e^{rT}$. An investor can enter into a long position on the forward and short the security at time $t = 0$. The value of the security is $S(0)$ which is invested at the risk-free rate compounded continuously. Thus the net cash flow at time $t = 0$ is $S(0) - S(0) = 0$. At $t = T$, the investor's cash balance is $S(0)e^{rT}$. The buyer pays $F_{0,T}$ as the delivery price of the forward in order to receive the security which is then used to unwind the short position. The cash flow at $t = T$ is therefore $-F_{0,T}$. Thus the net cash flow at $t = T$ is $S(0)e^{rT} - F_{0,T} > 0$. For no investment cost, the investor has made a positive profit. Hence if $F_{0,T} < S(0)e^{rT}$ Type B arbitrage is present.

Suppose $F_{0,T} > S(0)e^{rT}$. An investor can enter into a short forward contract with a delivery price of $F_{0,T}$ and maturity at time $t = T$ and borrow $S(0)$ at the risk-free rate to purchase the security at time $t = 0$. Thus the net cash flow at time $t = 0$ is $S(0) - S(0) = 0$. At $t = T$, the investor must repay the loan of $S(0)e^{rT}$ and will receive the delivery price of $F_{0,T}$. Thus the net cash flow at $t = T$ is $F_{0,T} - S(0)e^{rT} > 0$. Hence if $F_{0,T} > S(0)e^{rT}$ Type B arbitrage is present. Therefore the no arbitrage price of the forward contract on the non-dividend paying security is $F_{0,T} = S(0)e^{rT}$. $\qquad\square$

The reader should bear in mind that at the moment a forward contract has its delivery price determined (generally assumed to be at $t = 0$), the forward contract has no value. However, the value of the underlying asset may change after $t = 0$ and thus the value of the forward contract may at later times be positive, negative, or zero. Consider the following situation. A forward contract with maturity time $t = T$ is created at time $t = 0$ and has a delivery price of $F_{0,T} = S(0)e^{rT}$. If the forward had been created at $0 < t = \tau < T$, the delivery price would have been $F_{\tau,T} = S(\tau)e^{r(T-\tau)}$. Note that both forwards have the same delivery date. What was the value of the first forward contract at time $t = \tau$? At maturity the difference in the values of the forwards is the difference in the delivery prices for the underlying asset,

$$S(0)e^{rT} - S(\tau)e^{r(T-\tau)} = e^{rT}\left(S(0) - S(\tau)e^{-r\tau}\right).$$

The time $t = \tau$ present value of this difference is

$$e^{r(T-\tau)}\left(S(0) - S(\tau)e^{-r\tau}\right).$$

**Example 5.2.** Suppose the risk-free continuously compounded interest rate is 6% per annum. On February 1 the price of an asset is $100 and an investor enters into a long forward contract for the asset with a delivery date of June 1. On March 1 the price of the underlying asset is $110. What is the value of the forward contract?

**Solution.** The delivery price for a long forward created on February 1 is $100e^{0.06(4/12)} \approx \$102.02$. The delivery price for a long forward struck on March 1 is $110e^{0.06(3/12)} \approx \$111.66$. Thus the value of the 4-month forward contract on March 1 is approximately

$$(111.66 - 102.02)e^{-0.06(3/12)} \approx \$9.50.$$

Once a prepaid forward or forward contract has been priced and purchased, an investor will be interested in the profit from the transaction. The value of the security at time $t = T$ is a random variable and may differ from $\mathbb{E}\left(S(T)\right)$. The profit from the forward contract is defined as the difference between the price of the contract and the value of the security when $t = T$. This is the amount of money the investor would make if they immediately sold the security at time $t = T$, the time at which ownership is transferred to the investor. Mathematically this is expressed as

$$\text{profit} = S(T) - S(0)e^{rT}.$$

**Example 5.3.** Suppose a share of a non-dividend-paying security is currently trading for $25 and the risk-free interest rate is 4.65% per annum. What is the delivery price of a 2-month forward contract on the security? Plot the profit for the forward contract as a function of $S(2/12)$.

**Solution.** The price of a 2-month forward contract is

$$F_{0,2/12} = 25e^{0.0465(2/12)} \approx 25.1945.$$

A plot of the profit curve is shown in Fig. 5.1. A positive profit is made if at time $t = 2/12$ the stock is trading above $F_{0,2/12}$.

In the following section the pricing formulas developed above will be generalized to include the effects of transaction costs and assets that pay dividends.

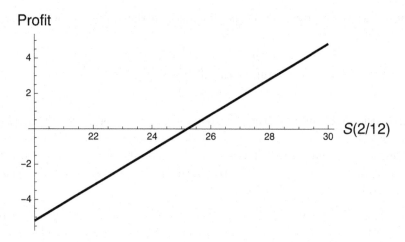

Fig. 5.1   The profit on a long forward contract is a linear function of the stock price at time $t = T$.

### Exercises

**5.2.1** Show that in the absence of arbitrage, the price paid for outright purchase of an asset should be $S(0)$, the price of the asset at the time of outright purchase.

**5.2.2** Suppose the continuously compounded risk-free interest rate is 5.05% per year. What is the cost to the buyer for a fully leveraged purchase of an asset worth now $17 and for which payment will be made in 1 month?

**5.2.3** Suppose the continuously compounded risk-free interest rate is 4.75% per year. What is the delivery price of a 3-month forward contract on a non-dividend paying security whose value currently is $23?

**5.2.4** Suppose a stock price is $45 and the continuously compounded risk-free interest rate is 4.25%. What is the delivery price of a 6-month forward contract for the stock, if the stock pays no dividends?

**5.2.5** Suppose the price of a non-dividend-paying security is $65 and the delivery price of a 6-month forward contract for the security is $68. What is the implied risk-free interest rate?

**5.2.6** A long forward contract with a maturity date in 8 months has a delivery price of $250. One month later the delivery price of an otherwise identical forward contract is $252. If the risk-free interest rate is 3% per annum compounded continuously, what is the value of the 8-month forward contract with 7 months to maturity?

**5.2.7** If a non-dividend-paying security is priced at \$1,200, the risk-free interest rate is 3.75%, and the delivery price of a 12-month forward contract is \$1,250, what arbitrage could an investor undertake?

**5.2.8** If a non-dividend-paying security is priced at \$1,200, the risk-free interest rate is 3.75%, and the delivery price of a 12-month forward contract is \$1,210, what arbitrage could an investor undertake?

**5.2.9** Suppose a security pays no dividends and the risk-free rate of continuously compounded interest is $r$. An investor has the choice between the following investments:

(i) Purchasing one share of the security at time $t = 0$ and selling it at time $t = T > 0$,
(ii) Entering into a long forward contract for the security at time $t = 0$ and lending the present value of the delivery price until time $t = T$.

Show that the payoffs of the two investments are the same.

**5.2.10** Suppose that $F_{0,T_1}$ and $F_{0,T_2}$ are forward contract delivery prices for contracts on the same stock with maturity dates of $T_1$ and $T_2$ respectively with $T_1 < T_2$. Show that

$$F_{0,T_2} = F_{0,T_1} e^{r(T_2 - T_1)},$$

where $r$ is the risk-free continuously compounded interest rate.

**5.2.11** Currency forwards are sometimes used by banks to eliminate risk due to uncertainties in future currency exchange rates when debts must be paid in a foreign currency. The currency forward allows the spot exchange rate to be "locked in" even though the debt will not be repaid until sometime in the future (generally one to 6 months hence [Cuthbertson and Nitzsche (2004)]). Suppose the spot exchange rate of euros for dollars is $S$ (*i.e.*, $S(0)$ euros will currently buy \$1). Let the annual continuously compounded interest rate for dollars be $r$ and the annual continuously compounded interest rate for euros be $r_e$. Let $F_{0,T}$ be the forward exchange rate (again in euros for dollars). Show that in the absence of arbitrage

$$F_{0,T} = S(0)e^{(r - r_e)T},$$

where $T$ is the time to maturity of the forward.

## 5.3   Dividends and Pricing

Dividends are paid to owners of shares of stock from the profits earned by the corporation issuing the shares. Not all shares earn dividends. Dividends on shares from individual corporations may be paid annually semi-annually, or quarterly. Occasionally it will be convenient to think of dividends on large and diverse collections of shares as paid continuously. Note that dividends are paid to the *shareholders*, not to the owners of prepaid forwards or forward contracts. When pricing a prepaid forward or forward contract, the effect dividends paid to shareholders between $t = 0$ and $t = T$ must be carefully considered, since the owners of forwards will not receive these disbursements. The delivery price of the forward must be decreased by the present value of the dividends paid during the interval $[0, T]$. If the risk-free interest rate is $r$, then the time-0 present value of an amount $D$ paid at time $t$ where $0 \le t \le T$ is $De^{-rt}$. Consequently if $n$ dividends in the amounts $\{D_1, D_2, \ldots, D_n\}$ are paid at times $\{t_1, t_2, \ldots, t_n\}$ in the interval $[0, T]$, the delivery price of a prepaid forward on a stock currently valued at $S(0)$ becomes

$$F_{0,T}^P = S(0) - \sum_{i=1}^{n} D_i e^{-rt_i}. \tag{5.2}$$

Using the idea that the delivery price of a forward contract is the future value of the prepaid forward, the price of the forward contract for a stock paying dividends at discrete times can be expressed as

$$F_{0,T} = S(0)e^{rT} - \sum_{i=1}^{n} D_i e^{r(T-t_i)}. \tag{5.3}$$

**Example 5.4.** Suppose the risk-free interest rate is 4.75%. A share of stock whose current value is $121 per share will pay a dividend in 6 months of $3 and another in 12 months of $4.

(i) What is the delivery price of a 1-year prepaid forward contract on the stock assuming that transfer of ownership will take place immediately after the second dividend is paid?

(ii) What is the delivery price of a 1-year forward contract on the stock assuming that transfer of ownership will take place immediately after the second dividend is paid?

**Solution.** The delivery price of the prepaid forward is

$$F_{0,1}^P = 121 - 3e^{-0.0475(6/12)} - 4e^{-0.0475(12/12)} \approx \$114.256.$$

The delivery price of a forward contract on the dividend paying stock is

$$F_{0,1} = 114.256e^{0.0475(12/12)} \approx \$119.814.$$

If a forward is being purchased on a portfolio of stocks paying dividends at many times, it may be convenient to think of the investment as paying dividends continuously. In this case let the annual dividend rate be denoted by $\delta$. Only simple modifications are needed in the pricing formulas above. The value of a prepaid forward on a stock currently worth $S(0)$ becomes

$$F_{0,T}^P = S(0)e^{-\delta T}. \tag{5.4}$$

The value of a forward contract on the same security is

$$F_{0,T} = S(0)e^{(r-\delta)T}. \tag{5.5}$$

**Example 5.5.** An investment valued at \$117 pays dividends continuously at the annual rate of 2.55%. The risk-free interest rate is 3.95%.

(i) What is the delivery price of a 4-month prepaid forward contract on the stock?
(ii) What is the delivery price of a 4-month forward contract on the stock?

**Solution.** The delivery price of a 4-month prepaid forward on the investment is

$$F_{0,4/12}^P = 117e^{-0.0255(4/12)} \approx \$116.01.$$

The delivery price of a 4-month forward contract on the investment is

$$F_{0,4/12} = 117e^{(0.0395-0.0255)(4/12)} \approx \$117.547.$$

A similar situation arises when pricing a forward contract on a currency exchange. Suppose \$$X(0)$ currently can be exchanged for 1€ (the two currencies mentioned here are irrelevant to the pricing principles used). Suppose the risk-free, continuously compounded, dollar-denominated interest rate is $r$ and the risk-free, continuously compounded, euro-denominated interest rate is $r_f$ (the subscript is used to indicate it is the foreign rate of interest). The dollar-denominated price of a forward contract to exchange \$$X(0)$ for 1€ at time $t = T$ is given by

$$F_{0,T} = X(0)\, e^{(r-r_f)T}. \tag{5.6}$$

Notice the similarity between this forward contract price and the price of a forward contract on a dividend-paying asset given in Eq. (5.5). The dividend rate of asset has been replaced by the foreign risk-free rate. In case the reader is not convinced by Eq. (5.6), the formula can be justified by the following argument.

If an investor would like to have 1€ $T$ units of time from now, they should invest $(1)e^{-r_f T}$ at time $t = 0$ at the foreign risk-free rate. In order to accomplish this investment $X(0)e^{-r_fT}$ dollars should be exchanged at time $t = 0$ for euros. Thus the price of this prepaid forward is

$$F_{0,T}^P = X(0)e^{-r_fT}.$$

The time $t = T$ value of this prepaid forward is the forward contract price shown in Eq. (5.6).

*Exercises*

**5.3.1** Suppose the risk-free interest rate is 3.65%. A share of stock whose current value is $97 per share will pay a dividend in 6 months of $2.50 and another in 12 months of $2.75. Find the delivery prices of a 1-year forward contract and a 1-year prepaid forward on the stock assuming that transfer of ownership will take place immediately after the second dividend is paid.

**5.3.2** A $80 stock pays a $2 dividend every 3 months, with the first dividend coming 3 months from today. The continuously compounded risk-free interest rate is 4.5%.

(a) What is the price of a prepaid forward contract that expires 12 months from today immediately after the fourth dividend payment?

(b) What is the price of a forward contract that expires 12 months from today immediately after the fourth dividend payment?

**5.3.3** A $75 stock pays a $2 dividend every 3 months, with the first dividend coming 3 months from today. The continuously compounded risk-free interest rate is 3.75%.

(a) What is the delivery price of a prepaid forward contract that matures 1 year from today, immediately after the fourth dividend is paid?

(b) What is the delivery price of a forward contract that matures 1 year from today, immediately after the fourth dividend is paid?

**5.3.4** The risk-free interest rate is 5% per annum compounded continuously and the dividend yield on a security is 2$ per annum. The current value of the security is $225. What is the delivery price of a 9-month forward contract for the security?

**5.3.5** A $75 stock pays a continuous dividend at the rate of 8%. The continuously compounded risk-free interest rate is 4.75%.

(a) What is the delivery price of a prepaid forward contract that expires 12 months from today?

(b) What is the delivery price of a forward contract that expires 12 months from today?

**5.3.6** An investment valued at $195 pays dividends continuously at the annual rate of 1.95%. The risk-free interest rate is 4.55%. Find the delivery prices of a 3-month prepaid forward and a 3-month forward contract on the investment.

**5.3.7** A $115 stock pays a 2% dividend continuously. The continuously compounded risk-free interest rate is 3.5%.

(a) What is the delivery price of a prepaid forward contract that matures 1 year from today?

(b) What is the delivery price of a forward contract that matures 1 year from today?

**5.3.8** The risk-free interest rate is 6% per annum compounded continuously. The delivery price for a 6-month forward contract on a security currently worth $197 is $200. What is the continuously compounded dividend yield for the security?

**5.3.9** Suppose the price of a stock is $85 and the continuously compounded interest rate is 6.25%. If the delivery price for a 12-month forward contract on the stock is $89, what is the continuous dividend yield on the stock?

**5.3.10** A security is currently priced at $1,000. The continuously compounded risk-free interest rate is 5.05% annually. The delivery price of a 6-month forward contract for the security is $990. If the security pays dividends continuously at rate $r$, find $r$ expressed as an annual percentage.

**5.3.11** Suppose a stock price is $45 and the continuously compounded risk-free interest rate is 4.25%.

(a) What is the delivery price of a 6-month forward contract for the stock, if the stock pays no dividends?

(b) If the stock pays a continuous dividend and a 6-month forward contract for the stock has a delivery price of $44, what is the continuous dividend yield for the stock?

**5.3.12** If the current price of a stock is $1,100 and the continuously compounded risk-free interest rate is 5%. In the market, there is a 9-month forward contract on the stock with a delivery price of $1,129.257.

(a) If the dividend yield will be $\delta = 0.005$, what arbitrage investment could an investor undertake?

(b) If the dividend yield will be $\delta = 0.03$, what arbitrage investment could an investor undertake?

**5.3.13** Suppose an investor has the choice of two investments:

**A:** Long position in a forward contract maturing at time $t = T > 0$ on a stock which pays dividends of amount $D_i$ at time $0 < t_i \leq T$ for $i = 1, 2, \ldots, n$ and lending the present value of the delivery price of the forward contract at time $t = 0$, with the loan to be repaid at time $t = T$. Selling the stock immediately after taking ownership at time $t = T$.
**B:** Long position on the stock at time $t = 0$ and selling the stock at time $t = T$.

Show that the two investments have the same payoff.
**5.3.14** A bond is currently selling for \$1,050. An investor owns a forward contract on this bond. The maturity date for the forward is 1 year from now and the delivery price of the contract is \$1,000. The bond pays an \$50 dividend every 6 months. One dividend will be paid six months from now and a second dividend will be paid 1 year from now just prior to the maturity of the forward contract. Find the current value of the forward contract if the interest rate for six months is 2.75% and the interest rate for 1 year is 3.25% compounded semiannually.
**5.3.15** Suppose 107.68¥ can be exchanged at present for \$1. Furthermore suppose the risk-free interest rate for yen savings is 3% and the risk-free interest rate for dollar savings is 4% (both rates compounded continuously). Find the price in dollars of a prepaid forward and a forward contract delivering ten billion yen in 3 months.
**5.3.16** Suppose the current \$/€ exchange rate is 1.23 and the 1-year forward \$/€ exchange rate is 1.25. The continuously compounded, risk-free interest rate for dollar-denominated saving and borrowing is 6% and the continuously compounded, risk-free rate for euro-denominated saving and borrowing is 4%. Describe an arbitrage opportunity available to an investor.

## 5.4    Incorporating Transaction and Storage Costs

The previous sections showed there are unique no arbitrage delivery prices for prepaid forwards and forward contracts. This ignores the possibility of there being transaction costs associated with the buying and selling of the security and the forward. This section generalizes the specifications surrounding the forward contract and determines an interval of no arbitrage delivery prices for the forward rather than a single no arbitrage price. This section uses of the following definitions and notation for the components of the transactions.

$S^a$: The time $t = 0$ ask price at which the security can be bought.

$S^b$: The time $t = 0$ bid price at which the security can be sold. In general $S^b < S^a$.

$r^b$: The continuously compounded interest rate at which money may be borrowed.

$r^l$: The continuously compounded interest rate at which money may be lent. In general $r^l < r^b$.

$k$: The cost per transaction for executing a purchase or sale.

One of the objectives of this section is to derive an interval of forward contract delivery prices of the form $[F^-, F^+]$ for which no arbitrage is possible when $F^- \leq F_{0,T} \leq F^+$. Outside of this interval arbitrage may be possible.

Suppose the forward contract has a delivery price of $F_{0,T}$. In the absence of arbitrage then

$$F_{0,T} \leq (S^a + 2k)e^{r^b T} \equiv F^+.$$

For the sake of contradiction assume that $F_{0,T} > F^+$. At time $t = 0$ the buyer may borrow amount $S^a + 2k$ at the rate $r_b$ to purchase the security and take a short position in the forward contract. Since a transaction cost of $k$ is incurred for both the purchase of the security and the short position of the forward contract, the amount $2k$, in addition to the ask price $S^a$ of the security at time $t = 0$, must be borrowed. As before, the payment for the forward contract will be made at time $t = T > 0$. Thus the net cash flow at time $t = 0$ is zero. At time $t = T$ the loan must be repaid in the amount of $(S^a + 2k)e^{r^b T}$ and the buyer receives $F_{0,T}$ for the forward. The net cash flow at time $t = T$ is therefore

$$F_{0,T} - (S^a + 2k)e^{r^b T} = F_{0,T} - F^+ > 0.$$

Hence when $F_{0,T} > F^+$ arbitrage results.

Likewise in the absence of arbitrage the following lower bound exists for the delivery price of the forward contract,

$$F_{0,T} \geq (S^b - 2k)e^{r^l T} \equiv F^-.$$

For the sake of contradiction assume that $F_{0,T} < F^-$. At time $t = 0$ the buyer can take a long position in the forward contract (for which $F_{0,T}$ will be paid at time $t = T > 0$) and short the security for $S^b$. A transaction cost of $k$ is paid at time $t = 0$ for the forward contract and another transaction cost of $k$ is incurred during the short sale the net proceeds from the sale are $S^b - 2k$. This amount is lent out at interest rate $r^l$ until time $t = T$. At time $t = T$ the buyer's cash balance is $(S^b - 2k)e^{r^l T}$. Also at this time the buyer pays $F_{0,T}$ for the forward contract and closes out the short position in the security. Thus the net cash flow at time $t = T$ is

$$(S^b - 2k)e^{r^l T} - F_{0,T} = F^- - F_{0,T} > 0.$$

Hence when $F_{0,T} < F^-$ arbitrage is possible. To summarize, for the situation when transaction costs are included in the pricing of a forward contract, the arbitrage-free forward contract delivery price must satisfy the inequality

$$(S^b - 2k)e^{r^l T} \leq F_{0,T} \leq (S^a + 2k)e^{r^b T}. \tag{5.7}$$

**Example 5.6.** Suppose the asking price for a certain stock is \$55 per share, the bid price is \$54.50 per share, the fee for buying or selling a share or a forward contract is \$1.50 per transaction, the continuously compounded lending rate is 2.5% per year, and the continuously compounded borrowing rate is 5.5% per year. Find the interval of no arbitrage delivery prices for a 3-month forward contract on the stock.

**Solution.** The delivery price of a 3-month forward contract on the stock would fall in the interval

$$(54.50 - 2(1.50))e^{0.025(3/12)} \leq F_{0,3/12} \leq (55 + 2(1.50))e^{0.055(3/12)}$$

$$\$51.7223 \leq F_{0,3/12} \leq \$58.8030.$$

Some assets may have storage costs associated with their ownership. For example, precious metals such as gold or silver may require storage in secure facilities which charge for this service. Storage costs can be thought of as negative cash flows. Only slight modifications to the pricing formulas already established are necessary. If $c$ is the time $t = 0$ present value of all the cost incurred in the storage of an asset during the interval $[0, T]$, then the party in the short forward position (the party who incurred the storage costs) is entitled to payment for these costs. Thus Eq. (5.1) can be re-written to incorporate the storage costs in the delivery price.

$$F_{0,T} = (S(0) + c)e^{rT}. \tag{5.8}$$

Specifically if storage fees $\{U_1, U_2, \ldots, U_n\}$ are assessed at times $0 \leq t_1 < t_2 < \cdots < t_n \leq T$, then

$$c = \sum_{i=1}^{n} U_i e^{-rt_i}.$$

**Example 5.7.** Morgan is taking a long position in a 6-month forward contract for 10 ounces of gold. The current price of an ounce of gold is $1,202.20. The storage costs for gold are $0.50 per ounce per month payable at the end of each month. The continuously compounded risk-free interest rate is 3.5%. What is the delivery price of this forward contract?

**Solution.** The delivery price for the forward contract is

$$F_{0,6/12} = 10\left(1202.20 + \sum_{i=1}^{6} 0.50e^{-0.035(i/12)}\right)e^{0.035(6/12)} \approx \$12,264.46.$$

Throughout the remainder of this book, unless specifically mentioned, the complications introduced by transaction costs, storage costs, and dividends will be ignored.

*Exercises*

**5.4.1** Suppose the asking price for a certain stock is $97.50 per share, the bid price is $97 per share, the fee for buying or selling a share or a forward contract is $0.50 per transaction, the continuously compounded lending rate is 3.5% per year, and the continuously compounded borrowing rate is 4.5% per year. Find the interval of no arbitrage delivery prices for a 4-month forward contract on the stock.

**5.4.2** Suppose the asking price for a certain stock is $75 per share, the bid price is $74 per share, the fee for buying or selling a share or a forward contract is $2 per transaction, the continuously compounded lending rate is 3% per year, and the continuously compounded borrowing rate is 4% per year. Find the interval of no arbitrage delivery prices for a 6-month forward contract on the stock.

**5.4.3** The current price of silver is $14.83. The storage cost for silver is $0.25 per ounce per quarter payable at the end of the quarter. If the risk-free interest rate is 4.2% per annum continuously compounded, find the delivery price for a 1-year forward contract on 1,000 ounces of silver.

**5.4.4** Suppose the price of copper is $4.10 per ounce. Copper purchased as an investment must be stored until it is sold. The storage cost must be paid monthly in advance and is $0.10 per ounce per month. Assume the risk-free interest rate is 3.5% compounded continuously. Find the price of a forward contract for an ounce of copper due to be delivered to the buyer in 6 months.

**5.4.5** The current price of silver is $14.83. The storage cost for silver is $0.25 per ounce per quarter payable at the beginning of the quarter. If the risk-free interest rate is 4.2% per annum continuously compounded, find the delivery price for a 1-year forward contract on 1,000 ounces of silver.

**5.4.6** Suppose $F_{0,T_1}$ and $F_{0,T_2}$ are the delivery prices of forward contracts on an asset with delivery dates of $T_1$ and $T_2$ respectively with $0 < T_1 < T_2$. Let the risk-free interest rate for borrowing and lending be $r$ and assume $c$ is the present value at time $t = T_1$ of the storage costs associated with the asset incurred during the interval $[T_1, T_2]$. Show that in an arbitrage-free setting

$$F_{0,T_2} \le (F_{0,T_1} + c)e^{r(T_2 - T_1)}.$$

**5.4.7** Suppose the storage cost per unit time for an asset is proportional to the value of the asset (with proportionality constant $\gamma$). Modify Eq. (5.1) to determine the delivery price for a forward contract on the asset. *Hint*: Treat the storage cost as a negative dividend yield.

**5.4.8** At the beginning of June, forward prices on copper were as listed in the table below. The storage cost of copper is $0.0053 per ounce per month payable at the beginning of the month. The forwards mature at the ends of the given months.

| Month | $F$ |
|-----------|--------|
| July | 4.0995 |
| August | 4.0945 |
| September | 4.1215 |
| December | 4.1235 |

Estimate the price of an ounce of copper on at the beginning of June and estimate the continuously compounded interest rate.

**5.4.9** Suppose at the beginning of June a company owns a copper mine which will produce the amounts of copper shown in the table below at the end of each listed month.

| Month | Copper Production (ounces) |
|-----------|----------------------------|
| July | 1,500 |
| August | 1,575 |
| September | 1,625 |
| December | 1,500 |

The cost to mine an ounce of copper is 1.25. Assume the risk-free interest rate is 3% compounded continuously. Using the forward prices for copper in Exercise 5.4.8, find the present value of the mining operation at the beginning of June.

## 5.5 Synthetic Forwards

The price and payoff of a financial instrument can sometimes be replicated by a combination of other financial instruments. In this situation the original instrument is said to have been "synthesized". In this section the method for replicating a forward contract, called a **synthetic forward** will be described. Suppose an investor wishes to have a long forward contract on a non-dividend-paying stock but no contracts are available. As an alternative, the investor borrows $S(0)$ (the current price of the stock) at the risk-free rate $r$ (compounded continuously) and purchases the stock at $t = 0$. The net cash flow of investor at $t = 0$ is zero, the same as the cash-flow of a forward contract on the stock at $t = 0$. At $t = T$, the stock will have value $S(T)$ and is sold, while the loan is repaid. The net cash flow at $t = T$ is thus

$$S(T) - S(0)e^{rT} = S(T) - F_{0,T}.$$

Using Eq. (5.1) this is the same as the payoff of a long forward contract on the stock. Thus by borrowing to purchase the stock, the investor has created a situation with the same $t = 0$ and $t = T$ cash flows as a long forward contract on the stock. This procedure has created a long synthetic forward contract.

If the stock pays dividends at a continuous rate $\delta$, then a synthetic forward can be created by borrowing $S(0)e^{-\delta T}$ at the continuously compounded, risk-free rate and purchasing $e^{-\delta T}$ shares of the stock at $t = 0$. Once again, the net cash flow of the investor at $t = 0$ is zero. Since the investor owns shares of the stock, the investor is entitled to the dividend payments, which are used to purchase additional shares of the stock continuously. At $t = T$, the investor will have a long position in $e^{-\delta T}e^{\delta T} = 1$ share worth $S(T)$. This share can be sold and the loan repaid. The net cash flow of the investor at $t = T$ is

$$S(T) - S(0)e^{-\delta T}e^{rT} = S(T) - S(0)e^{(r-\delta)T} = S(T) - F_{0,T},$$

the same as the cash flow of a long forward on a stock paying a continuous dividend according to Eq. (5.5).

If the stock pays discrete dividends $D_i$ at times $t_i$ $(i = 1, 2, \ldots, n)$ in the interval $(0, T]$ an investor can create a synthetic long forward contract by borrowing $S(0)$ at $t = 0$ and purchasing the stock. The $n$ dividend payments will be immediately invested at the risk-free rate upon receipt. Thus at $t = T$ the investor sells the stock for $S(T)$, withdraws the accumulated value of the dividend payments, and repays the loan. The net cash flow at $t = T$ is

$$S(T) + \sum_{i=1}^{n} D_i e^{r(T-t_i)} - S(0)e^{rT} = S(T) - F_{0,T},$$

using Eq. (5.3).

In each of these scenarios (non-dividend-paying stock, continuous dividends, and discrete dividends) the loan was repaid in a lump sum with continuously compounded interest at time $t = T$. Thus the loan is equivalent to a zero-coupon bond. Thus the synthetic long forward contract (and hence a forward contract) can be thought of as equivalent to a long position in the stock and a short position in a zero-coupon bond. Expressing the ownership of these financial instruments in an equation produces

$$\text{forward} = \text{stock} - \text{zero-coupon bond.}$$

Of course this equation can be manipulated to produce synthetic stock and synthetic zero-coupon bonds, namely

$$\text{stock} = \text{forward} + \text{zero-coupon bond}$$
$$\text{zero-coupon bond} = \text{stock} - \text{forward.}$$

*Exercises*

**5.5.1** Verify that a long forward contract and lending the present value of the forward contract price is equivalent to a share of stock. Be sure to consider the three cases of a non-dividend paying stock, a stock paying continuous dividends, and a stock paying discrete dividends.

**5.5.2** Verify that a long position in a stock and short position in a forward contract price is equivalent to a zero-coupon bond. Be sure to consider the three cases of a non-dividend paying stock, a stock paying continuous dividends, and a stock paying discrete dividends.

**5.5.3** A stock index has a current price of $1,200 and a continuous dividend yield of 1.5% per annum. The risk-free rate for lending and borrowing is 4% compounded continuously. If a 3-month forward price of $1,175 is observed in the market, what arbitrage opportunities can be created?

**5.5.4** A stock index has a current price of $1,200 and a continuous dividend yield of 1.5% per annum. The risk-free rate for lending and borrowing is 4% compounded continuously. If a 3-month forward price of $1,225 is observed in the market, what arbitrage opportunities can be created?

**5.5.5** A stock has a current price of $100 per share and will pay a dividend of $2 in 4 months (and no other dividends). The risk-free rate for lending and borrowing is 5% compounded continuously. If a 6-month forward price of $100 is observed in the market, what arbitrage opportunities can be created?

## 5.6 Futures

A forward contract is very "customizable" in that the terms of the contract can be arranged to the satisfaction of all the parties involved. The date of maturity of the forward, the amount of the asset to be delivered (securities, stocks, or commodities) at the maturity of the contract, and any necessary collateral to be held to reduce the risk of default by one or more of the parties may all be decided by the parties engaged in the contract. The parties may even decide that at maturity they will only exchange the net amount of profit earned by the parties on the transaction instead of actually selling or buying the underlying security or commodity.

**Futures** are similar to this last type of "cash-settled" forward contract with some additional differences. Futures are generally traded in a more structured exchange market. Futures have standardized maturity dates (typically a few months into the future) and standardized amounts of the underlying security or commodity. There are other important differences between forward and futures contracts. There is generally less risk that a party involved in a futures contract will default since daily adjustments to futures contracts take place and are managed by the clearinghouse associated with a futures exchange. The clearinghouse will require a deposit from both the party and the counter-party to the futures contract. This deposit is called a **margin**. The margin protects each party involved in the futures contract against default by the other party. The clearinghouse will then, based on subsequent changes in the futures price, require additional deposits to the margin so as to protect both parties from default. The request for the additional margin deposit is called a **margin call**. Bringing the margin up to the minimum level (sometimes called the **maintenance margin**) allows the futures position to continue. An investor who does not maintain the requested margin will have their positions liquidated.

The process of adjusting the financial amounts owed to the parties in the futures contract is called **marking-to-market**. In contrast, recall that forward contracts are settled on the date of maturity of the contract. A futures exchange will generally have rules governing the practice of trading depending on changes in the price of the contract traded. For example trading on a particular futures contract may be temporarily halted if a downward move in the price suddenly exceeds a specified threshold. The last difference is that due to the standardization and trading infrastructure provided by a futures exchange, futures are easily traded. Futures are described as more liquid than forward contracts. An investor wishing to rid themselves of a particular obligation implied by a futures contract may easily purchase an offsetting opposite contract with the same date of maturity as the original futures contract.

If the risk-free interest rate is constant for the life of the contract, the delivery price of a futures contract is the same as the delivery price of a forward contract. This will be assumed for the remainder of this chapter. The section will also present an extended example of the process of marking-to-market. For the purpose of the example assume an investor is purchasing a 10-day futures contract whose initial delivery price is \$1,000. The price may change daily until maturity. A volume of 1000 futures contracts will be purchased. The initial margin balance will be \$80 per futures contract, or in this case \$80,000. The clearinghouse will require that a minimum margin of 75% of the initial value of the futures contracts be maintained until maturity. The margin will earn interest at the risk-free rate of 10% per annum compounded continuously. The ultimate profit to the investor will be the difference between the final margin balance and the future value of the initial margin balance. To start, the investor deposits a margin of \$80,000. Suppose that on the next day the price of the futures contract has fallen to \$987.90. The change in price of one futures contract is $\Delta F = -12.10$. Multiplied by the 1,000 futures contracts the investor owns, the wealth of the investor has dropped by \$12,100. This loss will be taken from the margin balance. The initial margin balance has earned one day's interest, so its new post-loss balance is

$$80{,}000e^{0.10(1/365)} - 12{,}100 \approx \$67{,}921.92.$$

This current margin balance exceeds 75% of the initial margin balance so no margin call is issued.

After one more day, the price of a futures contract is now \$987.97, a change of $\Delta F = 0.07$ per contract from the previous day. This yields a

Table 5.1 The daily values of the futures price and the margin balance for a hypothetical 10-day futures contract. The dates NOV-16 and NOV-17 are assumed to be a weekend in which no futures trading took place. The number of futures contracts purchased by the investor was 1,000. The margin balance must be maintained at 75% of the initial margin balance. The column headed "Margin Balance" shows the margin balance after interest is credited and the daily change is added, but before the amount in the "Margin Call" column is added to the margin balance.

| Date | Futures Price | Daily Change | Cumulative Change | Margin Balance | Margin Call |
|------|-------|--------|------------|---------|------|
| NOV-11 | 1000.00 | | | 80,000.00 | |
| NOV-12 | 987.90 | −12,100 | −12,100.00 | 67,921.92 | |
| NOV-13 | 987.97 | 70 | −12,033.32 | 68,010.53 | |
| NOV-14 | 990.53 | 2,560 | −9,476.61 | 70,589.17 | |
| NOV-15 | 988.37 | −2,160 | −11,639.21 | 68,448.51 | |
| NOV-18 | 973.89 | −14,480 | −26,128.78 | 54,024.79 | 5,975.21 |
| NOV-19 | 968.70 | −5,190 | −31,325.94 | 54,826.44 | 5,173.56 |
| NOV-20 | 980.82 | 12,120 | −19,214.52 | 72,136.44 | |
| NOV-21 | 981.16 | 340 | −18,879.79 | 72,496.21 | |

1-day gain of $70 and an accumulated gain of

$$-12,100e^{0.10(1/365)} + 70 \approx -\$12,033.32.$$

Since the price of a futures contract increased, no margin call is made and the margin balance earns another day of interest to reach a value of

$$67,921.92e^{0.10(1/365)} + 70 \approx \$68,010.53.$$

The process continues in this manner through the first 4 days of the futures contract. The futures prices, daily gains/losses, cumulative gain/losses, and the margin balance are listed in Table 5.1.

At the end of the first business week (date NOV-15 in Table 5.1) the margin balance is $68,448.51. The next business or trading day is assumed to be NOV-18 and is notable for two reasons. First, 3 days have passed since the previous trading day and thus the margin balance and cumulative gain/loss earned 3 days of interest. Second, the value of a futures contract dropped by $\Delta F = -14.48$. Therefore the cumulative gain/loss is calculated as

$$-11,639.21e^{0.10(3/365)} - 14,480 \approx -\$26,128.78$$

while the margin balance has dropped to

$$68,448.51e^{0.10(3/365)} - 14,480 \approx \$54,024.79$$

which is below the minimum margin balance of $60,000. Hence a margin call in the amount of $60,000 - 54,024.79 = \$5,975.21$ is issued to restore the margin balance to its minimum level. One day later a drop of $\Delta F = -5.19$ in the price of a futures contract triggers another margin call. This daily process of marking-to-market continues until the futures contract matures. The daily values of the futures price and the margin balance are summarized in Table 5.1.

The profit to the holder of the long position in the futures contract is the cumulative gain/loss over the lifetime of the investment. In this example there is a cumulative loss of $18,879.79.

*Exercises*

**5.6.1** Suppose you can purchase a futures contract on 5,000 ounces of silver for $5.25 per ounce. The initial margin is $5,000 and the minimum margin which must be maintained is $4,500. Assume the risk-free interest rate is 0%. What change in the futures price will lead to a margin call?

**5.6.2** Suppose an investor buys 1,000 futures contracts for $875 each. The continuously compounded annual interest rate is 5.5% and the futures contracts will be marked to market weekly. The initial margin is equal to 25% of the value of the contracts purchased and margin equal to 80% of the initial margin must be maintained. What is the greatest price of the contract 1 week later which will trigger a margin call?

**5.6.3** A security is currently priced at $950 per share. An investor purchases a 6-month futures contract on 100 shares of the security. The continuously compounded risk-free interest rate is 6%. The initial margin on the futures contract is 12.5% and the futures position will be marked to market monthly. The maintenance margin is likewise 12.5%. What is the highest security price for which a margin call will be made after 1 month?

**5.6.4** An investor is purchasing a 7-day futures contract whose initial price is $208.99. The price of the contract changes daily following the path described in the table below.

| Day | Price | Day | Price |
|-----|--------|-----|--------|
| 0 | 208.99 | 4 | 218.95 |
| 1 | 208.85 | 5 | 219.08 |
| 2 | 209.33 | 6 | 223.02 |
| 3 | 210.71 | 7 | 224.84 |

A volume of 10,000 futures contracts will be purchased. The clearinghouse will require an initial margin of $20 per contract and the minimum margin will be 90% of the initial margin balance. The margin will earn interest at the risk-free rate of 12% per annum compounded continuously. Create a table for the daily accounting of marking-to-market for this futures contract similar to Table 5.1. Assume that futures trading takes place on weekends and holidays. What is the profit on the futures contract to the investor?

**5.6.5** An investor is purchasing a 11-day futures contract whose initial price is $775. The price of the contract changes daily following the path described in the table below.

| Day | Price | Day | Price |
|-----|--------|-----|--------|
| 1 | 774.67 | 6 | 735.64 |
| 2 | 779.39 | 7 | 741.59 |
| 3 | 778.42 | 8 | 759.88 |
| 4 | 749.56 | 9 | 766.25 |
| 5 | 742.87 | 10 | 805.36 |

A volume of 1,500 futures contracts will be purchased. The clearinghouse requires an initial margin of $100/contract and the minimum margin must be maintained at 90% of the initial margin during the life of the futures position. The margin will earn interest at the risk-free rate of 10% per annum compounded continuously. Create a table for the daily accounting of marking-to-market for this futures contract similar to Table 5.1. Assume that futures trading takes place on weekends and holidays. What is the profit/loss on the futures contract to the investor?

# Chapter 6

# Options

In the world of finance there are many types of financial instruments which go by the name of **derivatives**. One category of derivatives is known as **options**. At its simplest, an option is the right, but not the obligation, to buy or sell an asset such as a stock for an agreed upon price at some time in the future. The agreed upon price for buying or selling the asset is known as the **strike price**. Options come with a time limit at which (or prior to) they must be exercised or else they expire and become worthless. The deadline by which they must be exercised is known as the **exercise time**, **strike time**, or **expiry date**. Often the exercise time will simply be called **expiry**. In the remainder of the text the terms will be used interchangeably. An option to buy an asset in the future is called a **call option**. An option to sell is known as a **put option**. Types of options can also be distinguished by their handling of the expiry date. A **European option** can only be exercised at maturity, while an **American option** can be exercised at or before expiry. Of the two types, the European option is simpler to treat mathematically, and will be the focus of much of the rest of this book. However, in practice, American options are more commonly traded. There is a mathematical price to be paid in terms of complexity for the added flexibility of the American-style option.

Suppose stock in a certain company is selling today for $100 per share. For a number of reasons a investor may not want to buy this stock today, but may want to own it in 3 months. To reduce the risk of financial loss due to a potential large increase in the price of the stock during the next 3-month period, they buy a European call option on the stock with a 3-month strike time and a strike price of $110. At the expiry date, if the price of the stock is above $110 the investor has the right, as holder of the call option, to purchase the stock for $110. Even if the investor no longer wants to hold the stock, they will take the net profit generated by purchasing the stock of \$110 and immediately selling it at the stock's higher market price. Otherwise, if the value of the stock is below \$110 and the investor still wishes to buy the stock, they will let the call option expire without exercising it, and purchase the stock at its market price. Call options allow investors to protect themselves against paying an unexpectedly high price in the future for a stock which they are considering purchasing.

We have used the language of "buying" an option. Thus options themselves have a value (a price), just as the securities which underlie the options have a value. An important issue to consider is, how are values assigned to options? In light of the Arbitrage Theorem of Chapter 3, if the option is mis-priced relative to the security, arbitrage opportunities may be created. This chapter will explore some of the relationships between option and asset prices. Later Chapter 10 will derive the **Black–Scholes** partial differential equation together with the boundary and final conditions which govern the prices of European-style options.

## 6.1   Properties of Options

There are many relationships between the values of options, their underlying securities, and other market parameters. The maintenance of most of these relationships is necessary to eliminate the possibility of arbitrage. This section will cover some of these relationships and develop methods and strategies for proving that these relationships must hold in an arbitrage-free setting. The following notation for options and, variables, and parameters will be used throughout this chapter.

$C^a$: value of an American-style call option
$C^e$: value of a European-style call option
$K$: strike price of an option
$P^a$: value of an American-style put option
$P^e$: value of a European-style put option
$r$: continuously compounded, risk-free interest rate
$\delta$: continuously compounded, dividend yield rate
$S$: price of a share of a security
$T$: exercise time or expiry of an option
$t$: current time, generally with $0 \leq t \leq T$.

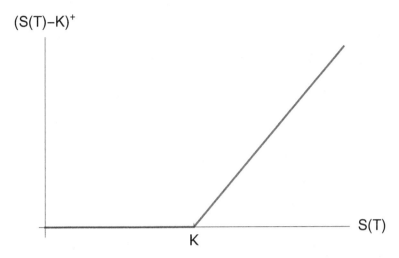

(S(T)–K)⁺

K

S(T)

Fig. 6.1    The piecewise linear curve representing the payoff of a call option.

The time $T$ payoff of a call option is 0 if $S(T) \leq K$ and equals $S(T) - K$ when $S(T) > K$. This piecewise linear payoff is illustrated in Fig. 6.1. In mathematical notation the payoff may be expressed concisely as follows.

$$(S(T) - K)^+ = \max\{S(T) - K, 0\} = \begin{cases} S(T) - K \text{ if } S(T) > K, \\ 0 \qquad\quad \text{ if } S(T) \leq K. \end{cases}$$

For a call option If $S(t) < K$ the option is said to be "out-of-the-money", if $S(t) > K$ the option is "in-the-money", and if $S(t) = K$ the option is "at-the-money".

The option values will be functions of other variables such as time and the price of the underlying asset and parameters such as the strike price and expiration date. When these dependencies are important the value of an option may be denoted as, for instance $C^e(S(t), t; K, T)$, to emphasize that this European call option depends on an underlying stock whose price at time $t$ is $S(t)$ and depends on the strike $K$ and expiry $T$. When those variables and parameters are omitted from the notation, they can be inferred from the context of the discussion. The interest rate $r$ is described as "risk-free". It is assumed that this is the rate of return for an investment which carries no risk. While the reader may wish to debate whether a complete absence of risk can ever be achieved, there are some investments which carry very little risk, for example, U.S. Treasury Bills.

For the initial exploration of option properties, assume that the security pays no dividends. Once the basic relationships have been established, the effects of dividends are straightforward to accommodate.

Consider the cost of an American option compared with a European option. In the absence of arbitrage, an American-style option must be worth at least as much as its European counterpart, *i.e.*, $C^a \geq C^e$ and $P^a \geq P^e$, assuming that the strike prices, strike times, and underlying securities are the same. On the contrary, suppose that an American-style option is worth less than its corresponding European-style option, *i.e.*, $C^a < C^e$. Taking the position of an informed investor, knowing that the American-style option has all the characteristics of the European option and in addition has the increased flexibility that it may be exercised early, no investor would purchase a European-style call option if it cost more than the American-style option. To formulate a more mathematical proof, suppose an investor sells the European-style call option and purchases the American-style option. Since $C^e - C^a > 0$, the investor may purchase a risk-free bond paying interest at rate $r$ compounded continuously. At the expiry date, the bond will have value $(C^e - C^a)e^{rT}$. If the owner of the European option wishes to exercise the option, the investor insures this is possible by exercising their own American option. If the owner of the European option allows it to expire unused, then the investor can do the same with the American option. Thus in both cases the investor makes a risk-free profit of $(C^e - C^a)e^{rT}$ which is an example of arbitrage.

It is also the case that $C^e \geq S(t) - Ke^{-r(T-t)}$ or arbitrage is present. The reader should practice interpreting these inequalities in financial terms. The last inequality can be understood as that a European call option on a non-dividend paying stock must be worth at least as much as the difference between the time $t$ of the stock and the time $t$ present value of the strike price associated with the option. To establish the inequality (and most of the option properties to come) use the technique of proof by contradiction. Assuming an absence of arbitrage and that $C^e < S(t) - Ke^{-r(T-t)}$, an investor can purchase the call option and sell the security at time $t$. Since $0 < Ke^{-r(T-t)} < S(t) - C^e$, the investor can invest $S(t) - C^e$ in a risk-free bond paying interest rate $r$. At the strike time $T$ the bond is worth $(S(t) - C^e)e^{r(T-t)}$ which is greater than $K$. Thus if the investor chooses to exercise the option (because $S(T) > K$), the bond can be cashed out, the security purchased for $K$, and there is still capital left over. In other words there is risk-free profit. On the other hand, if $S(T) < K$, the short position on the security is eliminated using the bond and purchasing the stock for $S(T)$. Again there will still be capital left over, a risk-free profit.

There is a strict arbitrage-free relationship between the value of a European call and put with the same strike price and expiry date written on the same underlying security. This is the important **Put-Call Parity Formula** of Eq. (6.1).

$$P^e(S(t), t; K, T) + S(t) = C^e(S(t), t; K, T) + Ke^{-r(T-t)}. \qquad (6.1)$$

In financial terms, the value of a European put and the stock equals the value of a European call and the present value of the strike price (with the assumptions that the underlying stocks for both options are the same, the stock pays no dividends, and the expiry dates of the options are the same). To see why this relationship must be true, imagine portfolio A represents the left-hand side of Eq. (6.1) while the right-hand side is represented by portfolio B. The Put-Call Parity formula implies that in an arbitrage-free setting these portfolios must have the same value.

Suppose portfolio A is worth less than portfolio B, *i.e.*,

$$P^e(S(t), t; K, T) + S(t) < C^e(S(t), t; K, T) + Ke^{-r(T-t)}. \qquad (6.2)$$

At time $t$ an investor can borrow at interest rate $r$ an amount equal to $P^e + S - C^e$. This would allow the investor to purchase the security, the European put option, and to sell the European call option. At the strike time of the two options, the investor must repay principal and interest in the amount of $(P^e + S - C^e)e^{r(T-t)}$. If the security is worth more than $K$ at time $T$, the put expires worthless and the call will be exercised by its owner. The investor must sell the security for $K$. Thus the net proceeds of this transaction are

$$K - (P^e(S(t), t; K, T) + S(t) - C^e(S(t), t; K, T))e^{r(T-t)} > 0. \qquad (6.3)$$

Since inequality (6.3) is equivalent to the one in (6.2). If the security is worth less than $K$ at time $T$, the call expires worthless and the put will be exercised by the investor. Again the investor will sell the security for $K$. The net proceeds of this transaction are the same as in the previous case. Thus there is a risk-free profit to be realized if portfolio A is worth less than portfolio B.

Now suppose portfolio A is worth more than portfolio B, *i.e.*,

$$P^e(S(t), t; K, T) + S(t) > C^e(S(t), t; K, T) + Ke^{-r(T-t)}. \qquad (6.4)$$

At time $t$ an investor can sell the security and the European put option and buy the call option. This generates an initial positive flow of capital in the amount of $S + P^e - C^e$. This amount will be invested in a risk-free bond earning interest at rate $r$ until the expiry date arrives. At that time the investor will have $(S + P^e - C^e)e^{r(T-t)}$. If the security is worth more than $K$ at time $T$, then the put option is worthless and investor will exercise the call option. The investor purchases the security for $K$ (thus canceling their short position). This leaves the investor with a net gain of

$$(P^e(S(t), t; K, T) + S(t) - C^e(S(t), t; K, T))e^{r(T-t)} - K > 0. \qquad (6.5)$$

Since inequality (6.5) is equivalent to the one in (6.4). If the security is worth less than $K$ at time $T$, the call option expires unused and the owner of the put option will exercise it. Thus the investor will clear their short position by buying the security for $K$ and their net gain is as before. Consequently, there exists an arbitrage opportunity if portfolio B is worth less than portfolio A. Therefore the two portfolios must have the same value, *i.e.*, the Put-Call Parity formula in Eq. (6.1) must hold.

### Exercises

**6.1.1** Show that in the absence of arbitrage $P^a \geq P^e$ where the underlying security, exercise time, and strike price for both options are the same.

**6.1.2** Consider a 6-month European call option with a strike price of $60 which costs $10. Draw a graph illustrating the net profit of the option for stock prices in the interval $[0, 100]$. Assume the continuously compounded risk-free interest rate is 5% per annum.

**6.1.3** Show that in the absence of arbitrage the value of a call option, either European or American, never exceeds the value of the underlying security.

**6.1.4** What is the minimum price of a European style call option with an exercise time of 3 months and a strike price of $26 for a security whose current value is $29 while the continuously compounded interest rate is 6%?

**6.1.5** Express in mathematical notation the payoff of a $K$-strike put option written on a stock whose price is $S(t)$.

**6.1.6** Show that $(S(T) - K)^+ - (K - S(T))^+ = S(T) - K$.

**6.1.7** Consider options (either American or European) which are identical except for the value of the underlying stock $S_1 < S_2$. Show that the following two inequalities hold.

$$C(S_1) \leq C(S_2),$$
$$P(S_1) \geq P(S_2).$$

**6.1.8** Consider options (American or European) which are identical except for the value of the underlying stock $S_1 < S_2$. If $0 < \gamma < 1$ show that the following two inequalities hold.

$$C(\gamma S_1 + (1-\gamma)S_2) \leq \gamma C(S_1) + (1-\gamma)C(S_2),$$
$$P(\gamma S_1 + (1-\gamma)S_2) \leq \gamma P(S_1) + (1-\gamma)P(S_2).$$

**6.1.9** Consider American options which are identical except for strike times $T_1 < T_2$. Show that the following two inequalities hold.

$$C^a(T_1) \leq C^a(T_2),$$
$$P^a(T_1) \leq P^a(T_2).$$

## 6.2   Including the Effects of Dividends

Now that the Put-Call Parity formula (6.1) for non-dividend paying stocks has been developed, incorporating the effects of dividends is a matter of understanding how dividends affect the value of the underlying security. This section will explore how discrete and continuous dividend payments alter the value of a security and modify the Put-Call Parity formula to account for these changes.

Suppose a stock will pay a dividend in the amount of $\delta S(t)$ (that is, a dividend proportional to the value of the stock $S(t)$) at time $t_d > 0$. What happens to the value of the stock at the instant the dividend is paid? Since something of value is leaving the corporation issuing the stock, the value of the company (and hence its stock) should reflect this loss. Consider the values of the stock immediately before and after the payment of the dividend denoted as

$$\lim_{t \to t_d^-} S(t) = S(t_d^-),$$

$$\lim_{t \to t_d^+} S(t) = S(t_d^+).$$

The before and after dividend stock values are related by the following equation.

$$S(t_d^+) = S(t_d^-)(1-\delta) \tag{6.6}$$

This equation states that the value of the stock must diminish by the amount of the dividend payment. A "no arbitrage" argument will establish this rigorously. If $S(t_d^+) > S(t_d^-)(1-\delta)$, then an investor may purchase the stock immediately before the dividend is paid for $S(t_d^-)$, collect a dividend in the amount of $\delta S(t_d^-)$, and sell the stock immediately after the dividend payment for $S(t_d^+)$, producing a net profit of

$$S(t_d^+) + \delta S(t_d^-) - S(t_d^-) = S(t_d^+) - S(t_d^-)(1 - \delta) > 0.$$

If $S(t_d^+) < S(t_d^-)(1 - \delta)$, then an investor may borrow the stock to short sell (in which case the investor must pay the dividend amount $\delta S(t_d^-)$). The investor sells the stock for $S(t_d^-)$ and pays the dividend $\delta S(t_d^-)$. The investor re-purchases the stock for $S(t_d^+)$. This produces a net profit of

$$(1 - \delta)S(t_d^-) - S(t_d^+) > 0.$$

In either case arbitrage is present.

Thus for a single discrete dividend payment the value of the stock drops by the amount of the dividend. This affects the left-hand side of the Put-Call Parity formula (6.1). We do not simply subtract the dividend payment from the left-hand side since this would ignore the time value of the dividend, instead we should subtract the present value of the dividend payment. This idea is generalized to a finite number of discrete dividend payments made between the sale of the options and expiry. If $n$ dividend payments of the form $\delta S(t_i^-)$ will be made at times $t_i^- > t$ for $i = 1, 2, \ldots, n$ then the Put-Call Parity formula for discrete dividend payments can be expressed as

$$P^e(S, t; K, T) + \left( S(t) - \delta \sum_{i=1}^{n} S(t_i^-)e^{-rt_i} \right) = C^e(S, t; K, T) + Ke^{-r(T-t)}.$$

(6.7)

In some situations it is mathematically convenient to think of the dividends as a continuous stream of payments at a rate $\delta$ per unit time. For example, a stock index is a financial instrument made up of hundreds or perhaps thousands of stocks, each of which may be paying dividends on its own schedule. In this context $\delta$ is called the **dividend yield**. If the dividend yield is constant then the dividend paid over a short time interval $[t, t + \Delta t]$ is approximately $\delta S(t)\Delta t$. Therefore

$$S(t + \Delta t) - S(t) = -\delta S(t)\,\Delta t$$

$$\frac{dS}{dt} = -\delta S(t) \text{ (as } \Delta t \to 0)$$

$$S(t) = S(0)e^{-\delta t}.$$

Consequently for European options on securities which pay dividends at a continuous, constant dividend yield $\delta$, the Put-Call Parity formula takes on the form

$$P^e(S(t), t; K, T) + S(t)e^{-\delta(T-t)} = C^e(S(t), t; K, T) + Ke^{-r(T-t)}. \quad (6.8)$$

The observant reader will notice that in each form of the Put-Call Parity formula, for non-dividend-paying securities (see Eq. (6.1)), securities paying discrete dividends (see Eq. (6.7)), and securities paying continuous dividends (see Eq. (6.8)), the formula for the price of a prepaid forward on the security appears. If needed, refer to Thm. 5.1, Eq. (5.2), and Eq. (5.4). Thus the three versions of the Put-Call Parity formula can be unified as in the following theorem.

**Theorem 6.1 (Put-Call Parity).** *A European put option and European call option written on the same underlying security with the same strike price $K$ and strike time $T$ must satisfy the following equation in the absence of arbitrage.*

$$P(S(t),t;K,T) + F^P_{t,T}(S(t)) = C(S(t),t;K,T) + Ke^{-r(T-t)}. \qquad (6.9)$$

*Variable $t$ represents the current time, $r$ is the risk-free continuously compounded interest rate, and $F^P_{t,T}(S(t))$ is a prepaid forward on the underlying security.*

### Exercises

**6.2.1** If a share of a non-dividend-paying security is currently selling for \$31, a 3-month European call option is \$3 with a strike price of \$31, and the risk-free interest rate is 10%, what is the arbitrage-free European put option price for the security?

**6.2.2** If a share of a non-dividend-paying security is currently selling for \$31, a 3-month European call option is \$3, a 3-month European put option is \$2.25, and the risk-free interest rate is 10%, what is the arbitrage-free strike price for the security?

**6.2.3** What is the minimum price of a European put option with an exercise time of 2 months and a strike price of \$14 for a non-dividend-paying stock whose value is \$11 while the continuously compounded interest rate is 7%?

**6.2.4** If a share of a non-dividend-paying security is currently selling for \$31, a 3-month European call option is \$3, a 3-month European put option is \$1, the strike price for both options is \$30, and the risk-free interest rate is 10%, what arbitrage opportunities are available to investors?

**6.2.5** A non-dividend-paying stock has a current value of \$36. A 4-month call option with a strike price of \$38 will cost an investor \$2.25. If the continuously compounded interest rate is 4.75%, find the price of the 4-month put option with a strike price of \$38.

**6.2.6** Show that the Put-Call Parity formula in Eq. (6.9) is equivalent to the equation,

$$C(S(t),t;K,T) - P(S(t),t;K,T) = (F_{t,T} - K)e^{-r(T-t)}.$$

**6.2.7** The current price of a non-dividend-paying stock is $50 and the risk-free interest rate is 6% compounded continuously. The price of a 45-strike European call option is $4.35 greater than the price of a 50-strike European call option. Both options expire in 6 months. Find the difference in the prices of 50-strike and 45-strike European put options on the stock.

**6.2.8** A 3-month European 70-strike call option on a stock is currently selling for $7. The current price of the stock is $71 and a dividend of $1 is expected to be paid in 1 month. No other dividends will be paid in the next 3 months. The risk-free interest rate is 5% continuously compounded. Find the price of a 3-month 70-strike European put option on the stock.

**6.2.9** Suppose an investor discovers the put option described in Exercise 6.2.8 has a market price of $6. Describe an arbitrage opportunity the investor may construct.

**6.2.10** A 4-month European 75-strike call option on a stock is currently selling for $10. The current price of the stock is $80 and the stock pays a continuous dividend at a yield of 1% per annum. The risk-free interest rate is 5% continuously compounded. Find the price of a 4-month 75-strike European put option on the stock.

**6.2.11** Suppose at time $t = 0$ one dollar can be exchanged for 1.1€. The dollar-denominated risk-free interest rate is 5% and the euro-denominated risk-free interest rate is 2%. The price of a 4-month European put to exchange $0.95 for one euro is $0.075. Use the Put-Call Parity formula Eq. (6.9) to find the price of the otherwise equivalent European call.

**6.2.12** Suppose the continuously compounded, risk-free interest rate is $r$, and $F_{0,T}$ is the price (determined at $t = 0$) of a futures contract. If $C^e$ and $P^e$ are the premiums for $K$-strike European call and put options on underlying the futures contract with expiry at $t = T$, develop a Put-Call Parity relationship for $F_{0,T}$, $K$, $C^e$, and $P^e$.

**6.2.13** Use Eq. (6.7) to show that for a European call option on a dividend-paying stock the following inequality holds.

$$C^e \geq S(t) - Ke^{-r(T-t)} - \mathrm{PV}(\mathrm{div}).$$

The expression $\mathrm{PV}(\mathrm{div})$ represents the time $t$ present value of the dividends paid.

## 6.3   Parity and American Options

This section will explore some of the parity relations obeyed by American-style options and comparisons between American-style and European-style

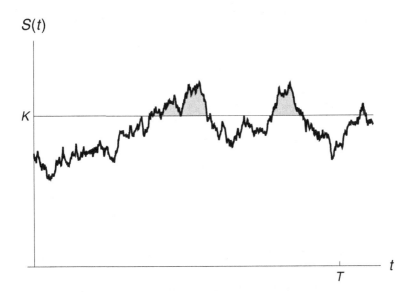

Fig. 6.2   The shaded region indicates the intervals before expiry in which the price of the underlying security $S(t)$ exceeds the strike price $K$. An American call option could be exercised profitably in these intervals, but a European call option could not, and would ultimately expire unused.

options. Earlier an arbitrage argument was used to show that American and European options, on the same underlying security and with the same strike price and expiry date, obey the following inequalities (see Sec. 6.1 and Exercise 6.1.1).

$$C^e(S(t), K, T) \leq C^a(S(t), K, T) \text{ and}$$
$$P^e(S(t), K, T) \leq P^a(S(t), K, T).$$

Intuitively these inequalities hold because the American options give the holder all the rights of the European options with the addition of the possibility of early exercise. There are market scenarios under which an American option may be exercised to yield a profit while the equivalent European option would expire out-of-the-money. See Fig. 6.2. Consider an American call versus a European call on the same underlying security with the same strike price $K$ and expiry time $T$. If the value of the security $S(t) > K + C^a$ for some time $0 \leq t < T$, the American option could be exercised and generate a positive profit. Since $t < T$ the European option could not be exercised. The European option would only generate a positive profit if $S(T) > K + C^e$ and there is no guarantee that will occur.

The European put and call prices also obey the Put-Call Parity formula expressed in Eq. (6.9). However, the American options do not obey the Put-Call Parity equation. American option prices do have bounds which will be explored in this section.

Assume that the American options are created at time $t$ when the price of the underlying security is $S(t)$. The risk-free interest rate is $r$ compounded continuously. If the strike price is $K$ then the values of the American put and call options with identical strike price and expiry $T > 0$ obey the following inequality.

$$C^a + K \geq S(t) + P^a. \tag{6.10}$$

To prove this, assume to the contrary that $C^a + K < S(t) + P^a$. In this case an investor could short the security, sell the put, and buy the call. This produces a cash flow of $S(t) + P^a - C^a$. If this amount is positive it can be invested at the risk-free rate, otherwise it is borrowed at the risk-free rate. If the holder of the American put chooses to exercise the put at time $t \leq \tau \leq T$, the investor can exercise the call option and purchase the security for $K$. At that time the investor's balance is

$$(S(t) + P^a - C^a)e^{r(\tau-t)} - K > Ke^{r(\tau-t)} - K \geq 0.$$

If the American put expires out-of-the-money, the investor will close their short position in the security at time $T$ by exercising the call option. In this case

$$(S(t) + P^a - C^a)e^{r(T-t)} - K > Ke^{r(T-t)} - K > 0.$$

Thus the investor receives a non-negative profit in either case, violating the principle of no arbitrage. Consequently the inequality in (6.10) holds.

Now it is also the case that

$$S(t) + P^a \geq C^a + Ke^{-r(T-t)}. \tag{6.11}$$

A proof by contradiction and a no arbitrage argument will establish this inequality as well. Suppose $S(t) + P^a < C^a + Ke^{-r(T-t)}$. An investor could sell an American call and buy the security and the American put. This generates a cash flow of $C^a - S(t) - P^a$. If necessary the investor will borrow funds at the risk-free rate $r$ compounded continuously. If the holder of the call decides to exercise it at any time $t \leq \tau \leq T$, the investor may sell the security for the strike price $K$. Thus at time $\tau$ the investor's asset balance is

$$(C^a - S(t) - P^a)e^{r(\tau-t)} + K = (C^a + Ke^{-r(\tau-t)} - S(t) - P^a)e^{r(\tau-t)}$$

$$\geq (C^a + Ke^{-r(T-t)} - S(t) - P^a)e^{r(\tau-t)},$$

since $r > 0$. By assumption $S(t) + P^a < C^a + Ke^{-r(T-t)}$, so the last expression above is positive. The investor has earned a risk-less positive profit, contradicting the no arbitrage assumption. Therefore the inequality in (6.11) is true.

By rearranging terms in Eqs. (6.10) and (6.11) and combining the two inequalities the following theorem is proved.

**Theorem 6.2.** *If the risk-free interest rate is $r$ compounded continuously, if $C^a$ and $P^a$ are the values of American call and put options respectively both with strike price $K$ and expiry $T$ written on a non-dividend-paying security, and if the value of the underlying security is $S(t)$, then*

$$S(t) - K \leq C^a - P^a \leq S(t) - Ke^{-r(T-t)}. \tag{6.12}$$

Perhaps one of the most unexpected results governing the value of American-style options is the equality holding between $C^e$ and $C^a$ for non-dividend-paying stocks. We have seen that $C^a \geq C^e$ for American and European call options with the same strike price, expiry time, and underlying security. We have also seen a simulation of an example for which the American call is in the money for a period before expiry and could be exercised to generate a positive profit while the European option would be out-of-the-money at expiry (Fig. 6.2). Since the American call, through possible early exercise, may generate a greater profit than its European counterpart, it is surprising that $C^a = C^e$. However, the usual type of no arbitrage proof will establish the equality. Suppose that $C^a > C^e$. Since the American call is (supposedly) worth more, an investor could sell the American call and buy a European call with the same strike price $K$, expiry date $T$, and underlying security. The net cash flow $C^a - C^e > 0$ would be invested at the risk-free rate $r$. Assuming the interest is compounded continuously, at any time $t \leq \tau \leq T$, the future value is $(C^a - C^e)e^{r(\tau-t)}$. If the holder of the American call chooses to exercise the option at some time $\tau \leq T$, the investor may sell short a share of the security for amount $K$ and add the proceeds to the amount invested at the risk-free rate. At time $T$ the investor must close out the short position in the security and may use the European option to do so. Upon settlement of the short position the holdings of the investor are

$$(C^a - C^e)e^{r(T-t)} + K(e^{r(T-\tau)} - 1) > 0.$$

If the American option is not exercised, the investor has holdings of $(C^a - C^e)e^{r(T-t)} > 0$ at expiry. A rational investor will only exercise the European option if profit is increased, else the European option will be allowed to expire unused. In either case the investor earns a risk-free positive profit. This completes the proof of the next theorem.

**Theorem 6.3.** *If $C^a$ and $C^e$ are the values of American and European call options respectively on the same non-dividend-paying security with identical strike prices and expiry times, then*

$$C^a(S(t), K, T) = C^e(S(t), K, T). \qquad (6.13)$$

Now that there is a means of determining the value of an American call on a non-dividend paying security, it can be used to establish a bound on the value of an American put on the same underlying security with the same expiry and strike price.

**Example 6.1.** The price of a non-dividend-paying security is currently $36, the risk-free interest rate is 5.5% compounded continuously, and the strike price of a 6-month American call option worth $2.03 is $37. Find a range of no arbitrage values for a 6-month American put on the same security with the same strike price.

**Solution.** The interval of no arbitrage prices can be found by making use of the inequality in (6.12).

$$36 - 37 \leq 2.03 - P^a \leq 36 - 37e^{-0.055(6/12)}$$
$$2.03 \leq P^a \leq 3.03.$$

The American put may be worth more than the European put (all parameters of the option being equal) due to the possibility of early exercise (even for non-dividend-paying stocks).

The preceding discussion assumes the underlying stock pays no dividends. In practice options are frequently written on stocks that do pay dividends. Typically individual stocks pay dividends one to four times per year. Stock indices formed by bundling a large number of stocks may be assumed to pay dividends continuously. Options are sometimes written on foreign currencies which pay a continuously compounded "dividend" in the form of the risk-free interest rate on the currency. The payment of a dividend may also trigger the early exercise of an American call option. Fortunately it is not difficult to generalize Eq. (6.12) to include dividends.

**Theorem 6.4.** *If the risk-free interest rate is $r$ compounded continuously, if $C^a$ and $P^a$ are the values of American call and put options respectively* *both with strike price K and expiry T written on a security whose value is S(t), then*

$$F_{t,T}^P(S) - K \leq C^a - P^a \leq S(t) - Ke^{-r(T-t)}, \qquad (6.14)$$

*where $F_{t,T}^P(S)$ is the price of a prepaid forward on the security.*

*Proof.* The proof will proceed as before by contradiction. Assuming $F_{t,T}^P(S) - K > C^a - P^a$ is equivalent to the inequality

$$K < F_{t,T}^P(S) + P^a - C^a.$$

An investor can sell the prepaid forward and an American put on the security while purchasing an American call. This generates a positive cash flow in excess of $K$ at time $t$. This cash flow is lent at the risk-free rate. Since the security has been borrowed in order to be shorted, it is the responsibility of the investor to pay the dividends to the buyer of the security, hence the need to invest not only the strike price but the present value of the dividends. During the life of the American put, the investor pays the buyer of the security the dividends due. If the holder of the American put chooses to exercise at time $\tau \in [t, T]$, the investor pays them $K$ for the security and receives the security which they use to close out the short position. The investor's portfolio now is worth

$$C^a + (F_{t,T}^P(S) + P^a - C^a)e^{r(\tau-t)} - K > C^a + Ke^{r(\tau-t)} - K > 0.$$

If the owner of the American put allows the option to expire, then the investor may exercise the American call at expiry to purchase the security for $K$ and close out their short position. At expiry the investor's portfolio is worth

$$(S(T) - K)^+ + (F_{t,T}^P(S) + P^a - C^a)e^{r(T-t)} - S(T)$$
$$> (S(T) - K)^+ + Ke^{r(T-t)} - S(T) = Ke^{r(T-t)} - K > 0.$$

In either case the investor receives a positive profit. Thus the inequality on the left of Eq. (6.14) has been established. The inequality on the right is the same as that in Eq. (6.12) and thus the same proof as in Thm. (6.2) holds. □

Comparing Eqs. (6.12) and (6.14) demonstrates that the payment of dividends may widen the separation in the prices of the American call and put options. When no dividend is paid the width of the interval potentially separating the two options is $K(1 - e^{-r(T-t)})$. Including the payment of dividends the width is increased to

$$S(t) - F_{t,T}^P(S) + K(1 - e^{-r(T-t)}).$$

The expression $S(t) - F_{t,T}^P(S)$ represents the dividend growth of the security during the interval $[t, T]$. This increased separation between the prices of the American call and put can be attributed to the increased uncertainty associated with whether the call option will be exercised prior to the dividend date, so as to receive the dividend.

## Exercises

**6.3.1** The price of a non-dividend-paying security is currently \$50, the risk-free interest rate is 6% compounded continuously, and the strike price of a 3-month American put option worth \$9.75 is \$51. Find the range of no arbitrage values of a 3-month American call on the same security with the same strike price.

**6.3.2** The price of a non-dividend-paying security is currently \$93, the risk-free interest rate is 5.74% compounded continuously, and the strike price of a 2-month American call option worth \$11.77 is \$90. Find the range of no arbitrage values of a 2-month American put on the same security with the same strike price.

**6.3.3** The price of a security is currently \$115, the risk-free interest rate is 3.75% compounded continuously, and the strike price of otherwise identical 6-month American call and put options is \$110. Find the maximum difference in the prices of the American call and put options if the security will pay a single dividend of 10% in 3 months time. Assume the price of the security remains constant during the lives of the options.

**6.3.4** The price of a security is currently \$98, the risk-free interest rate is 2.95% compounded continuously, and the strike price of otherwise identical 4-month American call and put options is \$100. Find the maximum difference in the prices of the American call and put options if the security pays a continuous stream of dividends at a rate of \$1 per month.

**6.3.5** A stock is currently selling for \$35 per share. The stock pays a continuous dividend at the rate of 3% per annum. The risk-free interest rate is 5% compounded continuously. A 9-month, \$36-strike American call option on the stock is sold for a premium of \$2.25. Find the range of no arbitrage prices for the associated American put.

**6.3.6** It was demonstrated that in the absence of arbitrage, the value of a European option on a non-dividend-paying stock satisfies the inequality $C^e \geq S - Ke^{-rT}$, where $K$ is the strike price, $r$ is the risk-free interest rate, and $T$ is the expiry date. Consider another type of option known as a perpetual call. The perpetual call can be exercised at any time and never expires. Show that the value of the perpetual call is $C = S$.

**6.3.7** Consider an American call option on a stock which will pay a dividend of $D$ at time $0 < t_d < T$ where $T$ is the expiry of the option. Show that if

$$D \leq K \left( 1 - e^{-r(T - t_d)} \right),$$

the option will not be exercised at time $t_d$. *Hint*: Consider the value of the option just before and just after $t_d$ using the result of Exercise 6.2.13.

## 6.4 Option Strategies

Earlier in this chapter some of the properties of put and call options were discussed. The most important of these is the Put-Call Parity formula, Eq. (6.9). In this section common uses of options will be explored to reinforce the idea that options function as a form of insurance against changes in the value of an asset (or the lack of change in the value of an asset). The strategies to follow will include cases of an investor being in a long as well as a short position in an asset and situations in which options are purchased from or sold to others.

The simplest case to consider is analogous to the purchase of homeowner's insurance. Suppose an investor has purchased a stock for price $S(0)$ and wishes to insure against the loss of value by the stock. The investor may purchase a put option which establishes the minimum price, the strike price $K$, for which the stock may be sold in the future. Homeowners purchase insurance on their dwellings to make certain that regardless of the occurrence of damages to the home due to natural events (within limits laid out in the insurance policy), the home maintains a minimum value, usually sufficient to provide replacement shelter in case the dwelling is no longer useable. The strategy of purchasing a put to insure a long position in an asset is called a **floor** since the asset holder is placing a "floor" under the value of the asset. This is apparent from the profit diagram of the strategy. Assume the investor borrows funds at the risk-free rate to finance the portfolio. If the asset is purchased for $S(0)$ and the put for $P$ while the strike price is $K$, the continuously compounded, risk-free interest rate is $r$, then at time $t$ with $0 \leq t \leq T$, the expiry time the value of the portfolio is

$$\max\{S(t), K\} - (S(0) + P)e^{rt}.$$

Since the investor purchased a put with strike price $K$, the stock can always be sold for a minimum of $K$. At a fixed time $t$ the graph of the profit resembles that in Fig. 6.3.

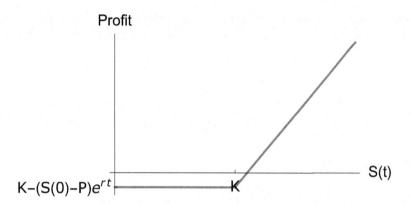

Fig. 6.3    The profit diagram for a portfolio consisting of long positions in an asset and a put option on that asset.

For the case of an investor employing the floor strategy, it is plausible that the investor believes the asset will increase in value and wants to insure against the possibility of it losing value. In contrast if the investor believes the asset will not increase in value, they may want to sell a call option on the asset to another investor. Since the investor who owns the asset is creating and selling the call option, this is sometimes called **writing a covered call**. If the asset is purchased for $S(0)$, the covered call with strike price $K$ is sold for $C$, and the continuously compounded, risk-free interest rate is $r$, then at time $t$ with $0 \leq t \leq T$ the profit from the covered call strategy is

$$\min\{S(t), K\} + (C - S(0))e^{rt}.$$

If the asset does not increase in value the call writer keeps the premium charged for the call and hence the call writer may earn a positive profit even when the asset does not appreciate in value. See Fig. 6.4.

The option strategies available to the investor in a short position in an asset mirror those of the investor in a long position. A short position in an asset must be cleared at a time in the future and the investor may want to insure against the price of the asset rising too high. In this case the investor may want to use an option strategy known as a **cap**, so called because a "cap" is placed on the amount the investor will have to pay to clear the short position. The cap consists of purchasing a call option on the asset. If the cash generated by the short sale of the asset is $S(0)$, the price of the call is $C$, the continuously compounded, risk-free interest rate is $r$, and the

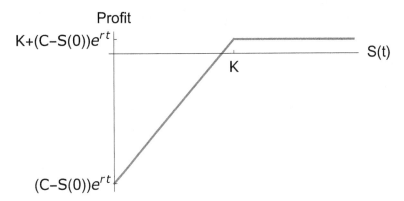

Fig. 6.4   The profit diagram for a portfolio consisting of a long position in an asset and short call.

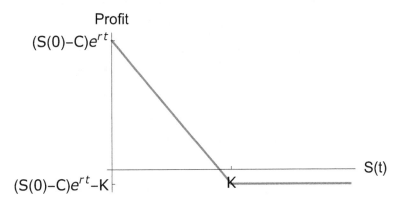

Fig. 6.5   The profit diagram for a portfolio consisting of a short position in an asset and a long call option on that asset.

strike price of the call is $K$, then at the time the call is exercised to clear the short position, the profit can be expressed as

$$(S(0) - C)e^{rt} - \min\{K, S(t)\}.$$

The net cash flow generated setting up the portfolio earns interest at the risk-free rate. The investor pays either the strike price or the time $t$ price for the asset (naturally whichever is lower) to close out the short position. Figure 6.5 illustrates the shape of the profit diagram.

An alternative available to the investor who has shorted an asset is to sell a put option. This strategy is known as writing a **covered put**. Since the investor has sold the put, they are obligated to buy the asset at the

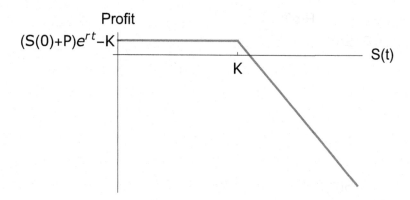

Fig. 6.6    The profit diagram for a portfolio consisting of a short position in an asset and short covered put.

strike price of the put, if the option is exercised. This may mean buying the asset at a loss if the current price is below the strike price (keep in mind the investor has shorted the stock and thus does not own it). The investor will have to buy the asset to close the short position in any case, thus the proceeds from the sale of the covered put may provide some extra, positive cash flow to offset the extra cost of the asset if it increases in value. If the put is not exercised, the investor earns the premium charged for the put. The profit of the covered put strategy is given by the following expression.

$$(S(0) + P)e^{rt} - \max\{S(t), K\}.$$

The proceeds from the short sale and the sale of the covered put earn interest at the risk-free rate. At time $t$ that the put is exercised, the investor purchases the stock for the strike price $K$ (in this case $S(t) < K$) or for the current price $S(t)$ (in this case the option expires unused). See Fig. 6.6.

The strategies considered so far have each involved an asset and a single option (either purchased or sold). More complicated strategies can be formulated with multiple options. The next category of strategies to be discussed involves two options of the same type (two calls or two puts) and are called **spreads**. When an investor purchases an option a premium is paid, *i.e.*, there is a cost. This cost decreases the future profit of the investment (but also lowers the risk). Investors can offset some of the cost of the purchased option by selling an otherwise identical option with a different strike price. In Exercise 6.4.1 the reader will show that calls are non-increasing functions of the strike price while puts are non-decreasing functions of the strike price. Thus, an investor may offset the cost of a call option with strike

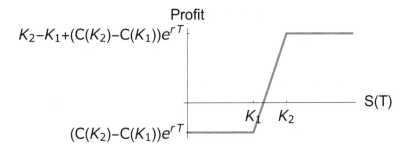

Fig. 6.7    The profit diagram for a typical bull spread.

price $K_1$ by selling a call option with strike price $K_2 > K_1$. Some out-of-pocket premium must still be paid, but the amount is reduced compared to the simple cap strategy. In the discussion to follow it will be assumed that the out-of-pocket costs are borrowed at the continuously compounded, risk-free interest rate. The reduced cost does potentially lower the profit inherent in the investment strategy.

An investor buying a call option has the belief that the underlying asset will increase in value before expiry. The investor views the call as a way to make a profit equal to the difference between the asset price (at the time of exercise) and the strike price of the option, minus the premium charged for the call. The investor may create a **bull spread** by selling a call option with a higher strike price, but otherwise identical to the one purchased. Let $C(K_1)$ and $C(K_2)$ be the premiums charged for the purchased and sold calls respectively. If $K_1 < K_2$ then the initial cash flow is $C(K_2) - C(K_1) \leq 0$. The profit will depend on where the asset price falls relative to the two strike prices. To keep the explanation simple, assume the options are of European type so that they must be exercised (if at all) at the same time. If $S(T) < K_1 < K_2$ then both options expire unused and the profit is $(C(K_2) - C(K_1))e^{rT}$. If $K_1 < S(T) < K_2$ then the purchased option is exercised and the sold option expires unused. In this case the profit is $(C(K_2) - C(K_1))e^{rT} + S(T) - K_1$. Finally, if $K_1 < K_2 < S(T)$ then both options will be exercised and the profit is $(C(K_2) - C(K_1))e^{rT} + K_2 - K_1$. The graph of the profit resembles the piecewise linear plot illustrated in Fig. 6.7. The investor limits their potential profit by selling the higher strike call.

A **bear spread** is constructed by purchasing a put with strike price $K_2$ and selling a put with strike price $K_1 < K_2$. An investor employing this option strategy may believe the underlying asset will decrease in value before expiry. Since $P(K_1) \leq P(K_2)$ the investor pays a premium to set up the bear spread; however, the premium is smaller than what would be paid to set up the floor option strategy. In much the same way as the profit for the bull spread was determined note that if $S(T) < K_1 < K_2$, both puts are exercised and the profit to the investor is $(P(K_1) - P(K_2))e^{rT} + K_2 - K_1$. The investor sells the asset for the higher strike price $K_2$ and must guarantee the owner of the lower strike option can sell the asset at price $K_1$. In effect the investor sells for $K_2$ and buys back for $K_1$. If $K_1 < S(T) < K_2$ the lower strike put is not exercised and the investor sells the asset for $K_2$. In this case the profit is $(P(K_1) - P(K_2))e^{rT} + K_2 - S(T)$. Finally, if $K_1 < K_2 < S(T)$ then both puts expire unused and the profit is $(P(K_1) - P(K_2))e^{rT} \leq 0$.

Investors can reduce risk by entering into a bull spread option strategy (if they believe asset prices will rise) or a bear spread strategy (if they believe asset prices will decline). Some investors may believe that asset prices will change, but are unable to commit to a direction of change. This type of investor believes that asset prices are volatile and that in the future the asset price is likely to be different than it is today. To reduce investment risk inherent in a volatile asset, an investor may wish to purchase a call and a put with the same expiration date. The call generates a profit if asset prices rise while the put generates a profit if the asset price falls. If the purchased call and put have the same strike price, generally near the current price of the asset so that both options are at-the-money, this strategy is known as a **straddle**. At-the-money options are generally more expensive since they are more likely to be exercised. Thus, a straddle may have an unacceptably high premium for some investors. A lower premium may be found when out-of-the-money calls and puts (again with the same expiry) are purchased. This strategy is known as a **strangle**. The trade-off with the strangle is that asset prices may have to move further to generate a profit than they would with a straddle.

**Example 6.2.** Suppose the current price of a security is \$100. An investor can purchase 3-month call and put options with a strike price of \$100 for a total of \$13.89 to create a straddle. As an alternative the investor can purchase a 3-month call with a strike price of \$110 for \$3.75 and a 3-month put with a strike price of \$90 for \$2.46. The premium cost to set up the strangle is \$6.21. The profit earned by these alternatives depends on the price of the security at the time of exercise. See Fig. 6.8. The profit generated by the straddle is higher than that produced by the strangle if

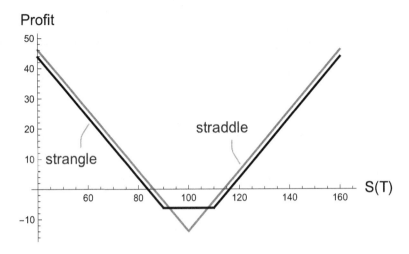

Fig. 6.8    Comparison of the profits generated by a straddle and a strangle.

the security price moves further from the common strike price of the straddle's put and call. However, the strangle generates a smaller loss than the straddle if the security price remains nearly the same over the lives of the options. The extra "insurance" provided by the strangle comes at the price of lower potential profit than that which could be earned with the straddle.

An investor holding contrary beliefs, namely that the price of an asset will exhibit less volatility than other investors predict may wish to write or sell a straddle. The profit from a written straddle has a graph which is a reflection of the profit curve of the purchased straddle of which an example is shown in Fig. 6.8. The straddle seller makes a positive profit if the asset's future price remains close to the strike price of the at-the-money written call and put options. However, the losses suffered by this option strategy can be large if the asset price changes significantly. In this case the straddle writer bet that asset prices would not change much, but in reality they did. Thus a cautious investor may wish to insure the written straddle against large losses by purchasing a strangle. The option strategy of combining a written straddle with a purchased strangle is known as a **butterfly spread**. As described the butterfly spread would involve two calls (one sold and one purchased) and two puts (again, one sold and one purchased). An equivalent payoff and profit can be achieved using two purchased calls (or puts) and two sold calls (or puts). Suppose an investor purchases two call options, one with strike price $K_1$ and one with strike price $K_3$ and sells two calls both

Table 6.1 Payoff from a butterfly spread created by buying two out-of-the-money calls and selling two at-the-money calls.

| Asset Price | Payoff from $K_1$-Strike Call | Payoff from $K_3$-strike Call | Payoff from $K_2$-Strike Call | Total Payoff |
|---|---|---|---|---|
| $S(T) < K_1$ | 0 | 0 | 0 | 0 |
| $K_1 < S(T) < K_2$ | $\lambda(S(T) - K_1)$ | 0 | 0 | $\lambda(S(T) - K_1)$ |
| $K_2 < S(T) < K_3$ | $\lambda(S(T) - K_1)$ | 0 | $-(S(T) - K_2)$ | $(1 - \lambda)(K_3 - S(T))$ |
| $K_3 < S(T)$ | $\lambda(S(T) - K_1)$ | $(1 - \lambda)(S(T) - K_3)$ | $-(S(T) - K_2)$ | 0 |

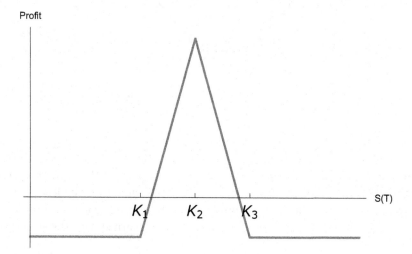

Fig. 6.9   Profit curve for a butterfly spread with $K_2 = (K_1 + K_3)/2$ implying $\lambda = 1/2$. In general the minimum profit/loss is $\lambda C(K_1) - C(K_2) + (1 - \lambda)C(K_3)$.

with strike price $K_2$. Assume that $K_1 < K_2 < K_3$ and generally that the $K_1$-strike and $K_3$-strike options are out-of-the-money. All four calls have the same expiration date. Define the quantity $\lambda$ as

$$\lambda = \frac{K_3 - K_2}{K_3 - K_1}.$$

This can be thought of as the proportion of the interval $[K_1, K_3]$ made up of the interval $[K_2, K_3]$. The butterfly spread is created by purchasing $\lambda$ $K_1$-strike European calls and $1 - \lambda$ $K_3$-strike European calls, and selling one $K_2$-strike European call. At expiry the payoff of the butterfly spread can be determined from Table 6.1.

The profit obtained from the butterfly spread is the payoff minus the net cost of the premium required to set up the option strategy. In general the profit curve resembles that shown in Fig. 6.9.

The opportunities and methods for combining options are limited only by the imaginations of the investors. The interested reader is urged to explore more option strategies in the texts by Hull (2000) and McDonald (2013).

### Exercises

**6.4.1** Suppose there are options on the same stock with two different strike prices, $K_1 < K_2$.

(a) Show that $C(K_1) \geq C(K_2)$.
(b) Show that $P(K_1) \leq P(K_2)$.
(c) Show that $C(K_1) - C(K_2) \leq K_2 - K_1$.
(d) Show that $P(K_2) - P(K_1) \leq K_2 - K_1$.

**6.4.2** Consider three call options $C(K_1)$, $C(K_2)$, and $C(K_3)$ and three put options $P(K_1)$, $P(K_2)$, and $P(K_3)$ with corresponding strike prices $K_1 < K_2 < K_3$.

(a) Show that $\dfrac{C(K_2) - C(K_1)}{K_2 - K_1} \leq \dfrac{C(K_3) - C(K_2)}{K_3 - K_2}$.

(b) Show that $\dfrac{P(K_2) - P(K_1)}{K_2 - K_1} \leq \dfrac{P(K_3) - P(K_2)}{K_3 - K_2}$.

**6.4.3** Suppose there are call and put options with strike prices $K_1 < K_2 < K_3$. Show that there exists a number $0 < \gamma < 1$ such that the following two inequalities hold.

$$C(K_2) \leq \gamma C(K_1) + (1 - \gamma)C(K_3).$$
$$P(K_2) \leq \gamma P(K_1) + (1 - \gamma)P(K_3).$$

Use the results of Exercise 6.4.2.

**6.4.4** Suppose you observe call options with the characteristics listed in the table below.

| C | K |
|------|-----|
| 4.57 | 100 |
| 3.80 | 105 |
| 2.13 | 115 |

Does the possibility of arbitrage exist?

**6.4.5** Suppose the continuously compounded, risk-free interest rate is 3.25%, an investor borrows money to purchase a stock for $500 and a six-month American put option for $40. The strike price of the put is $495.

(a) If after 2 months the price of the stock is $450 and the investor decides to exercise the put option, determine the profit for this floor strategy.
(b) If after 2 months the price of the stock is $550 and the investor decides to sell the stock, determine the profit for this floor strategy.

**6.4.6** Suppose the continuously compounded, risk-free interest rate is 3.25%, an investor borrows money to purchase a stock for $498 and sells a 6-month American covered call for $35. The strike price of the call is $500.

(a) At expiry the price of the stock is $490 and the investor decides to sells the stock. Determine the profit for this covered call strategy.
(b) At expiry the price of the stock is $510 and the owner of the call decides to exercise it. Determine the profit to the option writer for this covered call strategy.

**6.4.7** Suppose the continuously compounded, risk-free interest rate is 2.95%, an investor shorts a stock for $525 and purchases a 4-month American call option for $50. The strike price of the call is $530.

(a) If after 2 months the price of the stock is $500 and the investor decides to close the short position, determine the profit for this cap strategy.
(b) If after 2 months the price of the stock is $555 and the investor decides to close the short position, determine the profit for this cap strategy.

**6.4.8** Suppose the continuously compounded, risk-free interest rate is 2.95%, an investor shorts a stock for $475 and sells a 3-month American covered put for $45. The strike price of the put is $495.

(a) At expiry the price of the stock is $485 and the investor clears the short position. Determine the profit for this covered put strategy.
(b) At expiry the price of the stock is $515 and the owner of the put decides to exercise it. Determine the profit to the option writer for this covered put strategy.

**6.4.9** Suppose the continuously compounded, risk-free interest rate is 3.75%, an investor creates a bull spread on a stock by purchasing a 2-month European call option with a strike price of $100 for $7.57 and selling a 2-month European call option with a strike price of $110 for $4.75. Find the profit to the investor if at expiry the stock is worth

(a) $98,
(b) $107,
(c) $115.

**6.4.10** Suppose the continuously compounded, risk-free interest rate is 4.25%, an investor creates a bear spread on a stock by purchasing a 3-month European put option with a strike price of $425 for $15.75 and selling a 3-month European put option with a strike price of $400 for $10.25. Find the profit to the investor if at expiry the stock is worth

(a) $375,
(b) $410,
(c) $450.

**6.4.11** A stock is currently trading for $50. An investor unsure about the future direction of the stock price may purchase the following options.

| Option Type | Strike Price | Premium |
|---|---|---|
| Call | $50 | $3.64 |
| Call | $55 | $1.70 |
| Put | $50 | $2.72 |
| Put | $45 | $0.94 |

(a) What is the premium cost of a straddle?
(b) What is the premium cost of a strangle?
(c) In what stock price intervals does the straddle produce a higher profit than the strangle?

**6.4.12** An investor creates a butterfly spread using puts. Puts with strike prices of $45 and $55 are purchased while two puts with strike prices of $50 are sold. The premiums (per option) for the puts are shown in the table below.

| Strike Price | Premium |
|---|---|
| $45 | $0.94 |
| $50 | $2.72 |
| $55 | $5.68 |

Determine the profit to the investor if the underlying asset at expiry has the following values.

(a) $40,
(b) $47,
(c) $52,
(d) $57.

**6.4.13** A butterfly spread can be created by buying two out-of-the-money calls and selling two at-the-money calls or by buying a strangle and selling a straddle. Consider buying a strangle consisting of a $K_1$-strike put and $K_3$-strike call and selling a straddle with strike price $K_2$ where $K_1 < K_2 < K_3$ and $\lambda = (K_3 - K_2)/(K_3 - K_1) = 1/2$. Show that the payoff is similar to the chart shown in Table 6.1.

Chapter 7

# Approximating Option Prices Using Binomial Trees

One of the objectives of this text is the development and justification of the mathematical machinery to support formulas for the pricing of put and call options. While some formulas producing exact values will appear in Chapter 10, enough background has been presented to develop a useful method for approximating the values of options, namely **binomial trees**. This method will be explored in this chapter to give the reader an elementary numerical method for approximating the prices of a variety of financial options and to excite the reader about the possibility of exact formulas for the prices. The number of types of options for which prices are needed exceeds the number for which applied mathematicians have developed exact, closed form formulas, thus the binomial tree approach developed here will also be the method by which some of the exotic options are priced later in Chapter 14.

The binomial tree approximation method for pricing options depends on the notions of present value (see Chapter 1.4), binomial random variables (see Sec. 2.6), and the payoff expressions for options (see Chapter 6). This chapter also serves as a bridge between the material on discrete random variables and the later material on continuous normal and lognormal random variables (see Sec. 8.4 and Sec. 8.5).

## 7.1 One Period Binomial Model

The derivation of the **binomial model** was initially developed by Cox, Ross, and Rubinstein [Cox et al (1979)] and is sometimes referred to as the **Cox-Ross-Rubinstein model**. The model takes its name from the assumption that an asset currently worth $S(0)$ can change to only one of two prices $S(1) = S_u$ or $S(1) = S_d$ over a single, indivisible unit of time. Generally $S_d < S(0) < S_u$.

To illustrate the simplest case, assume the risk-free interest rate is $r$ compounded continuously, one unit of time passes, and the asset is a non-dividend-paying stock. An at-the-money, one period European call option will have a payoff of either $(S_u - S(0))^+ = S_u - S(0)$ or $(S_d - S(0))^+ = 0$. The call can be priced using an arbitrage argument. Suppose an investor may choose between two portfolios. The investor may purchase the European call (portfolio one) or the investor may borrow $B$ at the risk-free rate and purchase $\Delta$ shares of the stock and (portfolio two). After one period the investor will either exercise the call option (if portfolio one is chosen and the call is in-the-money) or the investor will repay the loan and sell the shares of stock held (if portfolio two is chosen). At time $t = 1$ the payoff of portfolio one is

$$(S(1) - K)^+ = \begin{cases} S_u - S(0) & \text{if } S(1) = S_u, \\ 0 & \text{if } S(1) = S_d. \end{cases}$$

At time $t = 1$ the payoff of portfolio two is

$$(\Delta)S(1) - Be^r = \begin{cases} (\Delta)S_u - Be^r & \text{if } S(1) = S_u, \\ (\Delta)S_d - Be^r & \text{if } S(1) = S_d. \end{cases}$$

If the payoffs for the two portfolios are the same then the following set of equations

$$S_u - S(0) = (\Delta)S_u - Be^r,$$
$$0 = (\Delta)S_d - Be^r.$$

can be solved to reveal that

$$\Delta = \frac{S_u - S(0)}{S_u - S_d},$$
$$B = \frac{S_d e^{-r}(S_u - S(0))}{S_u - S_d}.$$

Under the mild assumptions imposed so far $0 < \Delta < 1$ and $B > 0$. Thus it is necessary to assume that fractions of a share of stock can be purchased. Otherwise the purchases can be scaled up so that an integer number of shares and options are purchased.

Since the portfolios now have the same payoffs after one period, they must have the same costs at $t = 0$. The time $t = 0$ cost of portfolio two is $(\Delta)S(0) - B$, so the cost of the European call option is

$$C^e = e^{-r}\frac{S_u - S(0)}{S_u - S_d}(e^r S(0) - S_d).$$

Note that determination of the value of the call option did not require knowing which of the two prices the underlying stock would take after one period.

**Example 7.1.** Suppose the risk-free interest rate is 5% per annum compounded continuously. A non-dividend-paying stock currently sells for $50/share and in 1 year will either sell for $55/share or $45/share. Find the price of a $51-strike European call option on the stock which expires in 1 year using a one-period binomial model.

**Solution.** Consider the portfolio consisting of $\Delta$ shares of the stock and borrowing in the amount of $B$ which replicates the payoff of the European call in 1 year. This implies

$$0 = (\Delta)45 - Be^{0.05},$$
$$4 = (\Delta)55 - Be^{0.05},$$

which has solution $\Delta = 0.4$ and $B = 17.1221$. Thus the price of the European call option is

$$C^e = (0.4)(50) - 17.1221 = \$2.8779.$$

After this brief introduction the binomial model can be generalized to approximate the price of other options with a European style of exercise. The following assumptions will be made.

- Strike price of the option is $K$.
- Exercise time of the option is $T$.
- Initial price of the asset is $S$.
- At time $T$ the asset will be worth $uS$ where $u > 1$ or will be worth $dS$ where $0 < d < 1$.
- The asset pays a continuous dividend at a rate $\delta$.
- Continuously compounded, risk-free interest rate is $r$.

The value of the option at $t = 0$ will be denoted $V$ and will have a payoff at time $t = T$ denoted as $V_u$ (if the value of the asset is $uS$) or $V_d$ (if the value of the asset is $dS$). The central idea behind using the binomial tree approach to price the option is that the value of the option, in the absence of arbitrage, should be the same as the value of the **replicating portfolio**. This portfolio will have the same payoff as the option at time $t = T$. The replicating portfolio will consist of $\Delta$ shares of the underlying asset and borrowing amount $B$ at the risk-free rate. To incorporate the payment of dividends by the asset, it will be assumed that all dividends are reinvested in additional shares of the asset. Setting up and solving the two equations for the possible payoffs of the portfolios:

$$(\Delta)dSe^{\delta T} - Be^{rT} = V_d,$$
$$(\Delta)uSe^{\delta T} - Be^{rT} = V_u,$$

produces

$$\Delta = e^{-\delta T}\frac{V_u - V_d}{S(u - d)},$$

$$B = e^{-rT}\frac{dV_u - uV_d}{u - d}.$$

Therefore the time $t = 0$ price of the option is

$$V = e^{-\delta T}\frac{V_u - V_d}{u - d} - e^{-rT}\frac{dV_u - uV_d}{u - d}. \tag{7.1}$$

**Example 7.2.** Stock in XYZ Corporation is currently selling for \$65/share. The stock pays a continuous dividend at a rate of 2% per year. The risk-free continuously compounded interest rate is 4.5%. One year from now stock in XYZ will sell for either \$62/share or \$68/share. Determine the price of a 1-year, \$63-strike European put option on XYZ stock.

**Solution.** Let $\delta = 0.02$, $T = 1$, $V_u = (63 - 68)^+ = 0$, $V_d = (63 - 62)^+ = 1$, $u = 68/65$, $d = 62/65$, and $r = 0.045$. Substituting these values in Eq. (7.1) yields

$$P^e = e^{-0.02(1)}\frac{0 - 1}{\frac{68}{65} - \frac{62}{65}} - e^{-0.045(1)}\frac{\frac{62}{65}(0) - \frac{68}{65}(1)}{\frac{68}{65} - \frac{62}{65}} = \$0.2158.$$

The upcoming exercises will give the reader more practice estimating option prices using the one-period binomial model. In the next section, multi-period binomial models will be explored as a generalization and numerically more accurate means of estimating option prices.

*Exercises*

**7.1.1** The current price of a non-dividend-paying stock is \$40/share. At the end of 3 months the stock price will either be \$42/share or \$37/share. The continuously compounded, risk-free interest rate is 3% per annum. Find the price of a 3-month, \$41-strike European call option on the stock.

**7.1.2** The current price of a non-dividend-paying stock is \$60/share. At the end of 4 months the stock price will either be \$65/share or \$50/share. The continuously compounded, risk-free interest rate is 5% per annum. The market price of a 4-month, \$60-strike European put option on the stock is observed to be \$1.75. Construct an arbitrage opportunity for an investor.

**7.1.3** The current price of a stock is \$120/share. The stock pays dividends continuously at a rate of 1%. At the end of 3 months the stock price will either be \$130/share or \$115/share. The continuously compounded, risk-free interest rate is 5% per annum.

(a) Find the price of a 3-month, \$125-strike European call option on the stock.
(b) Find the price of a 3-month, \$125-strike European put option on the stock.
(c) Show the option prices satisfy the Put-Call Parity formula of Eq. (6.9).

**7.1.4** The current price of a non-dividend-paying stock is \$75/share. At the end of 4 months the stock price will either be \$95/share or \$60/share. The continuously compounded, risk-free interest rate is 4% per annum. A 4-month European strangle on the stock will have a payoff determined as

$$\text{payoff} = \begin{cases} 70 - S(4/12) & \text{if } S(4/12) < 70 \\ 0 & \text{if } 70 \le S(4/12) \le 90 \\ S(4/12) - 90 & \text{if } S(4/12) > 90. \end{cases}$$

Determine the time $t = 0$ price of the strangle.

**7.1.5** The current price of a stock is \$125/share. At the end of 6 months the stock price will either be \$145/share or \$100/share. The stock pays a continuous dividend at the rate of 2%. The continuously compounded, risk-free interest rate is 5% per annum. A 6-month European butterfly spread on the stock will have a payoff determined as

$$\text{payoff} = \begin{cases} 0 & \text{if } S(4/12) \le 105 \\ S(6/12) - 105 & \text{if } 105 < S(6/12) < 125 \\ 145 - S(6/12) & \text{if } 125 < S(6/12) < 150 \\ 0 & \text{if } S(6/12) > 150. \end{cases}$$

Determine the time $t = 0$ price of the butterfly spread.

**7.1.6** The current price of a stock is \$125/share. At the end of 6 months the stock price will either be \$145/share or \$100/share. The stock pays a continuous dividend at the rate of 2%. The continuously compounded, risk-free interest rate is 5% per annum. A European option will pay $S(6/12)$ if $S(6/12) > 140$ or 0 otherwise. Find the price of the option at $t = 0$.

**7.1.7** The current price of a stock is \$100/share. At the end of 5 months the stock price will either be \$120/share or \$90/share. The stock pays a continuous dividend at the rate of 3%. The continuously compounded, risk-free interest rate is 6% per annum. A European option will pay $(S(5/12))^2$ if $S(5/12) > 100$ or 0 otherwise. Find the price of the option at $t = 0$.

**7.1.8** Brooklyn is modeling the changes in the price of a 1-year futures contract using a one-period binomial tree. Suppose for this tree, $u/d = 5/4$, the risk-free interest rate is 6% compounded continuously, and the premium for an at-the-money European call for the futures contract is \$5.24827. The initial price of the futures contract is \$100. What is $u$?

**7.1.9** Charlie is modeling the changes in the price of a 1-year futures contract using a one-period binomial tree. Suppose for this tree, $u/d = 4/3$, the risk-free interest rate is 6% compounded continuously, and the premium for an at-the-money European put for the futures contract is \$17.0738. The initial price of the futures contract is \$250. What is $d$?

## 7.2   Multi-Period Binomial Models

One notational simplification can be made to Eq. (7.1). Grouping the terms involving the two payoffs of the European option enables the equation to be re-written as follows.

$$V = e^{-rT} \left( \frac{e^{(r-\delta)T} - d}{u - d} V_u + \frac{u - e^{(r-\delta)T}}{u - d} V_d \right).$$

Define the quantity $p^* = (e^{(r-\delta)T} - d)/(u - d)$, then

$$V = e^{-rT}((p^*)V_u + (1 - p^*)V_d). \tag{7.2}$$

Under the mild assumption that $d < e^{(r-\delta)T} < u$, then $0 < p^* < 1$ and $p^*$ can be thought of as a probability. When $0 < p^* < 1$, $p^*$ is called a **risk-neutral probability**. This should *not* be thought of as the probability that the asset whose current price is $S$ will be worth $uS$ at time $t = T$. In fact, no mention of the probability of the "up" or "down" movement of the asset price has been mentioned or needed thus far. The quantity $p^*$ acts like a probability and gives a new interpretation to the value of the European option. The value of a European option is the present value of the risk-neutral expected value of the payoff of the option.

Rather than taking one indivisible time step of length $T$, suppose the interval $[0, T]$ is divided into $n$ subintervals of length $h = T/n$. If the value of an asset is $S(ih)$ at time $t = ih$ then the asset can be worth $uS(ih)$ or $dS(ih)$ at time $t = (i+1)h$ for $i = 0, 1, \ldots, n-1$. If $u(dS(ih)) = d(uS(ih))$, in other words if the price of the asset after a down step followed by an up step is the same as the price of the asset after an up step followed by a down step, the graph of prices forms a **binomial lattice** or **recombining tree**. If $u(dS(ih)) \neq d(uS(ih))$ the graph of prices is described as a **binomial tree** or **non-recombining tree**. Figure 7.1 illustrates a binomial lattice

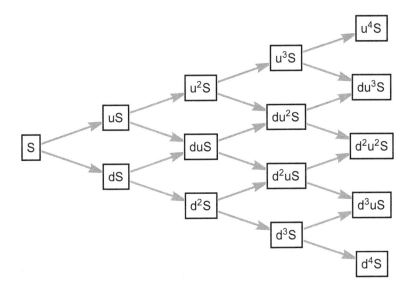

Fig. 7.1 A binomial lattice with four discrete time steps.

with four time steps. The risk-neutral probability of taking the up step from a vertex in the lattice is

$$p^* = \frac{e^{(r-\delta)h} - d}{u - d}. \tag{7.3}$$

Pricing a European option using a multi-period binomial lattice (or tree) is very similar to the to one period model. The time $t = 0$ price of the option will be the present value of the risk-neutral expected value of the option's payoff at time $t = T$. If the price movements of the asset are modeled using a recombining tree, then the probability of taking a path from initial value $S$ to final value $d^{n-k}u^k S$ is

$$\binom{n}{k}(p^*)^k(1 - p^*)^{n-k}.$$

For example the time $t = 0$ price for a $K$-strike European call option is

$$C^e = e^{-rT}\sum_{k=0}^{n}\binom{n}{k}(p^*)^k(1 - p^*)^{n-k}(d^{n-k}u^k S - K)^+$$

$$= e^{-rT}\,\mathbb{E}^*\left((S(T) - K)^+\right),$$

where $\mathbb{E}^*\left(X\right)$ denotes the risk-neutral expectation of $X$.

**Example 7.3.** Approximate the value of a 5-month $60-strike European call option on a non-dividend-paying stock for which current price of the stock is $62 per share and the continuously compounded risk-free interest rate is 10% per year. At the end of each month the price of the stock will either increase by a factor of 1.06 or decrease by a factor of 0.94. Use a recombining tree with five time steps.

**Solution.** To approximate the value of the option we must create a recombining tree of values of the security. The time step is 1 month, so $h = 1/12$. The risk-neutral probability of an up step is

$$p^* = \frac{e^{(0.10-0)(1/12)} - 0.94}{1.06 - 0.94} = 0.569735.$$

The lattice of security values is shown in Fig. 7.2. Since the option is of European type only the values for $S$ in the right-most column are important for the determination of the value of the option. Therefore,

$$C^e = e^{-0.10(5/12)} \sum_{k=0}^{5} \binom{5}{k} (p^*)^k (1 - p^*)^{5-k} (u^k d^{5-k} 62 - 60)^+$$

$$= e^{-0.10(5/12)} \left[ \binom{5}{3} (0.569735)^3 (0.430265)^2 (5.24767) \right.$$

$$+ \binom{5}{4} (0.569735)^4 (0.430265)^1 (13.5772)$$

$$\left. + \binom{5}{5} (0.569735)^5 (0.430265)^0 (22.97) \right]$$

$$= \$5.9979.$$

### Exercises

**7.2.1** Verify that, in the absence of arbitrage, the expression for $p^*$ in Eq. (7.3) lies in the interval $[0, 1]$ and thus is a valid value for a probability.

**7.2.2** Show that $1 - p^* = \dfrac{u - e^{(r-\delta)T}}{u - d}$.

**7.2.3** A share of a certain stock currently sells for $105. At the end of a month the price of the stock will either increase by 3% or decrease by 3%. The stock pays a continuously compounded dividend at a yield of 2%. The risk-free continuously compounded interest rate is 7%. Use a binomial model with two time steps (of 1 month each) to determine the price of a 2-month, at-the-money European put option for the stock.

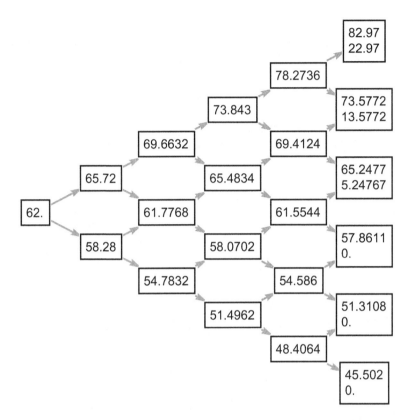

Fig. 7.2　A binomial lattice of security prices for Example 7.3. The number on the second row of each terminating leaf of the lattice is the time $t = 5/12$ payoff of the $60-strike European call option.

**7.2.4** A non-dividend-paying stock has a current price of $150/share. At the end of a month, the price of the stock will either increase by a factor of 1.15 or decrease by a factor of 0.86. The continuously compounded, risk-free interest rate is 5%. Estimate the price of a 3-month, $155-strike European call option using a three-period binomial model.

**7.2.5** A non-dividend-paying stock has a current price of $125/share. At the end of a month, the price of the stock will either increase by a factor of 1.11 or decrease by a factor of 0.89. The continuously compounded, risk-free interest rate is 7%. Estimate the price of a 3-month, $125-strike European straddle using a three-period binomial model.

**7.2.6** A non-dividend-paying stock has a current price of $200/share. At the end of a month, the price of the stock will either increase by a factor of $u$ or decrease by a factor of $1/u$ where $u > 1.01$. The continuously compounded, risk-free interest rate is 5%. The price of a 2-month, at-the money European call option is \$3.75. Determine the factor $u$.

## 7.3   Estimating Increase/Decrease Factors

Important components of the binomial model are the increasing and decreasing factors by which the asset price changes during each time step. This section will discuss a method for estimating $u$ and $d$ from historical pricing data and elementary descriptive statistics. Suppose a sequence of historical prices $\{S(k)\}_{k=0}^{n}$ is collected for an asset. The expression $S(k)$ represents the price of the asset at time $t = kh$ where $h$ is the length of time between observations. It is assumed that the prices are observed at regularly spaced intervals of time. Many internet search engines can provide a list of closing prices for stocks trading on an exchange. The closing price of a stock is the last trading price for the stock on a given day at the exchange. Publicly traded companies often include a record of their stock prices on the company web site.

Under the assumption that consecutive prices are related through the exponential model,

$$S(k+1) = S(k)e^{R_k} \text{ for } k = 0, 1, \ldots, n-1,$$

define the sequence

$$R_k = \ln \frac{S(k+1)}{S(k)},$$

for $k = 0, 1, \ldots, n-1$. The quantity $R_k$ can be interpreted as the continuously compounded return of the stock during the time interval $[kh, (k+1)h]$. Let $\bar{r}$ be the mean of this sequence.

$$\bar{r} = \frac{1}{n} \sum_{k=0}^{n-1} R_k = \frac{1}{n} \ln \frac{S(n)}{S(0)}.$$

Define $s_R^2$ to be the **sample variance** of $\{R_k\}_{k=0}^{n}$.

$$s_R^2 = \frac{1}{n-1} \sum_{k=0}^{n-1} (\bar{r} - R_k)^2. \tag{7.4}$$

Note that the formula for the variance of sample data is different from the formula for the variance of a discrete random variable given in Eq. (2.13). The chief difference is the expression $n - 1$, known as the **degrees of**

**freedom.** It can be thought of as the number of observations which may vary after the mean of the observations has been fixed. Using the definitions of $\bar{r}$ and $R_k$, the sample variance of the observed returns can also be written as

$$s_R^2 = \frac{1}{n-1} \sum_{k=0}^{n-1} \left( \frac{1}{n} \ln \frac{S(n)}{S(0)} - \ln \frac{S(k+1)}{S_k} \right)^2.$$

The sample variance of the returns is an estimate of the quantity $\sigma^2 h$ where $\sigma$ is called the **volatility** of the returns. Therefore the volatility can be estimated from the historical returns as $\sigma \approx s_R / \sqrt{h}$.

Once the volatility $\sigma$ is estimated, the increase and decrease factors $u$ and $d$ can be determined. Recall that the price of a forward maturing $h$ units of time from now on an asset paying a continuous dividend can be expressed as

$$F_{t,t+h} = S(t)e^{(r-\delta)h}.$$

If there is no uncertainty in the time $t + h$ price of the asset then $S(t + h) = F_{t,t+h}$. One way to incorporate the uncertainty in the future price of the asset is to include the volatility. If the annualized volatility is $\sigma$, then the volatility per interval of length $h$ is $\sigma\sqrt{h}$. This assumes that returns generated by the asset in non-overlapping intervals of time are independent and identically distributed and that the variance in the return over interval $[t, t + h]$ is proportional to the length of the time interval, $h$. Thus an estimate of the increase factor $u$ would be that the return of the asset is a standard deviation above the return reflected in the forward price. Likewise an estimate of the decrease factor $d$ would be that the return is one standard deviation below the return inherent in the forward price. This can be expressed as the following pair of equations.

$$uS(t) = F_{t,t+h}e^{\sigma\sqrt{h}},$$
$$dS(t) = F_{t,t+h}e^{-\sigma\sqrt{h}}.$$

This implies that

$$u = \frac{F_{t,t+h}}{S(t)}e^{\sigma\sqrt{h}}, \tag{7.5}$$

$$d = \frac{F_{t,t+h}}{S(t)}e^{-\sigma\sqrt{h}}. \tag{7.6}$$

This implies the risk-neutral probability of an up step in the binomial tree can be written as

$$p^* = \frac{1}{1 + e^{\sigma\sqrt{h}}}. \tag{7.7}$$

*Exercises*

**7.3.1** Suppose a stock has an annualized volatility of $\sigma$ and the risk-free, continuously compounded interest rate is $r$. A binomial model of the movements of the stock price is to be constructed with a time step of $h$. Show that

$$u = e^{(r-\delta)h+\sigma\sqrt{h}},$$

$$d = e^{(r-\delta)h-\sigma\sqrt{h}},$$

where $\delta$ is the continuously compounded dividend yield for the stock.

**7.3.2** Use Eqs. (7.5) and (7.6) to derive the expression for the risk-neutral probability $p^*$ given in Eq. (7.7).

**7.3.3** A share of stock was sold for the following on the last day of each month (assume every month has 30 days). Find the monthly volatility and the annualized volatility for the share price.

| Month | Price | Month | Price |
|-------|-------|-------|-------|
| 1 | 78.41 | 7 | 100.79 |
| 2 | 109.14 | 8 | 104.10 |
| 3 | 96.08 | 9 | 67.79 |
| 4 | 144.71 | 10 | 152.56 |
| 5 | 119.80 | 11 | 72.59 |
| 6 | 80.57 | 12 | 115.02 |

**7.3.4** The current price of non-dividend-paying stock is $60. The annualized volatility of the stock is 30%. The risk-free, continuously compounded interest rate is 5%. Estimate the price of a 1-year, $61-strike European call option for the stock using a one period binomial tree.

**7.3.5** The current price of non-dividend-paying stock is $50. The annualized volatility of the stock is 25%. The risk-free, continuously compounded interest rate is 8%. Estimate the price of a 1-year, $55-strike European put option for the stock using a two-period binomial tree.

**7.3.6** Suppose the current \$/€ exchange rate is 1.25 and the annualized volatility in the exchange rate is 15%. The continuously compounded, risk-free interest rate for dollar-denominated saving and borrowing is 6% and the continuously compounded, risk-free rate for euro-denominated saving and borrowing is 2%. Use a two-period binomial lattice model to estimate the price of a 1-year, 1.275-strike European call on the exchange rate.

## 7.4   American Options

The reader will recall that one of the most apparent differences between European-style and American-style options is that the latter have the right of early exercise. In this section the binomial pricing model will be used to estimate the prices of American call and put options. This requires only a small modification of the binomial pricing model used for European-style options.

To illustrate the method consider a $K$-strike American put option on a stock whose initial price is $S(0)$. The expiry of the option is at $t = T$ and there are $n$ time steps in the binomial model. During a time step of duration $h = T/n$ the price of the stock may increase by a factor $u > 1$ or decrease by a factor $0 < d < 1$. As before the risk-free interest rate is denoted $r$ and is compounded continuously. The risk-neutral probability of an upward movement in the stock price is $p^*$. Since the style of exercise is American the option holder may decide to exercise the option at any time step $t \in \{0, h, 2h, \ldots, T\}$ in the model. A rational investor will decide to exercise the American put at time $t$ only if the immediate payoff of exercise exceeds the time $t$ present value of the risk-neutral expected value of the option if it is held for another time step. Let the payoff from immediate exercise of the put be $(K - S(t))^+$, known as the **intrinsic value** of the put. Then early exercise of the American put can take place when

$$K - S(t)$$
$$> e^{-rh}\left[p^* P^a(uS(t), t + h; K, T) + (1 - p^*)P^a(dS(t), t + h; K, T)\right].$$

Figure 7.3 illustrates the relationships between the stock price, intrinsic value of the American Put, and the premium for the American put. An American-style option is always worth at least as much as its current intrinsic value. The value of the American put at time $t$ is therefore

$$
\begin{aligned}
&P^a(S(t), t; K, T)\\
&= \max\{K - S(t), e^{-rh}\left[p^* P^a(uS(t), t + h; K, T)\right.\\
&\quad + \left. (1 - p^*)P^a(dS(t), t + h; K, T)\right]\}.
\end{aligned}
\tag{7.8}
$$

The following example will illustrate the steps used to price an American put using a binomial model.

**Example 7.4.** Suppose the current price of a non-dividend-paying stock is \$75, the risk-free, continuously compounded interest rate is 4.5%, and the volatility of the stock price is 20%. Use a binomial model with two

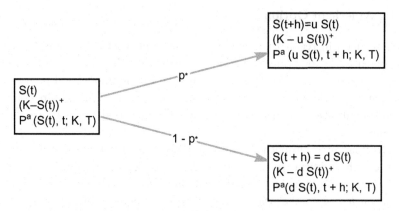

Fig. 7.3  A single step in the binomial tree model of an American put option. The top quantity in each vertex is the current price of the underlying stock. The middle quantity is the intrinsic value of the option. The bottom quantity is the risk-neutral premium for the option.

time steps to estimate the premium for a 6-month, \$72-strike American put option on the stock.

**Solution.** If $h = 3/12$ then the upward and downward movement factors and the risk-neutral probability of an upward movement are respectively,

$$u = e^{(0.045-0)(3/12)+0.20\sqrt{3/12}} = 1.09008,$$

$$d = e^{(0.045-0)(3/12)-0.20\sqrt{3/12}} = 0.93824,$$

$$p^* = \frac{1}{1 + e^{0.20\sqrt{3/12}}} = 0.481259.$$

Figure 7.4 shows the stock prices, option intrinsic values, and the option premium at each vertex in the binomial lattice. The pricing algorithm begins at time $t = 6/12$. The intrinsic value of the American option at expiry is the same as the value of the option at expiry. Since the two upper vertices indicate the value of the American put is 0 then the time $t = 3/12$ present value of the risk-neutral expected value of the American put is also 0. The intrinsic value of the put at the upper vertex at time $t = 3/12$ is also 0, thus the premium of the American put at that vertex is 0. The premiums of the American put at the two lower vertices of the lattice at time $t = 6/12$ are respectively 0 and 5.97799. Thus the time $t = 3/12$ present value of the risk-neutral expected value of the American put is

$$e^{-0.045(3/12)}\left[(0.481259)(0) + (0.518741)(5.97799)\right] = 3.06634.$$

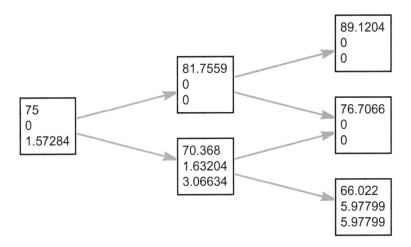

Fig. 7.4   A two-step binomial lattice used to estimate the premium of an American put option described in Example 7.4.

Since this value exceeds the intrinsic value of the option at the lower node of the lattice at time $t = 3/12$, the option would not be exercised at that time and the premium for the option would be

$$P^a(70.368, 3/12; 72, 6/12) = \max\{0, 3.06634\} = 3.06634.$$

The time $t = 0$ premium of the American put is found in a similar way. The prices of the American put at the two vertices found at time $t = 3/12$ are respectively 0 and 3.06634. The time $t = 0$ present value of the risk-neutral expected value of the American put is

$$e^{-0.045(3/12)}[(0.481259)(0) + (0.518741)(3.06634)] = 1.57284.$$

Since this value exceeds the intrinsic value of the option at the lower node of the lattice at time $t = 0$, the option would not be exercised at time $t = 0$ and the premium charged for the option would be

$$P^a(75, 0; 72, 6/12) = \max\{0, 1.57284\} = \$1.57284.$$

This approach can be taken to pricing American put and call options on other underlying assets such as currency exchange rates and futures contracts. These applications will be explored in the exercises.

Before exploring some of the properties of the binomial lattice formula for the price of an American option, consider the following example of a three-step binomial model for the pricing of an American call on a stock which pays a continuous dividend.

**Example 7.5.** Suppose the current price of a security is \$32, the risk-free interest rate is 10% compounded continuously, and the volatility of the return for the security is 20%. The security pays a continuous dividend at a rate of 3% per annum. Find the price of a 3-month, \$34-strike American call on the security using a binomial model with three steps.

**Solution.** The price of a 3-month American call with a strike price of \$34 on the security can be found as outlined below. The length of a time step will be $h = 1/12$. The formulas for the parameters $u$ and $d$ given in Exercise 7.3.1 produce the following "up" and "down" factors.

$$u = e^{(0.10-0.03)(1/12)+0.20\sqrt{1/12}} \approx 1.06563,$$

$$d = e^{(0.10-0.03)(1/12)-0.20\sqrt{1/12}} \approx 0.94942.$$

The risk-neutral probability of an increase in the price of the security occurring between time steps is given by the formula in Eq. (7.7).

$$p = \frac{1}{1 + e^{0.20\sqrt{1/12}}} \approx 0.48557.$$

The time until expiry will be divided into three single-month time steps and a binomial lattice of security prices will be created. The binomial lattice of security values is shown in Fig. 7.5. The number at the top of each node is the value of the underlying stock at that node in the tree. The tree summarizes the immediate payoffs of the call at each node. These are the middle values in each node. Most of the work is involved in finding the value of the American call at each node in the tree. For the four leaves of the tree (the right-most nodes), the value of the call is its payoff value, the same as its intrinsic value. Working backwards in time from the leaves, at $t = 2/12$ the price of the American call is the present value of the risk-neutral expected value at $t = 2/12$. These are as follows.

$$e^{-0.10(1/12)} \left[ (0.48557)(4.72328) + (1 - 0.48557)(0.5004) \right] = 2.52974,$$

$$e^{-0.10(1/12)} \left[ (0.48557)(0.5004) + (1 - 0.48557)(0) \right] = 0.24096,$$

$$e^{-0.10(1/12)} \left[ (0.48557)(0) + (1 - 0.48557)(0) \right] = 0.$$

Since the value of the option at $t = 2/12$ is greater than the immediate payoff at $t = 2/12$ the call would not be exercised at that time. At $t = 1/12$ the value of the call is calculated as,

$$e^{-0.10(1/12)} \left[ (0.48557)(2.52974) + (1 - 0.48557)(0.24096) \right] = 1.34110,$$

$$e^{-0.10(1/12)} \left[ (0.48557)(0.24086) + (1 - 0.48557)(0) \right] = 0.11603.$$

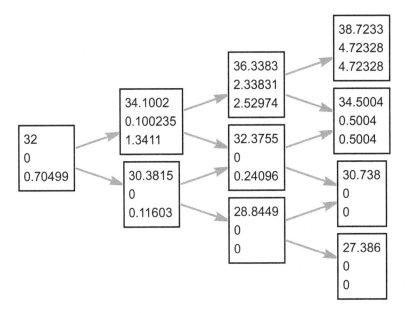

Fig. 7.5 A three-step discrete approximation to the evolution of the value of a security. The top value in each node is the value of the security. The middle number is the payoff of the call if it were exercised immediately (the intrinsic value of the option). The bottom number is the price of the option at that step in time.

Once again these values exceed the intrinsic value of the option, so the call would not be exercised early. Lastly at $t = 0$ the value of the American call is

$$C^a = e^{-0.10(1/12)} \left[ (0.48557)(1.34110) + (1 - 0.48557)(0.11603) \right]$$
$$= \$0.70499.$$

The pricing algorithm for an American-style put option given in Eq. (7.8) is simple to implement in a spreadsheet or computer program. This simplicity may hide some of the mathematical properties of the value of the option. In the remainder of this section, brief descriptions and justifications of some of the properties of the put option will be presented. These results will be helpful in deciding the optimal time to exercise an American put option.

**Lemma 7.1.** *Suppose the value of the security underlying an American put option follows a path through the n-step binomial lattice such that for some $i \in \{0, 1, \ldots, n-1\}$ we have $(K - dS(t_i))^+ = 0$, then $(K - S(t_i))^+ = 0$.*

*Proof.* Since $d < 1 < u$ then $dS(t_i) < S(t_i) < uS(t_i)$ which implies

$$(K - uS(t_i)) < (K - S(t_i)) < (K - dS(t_i)),$$

from which it follows that

$$(K - uS(t_i))^+ \leq (K - S(t_i))^+ \leq (K - dS(t_i))^+.$$

If $(K - dS(t_i))^+ = 0$ then $(K - S(t_i))^+ = 0$. $\qquad\square$

This lemma can be thought of as stating that if the intrinsic value of an American put at the next downward movement of the price of the underlying security would be zero, then the current intrinsic value of the option is zero as well. Likewise the intrinsic value of the American put at higher values of the security price would be zero. The following lemma reveals the effect this has on the value of the option.

**Lemma 7.2.** *Suppose the value of the security underlying an American put option follows a path through the n-step binomial lattice such that for some $i \in \{0, 1, \ldots, n-1\}$ the expression $(K - dS(t_i))^+ = 0$, then $P^a(t_i) = e^{-r\Delta t} P^a(t_{i+1})$.*

*Proof.* By Lemma 7.1 $(K - S(t_i))^+ = 0$ and by Eq. (7.8),

$$
\begin{aligned}
P^a(t_i) &= \max_{j \in \{i, i+1, \ldots, n, \infty\}} \left\{ e^{-r(j-i)\Delta t} \, \mathbb{E}^* \left( (K - S(t_j))^+ \right) \right\} \\
&= \max_{j \in \{i+1, i+2, \ldots, n, \infty\}} \left\{ e^{-r(j-i)\Delta t} \, \mathbb{E}^* \left( (K - S(t_j)^+ \right) \right\} \\
&= e^{-r\Delta t} \max_{j \in \{i+1, i+2, \ldots, n, \infty\}} \left\{ e^{-r(j-[i+1])\Delta t} \, \mathbb{E}^* \left( (K - S(t_j))^+ \right) \right\} \\
&= e^{-r\Delta t} P^a(t_{i+1}).
\end{aligned}
$$

$\qquad\square$

Now consider an American put that expires in the money.

**Lemma 7.3.** *Suppose the value of the security underlying an American put option follows a path through the n-step binomial lattice such that for some $i \in \{0, 1, \ldots, n-1\}$ the expression $(K - uS(t_i))^+ > 0$, then $(K - S(t_i))^+ > e^{-r\Delta t} \mathbb{E}^* \left( (K - S(t_{i+1}))^+ \right)$.*

*Proof.* See Exercise 7.4.7. $\qquad\square$

## Exercises

**7.4.1** Create a two-period binomial lattice to determine the premium for a 1-year, \$108-strike American put option on a stock currently priced at \$105. The stock has a volatility of 50% and pays a continuous dividend at a yield of 1%. The risk-free, continuously compounded interest rate is 6%. Would early exercise of the option be optimal?

**7.4.2** Create a two-period binomial lattice to determine the premium for a 1-year, \$110-strike American call option on a stock currently priced at \$107. The stock has a volatility of 40% and pays a continuous dividend at a yield of 3%. The risk-free, continuously compounded interest rate is 9%. Would early exercise of this option be optimal?

**7.4.3** Create a three-period binomial lattice to determine the premium for a 1-year, \$150-strike American call option on a stock currently priced at \$150. The stock has a volatility of 45% and pays a continuous dividend at a yield of 3%. The risk-free, continuously compounded interest rate is 7%.

**7.4.4** Suppose the current \$/€ exchange rate is 1.10 and the annualized volatility in the exchange rate is 25%. The continuously compounded, risk-free interest rate for dollar-denominated saving and borrowing is 6% and the continuously compounded, risk-free rate for euro-denominated saving and borrowing is 2%. Use a three-period binomial lattice model to estimate the price of a 1-year, 1.15-strike American call on the exchange rate.

**7.4.5** An American call option has a crude oil futures contract as its underlying asset. The current 1-year crude oil futures price is \$60/barrel, the strike price of the option is \$65/barrel, and the continuously compounded, risk-free interest rate is 5%. The volatility of the futures price is 50%. Find the premium for the call option using a binomial lattice with two time steps.

**7.4.6** The price of a non-dividend-paying security is \$56, the risk-free interest rate is 12% compounded continuously, and the volatility of the security is 25%. Find the price of a 2-month American put on the security with a strike price of \$58 using a binomial tree with two time steps.

**7.4.7** Complete the proof of Lemma 7.3.

The next chapter will introduce the notion of continuous random variables, specifically the normal and lognormal random variables. They will allow closed form formulas to be developed for European put and call option prices. The properties of the normal and lognormal random variable will also be used in the later discussion of stochastic calculus.

Chapter 8

# Normal Random Variables and Probability

Whereas in Chapter 2 random variables could take on only a finite number of values taken from a set with gaps between the values, in the present chapter, **continuous** random variables will be described. A continuous random variable can take on an infinite number of different values from a range without gaps. Calculus-based methods for determining the expected value and variance of a continuous random variable will be described. Two commonly used types of continuous random variables in the study of financial mathematics are the **normal** and **lognormal** random variable. When many random factors and influences come together (as in the complex situation of a financial market), the aggregate influence can be modeled by a normal random variable. Empirical examination of stock market data will detect the presence of normal randomness in the fluctuations of stock prices.

## 8.1 Continuous Random Variables

The interpretation of probability must change to understand the difference between discrete and continuous random variables. Consider the interval $[0, 1]$. If the sample space of outcomes is only the integers contained in this interval, then only the discrete set $\{0, 1\}$ need be considered. The probability of selecting either integer from this set is $1/2$ and thus the notion of discrete probability, random variables, and distributions is sufficient. If the sample space of outcomes of an experiment consists of selecting, with equal likelihood, a number from the interval $[0, 1]$ of the form $k/10$ where $k \in \{0, 1, \ldots, 10\}$, then once again the notion of discrete probability and random variables is sufficient. The $\mathbb{P}\left(X = k/10\right) = 1/11$ for $k = 0, 1, \ldots, 10$. Continuing in this way, for a fixed positive integer $n$

if the outcomes are the real numbers in $[0, 1]$ of the form $k/n$ where $k \in \{0, 1, \ldots, n\}$, then $\mathbb{P}(X = k/n) = \frac{1}{n+1}$. So long as $n$ is finite the familiar concepts of discrete probability are all that are needed. What happens as $n \to \infty$? The limiting case can be thought of as **continuous probability**. The description continuous is used because in the limit, the gaps between the outcomes in the sample space disappear. However, using the discrete notion of probability, the likelihood of choosing a particular number from the interval $[0, 1]$ becomes

$$\lim_{n \to \infty} \frac{1}{n+1} = 0.$$

This is correct but brings to mind a paradox. There is nothing wrong with the notion of the limit or the ideas of discrete probability, but does this imply that the probability of choosing any number from $[0, 1]$ is 0? Surely some number must be chosen. The paradox arises from applying the notions of discrete random variables to continuous random variables. What is needed is a change in the notion of probability when dealing with continuous random variables. Instead of determining the probability that a continuous random variable equals some real number, the meaningful determination is the likelihood that the random variable lies in a set, usually an interval or finite union of intervals on the real line. These notions are defined next.

A random variable $X$ has a **continuous distribution** if there exists a non-negative function $f_X : \mathbb{R} \to \mathbb{R}$ such that for an interval $[a, b]$ the

$$\mathbb{P}(a < X \leq b) = \int_a^b f_X(x)\, dx. \tag{8.1}$$

The function $f_X$, which is known as the **probability distribution function** or **probability density function**, must in addition to satisfying $f_X(x) \geq 0$ on $\mathbb{R}$, have the following property,

$$\int_{-\infty}^{\infty} f_X(x)\, dx = 1. \tag{8.2}$$

These properties are the analogues of properties of discrete random variables and their distributions. The probabilities of exact values of the random variable are non-negative. In fact, as explained above $\mathbb{P}(X = x) = 0$ for a continuous random variable provided $f_X(t)$ is continuous at $t = x$. By contrast, $\mathbb{P}(X = x)$ can be greater than zero for a discrete random variable. The totality, expressed now as an integral over the real number line, of the values of the probability distribution function must be one. Interpreted graphically, if $X$ represents a continuous random variable then

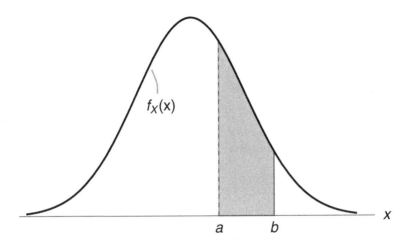

Fig. 8.1   An example of a probability distribution function. The shaded area represents the $\mathbb{P}\,(a < X \le b)$.

$\mathbb{P}\,(a < X \le b)$ represents the area of the region bounded by the graph of the probability distribution function, the $x$-axis, and the lines $x = a$ and $x = b$. Note that $f_X(x)$ is not required to be continuous itself, though in many cases it will be. See Fig. 8.1. The total area under the graph of the probability density function is one.

Perhaps the most elementary of the continuous random variables is the **uniformly distributed continuous random variable**. A continuous random variable $X$ is uniformly distributed in the interval $[a, b]$ (with $a < b$) if the probability that $X$ belongs to any subinterval of $[a, b]$ is equal to the length of the subinterval divided by $b - a$. This is frequently denoted $X \sim \mathcal{U}(a, b)$. From the definition of the continuous uniform random variable the probability density function for $X$ can be determined. Since the length of the subinterval is proportional to the probability that $X$ lies in the subinterval then the probability density function $f(x)$ must be constant. Suppose this constant is $k$. From the property expressed in Eq. (8.2) it must be the case that

$$1 = \int_{-\infty}^{\infty} f(x)\,dx = \int_{a}^{b} k\,dx = \int_{a}^{b} \frac{1}{b - a}\,dx.$$

Assume here the probability distribution function vanishes outside of the interval $[a, b]$. This implies the improper integral in Eq. (8.2) converges to a finite value. Thus the probability density function for a continuously, uniformly distributed random variable on interval $[a, b]$ is the piecewise-defined function

$$f_X(x) = \begin{cases} \frac{1}{b-a} & \text{if } a \leq x \leq b, \\ 0 & \text{otherwise.} \end{cases}$$

**Example 8.1.** Random variable $X$ is continuously uniformly randomly distributed in the interval $[-1, 12]$. Find the probability that $2 \leq X \leq 7$.

**Solution.** According to the previous discussion of uniformly distributed random variables, the probability density function of $X$ is

$$f_X(x) = \begin{cases} \frac{1}{13} & \text{if } -1 \leq x \leq 12, \\ 0 & \text{otherwise.} \end{cases}$$

Thus

$$\mathbb{P}\left(2 \leq X \leq 7\right) = \int_2^7 f_X(x)\, dx = \int_2^7 \frac{1}{13}\, dx = \frac{5}{13}.$$

If $f_X(x)$ is the probability density function for continuous random variable $X$, then the **cumulative distribution function** (or **CDF**) is denoted $F$ and can be interpreted as the probability that $X$ is less than or equal to a given value.

$$F(x) = \int_{-\infty}^x f_X(t)\, dt = \mathbb{P}\left(X \leq x\right). \tag{8.3}$$

If $f_X(x)$ is a valid probability density function, then the improper integral converges and the probability is well defined. When $f_X$ is continuous then $F'(x) = f_X(x)$ according to the Fundamental Theorem of Calculus [OpenStax (2018), Sec. 5.3]. For example the CDF for the random variable in the previous example can be expressed as

$$F(x) = \begin{cases} 0 & \text{if } x < -1, \\ (x+1)/13 & \text{if } -1 \leq x \leq 12, \\ 1 & \text{if } 12 < x. \end{cases}$$

**Theorem 8.1.** *Let $f_X(x)$ be the probability density function of a continuous random variable and let $F(x)$ be the corresponding cumulative distribution function, then the following properties hold.*

*(i) $0 \leq F(x) \leq 1$ for all $x \in \mathbb{R}$.*
*(i) $\lim_{x \to -\infty} F(x) = 0$ and $\lim_{x \to \infty} F(x) = 1$.*
*(iii) If $a < b$ then $F(a) \leq F(b)$, i.e., $F$ is a non-decreasing function.*
*(iv) $\mathbb{P}\left(a < X \leq b\right) = F(b) - F(a)$.*

*(v) F is continuous.*

*(vi)* $F'(x) = f_X(x)$ *whenever the derivative exists.*

*Proof.*

(i) Since $F(x) = \mathbb{P}\left(X \leq x\right)$ then $0 \leq F(x) \leq 1$ for all real numbers $x$.

(ii) Let $\{x_n\}_{n=1}^{\infty}$ be any strictly increasing sequence of real numbers such that $x_n \to \infty$ as $n \to \infty$.

$$1 = \int_{-\infty}^{\infty} f_X(t)\, dt$$

$$= \int_{-\infty}^{x_1} f_X(t)\, dt + \lim_{n \to \infty} \sum_{k=1}^{n-1} \int_{x_k}^{x_{k+1}} f_X(t)\, dt$$

$$= \lim_{n \to \infty} \left( \int_{-\infty}^{x_1} f_X(t)\, dt + \sum_{k=1}^{n-1} \int_{x_k}^{x_{k+1}} f_X(t)\, dt \right)$$

$$= \lim_{n \to \infty} \int_{-\infty}^{x_n} f_X(t)\, dt = \lim_{n \to \infty} F(x_n).$$

Since this result holds for any strictly increasing sequence, then it holds for all $x$ as $x \to \infty$. As a corollary to this result,

$$\int_{-\infty}^{-x} f_X(t)\, dt + \int_{-x}^{\infty} f_X(t)\, dt = 1$$

$$\int_{-x}^{\infty} f_X(t)\, dt = 1 - F(-x)$$

$$\lim_{x \to \infty} \int_{-x}^{\infty} f_X(t)\, dt = \lim_{x \to \infty} (1 - F(-x))$$

$$\lim_{x \to \infty} F(-x) = 0.$$

Consequently $\lim_{x \to -\infty} F(x) = 0$.

(iii) Suppose $a < b$ then since $f_X(t) \geq 0$ for all $t$,

$$\int_{-\infty}^{a} F_X(t)\, dt = F(a) \leq F(b) = \int_{-\infty}^{b} F_X(t)\, dt.$$

(iv) Note that

$$\mathbb{P}(a < X \leq b) = \int_{a}^{b} f_X(t)\, dt = \int_{-\infty}^{b} f_X(t)\, dt - \int_{-\infty}^{a} f_X(t)\, dt = F(b) - F(a).$$

(v) Let $a$ be a fixed real number and let $\{a_n\}_{n=1}^{\infty}$ be a decreasing sequence with $\lim_{n \to \infty} a_n = a$.

$$\lim_{n \to \infty} F(a_n) = \lim_{n \to \infty} \int_{-\infty}^{a_n} f_X(t)\, dt = \int_{-\infty}^{a} f_X(t)\, dt = F(a).$$

Once again, since the decreasing sequence is arbitrary then $\lim_{x \to a^+} F(x) = F(a)$. To determine the limit from the left, let $\{\alpha_n\}_{n=1}^{\infty}$ be an increasing sequence with $\lim_{n \to \infty} \alpha_n = a$.

$$\mathbb{P}\left(X < a\right) = \lim_{n \to \infty} \mathbb{P}\left(X \le \alpha_n\right) = \lim_{n \to \infty} F(\alpha_n) := F(a^-).$$

Therefore $\lim_{x \to a^-} F(x) = F(a^-)$. Thus

$$0 = \mathbb{P}\left(X = a\right) = \mathbb{P}\left(X \le a\right) - \mathbb{P}\left(X < a\right) = F(a) - F(a^-),$$

which implies $\lim_{x \to a^-} F(x) = F(a)$ and $F$ is continuous at $a$ (an arbitrary real number).

(vi) The derivative property follows from the Fundamental Theorem of Calculus. $\qquad\square$

In some instances the probability density function for a continuous random variable $X$ is known and the PDF for another variable $Y$ defined as a function of $X$ is needed. The relationship between the probability density function and the cumulative distribution function is often helpful in these cases. Suppose $X$ is a continuous random variable with probability density function $f_X(x)$ and suppose $r$ is a function such that $Y = r(X)$. In other words $Y$ is a random variable defined as a function of the random variable $X$. If $a < X < b$ and $r(x)$ is continuous and strictly increasing on the interval $(a, b)$, then $r$ maps interval $(a, b)$ onto an interval $(\alpha, \beta)$. Furthermore for each $y \in (\alpha, \beta)$ there exists a unique $x \in (a, b)$ such that $r(x) = y$. Hence $r$ has an inverse and $x = r^{-1}(y)$. Since $r$ is assumed to be continuous and strictly increasing then $r^{-1}$ is also continuous and strictly increasing on $(\alpha, \beta)$. Suppose $y \in (\alpha, \beta)$ then

$$\mathbb{P}\left(Y \le y\right) = \mathbb{P}\left(r(X) \le y\right) = \mathbb{P}\left(X \le r^{-1}(y)\right) = F(r^{-1}(y)),$$

where $F$ is the cumulative distribution function for random variable $X$. Consequently the cumulative distribution function for random variable $Y$ is $G(y) = F(r^{-1}(y))$. If $r^{-1}(y)$ is differentiable on $(\alpha, \beta)$ then by Thm. 8.1, $Y$ is a continuous random variable with probability density function

$$g_Y(y) = \frac{d}{dy}\left[G(y)\right] = \frac{d}{dy}\left[F(r^{-1}(y))\right] = f_X(r^{-1}(y))\frac{d}{dy}\left[r^{-1}(y)\right]$$

$$= \frac{1}{r'(r^{-1}(y))} f_X(r^{-1}(y)).$$

A similar statement can be made if $r$ is continuous and strictly decreasing on $(a, b)$. The inverse function $r^{-1}$ is continuous and strictly decreasing on and interval $(\alpha, \beta)$.

$$\mathbb{P}\left(Y \le y\right) = \mathbb{P}\left(r(X) \le y\right) = \mathbb{P}\left(X \ge r^{-1}(y)\right) = 1 - F(r^{-1}(y)).$$

If $r^{-1}(y)$ is differentiable on $(\alpha, \beta)$ then the probability density function for $Y$ can be written as

$$g_Y(y) = \frac{d}{dy}\left[1 - F(r^{-1}(y))\right] = \frac{-1}{r'(r^{-1}(y))} f_X(r^{-1}(y)).$$

The two cases can be summarized by stating that if $r$ is differentiable and monotone (increasing or decreasing) on $(a, b)$ and $Y = r(X)$, then the PDF for random variable $Y$ can be expressed as

$$g_Y(y) = \left|\frac{1}{r'(r^{-1}(y))}\right| f_X(r^{-1}(y)). \tag{8.4}$$

For example if $Y = aX + b$ where $a \ne 0$ and $b$ are constants, then the probability density function of $Y$ can be expressed as

$$g_Y(y) = \frac{1}{|a|} f_X\left(\frac{y - b}{a}\right).$$

### Exercises

**8.1.1** Random variable $X$ is continuously uniformly distributed in the interval $[-4, 1]$. Find $\mathbb{P}\left(X \ge 0\right)$.

**8.1.2** Suppose the probability distribution function of a continuous random variable $X$ is

$$f_X(x) = \frac{C}{1 + x^2}.$$

(a) Find the value of the constant $C$.
(b) Find $\mathbb{P}\left(X > 1\right)$.

**8.1.3** Random variable $X$ is continuously distributed on the interval $(1, \infty)$, with probability distribution function

$$f_X(x) = \begin{cases} C/x^3 & \text{if } 1 \le x, \\ 0 & \text{otherwise.} \end{cases}$$

Determine the value of $C$.

**8.1.4** Show using properties of the definite integral that for a continuous random variable $X$ with probability distribution function $f_X(x)$,

$$\mathbb{P}\left(X \ge a\right) = 1 - \mathbb{P}\left(X < a\right).$$

**8.1.5** Suppose $F(x)$ is a cumulative distribution function for a continuous random variable $X$. Show that $\mathbb{P}\left(a < X < b\right) = F(b^-) - F(a)$ when $b > a$.

**8.1.6** Find the cumulative distribution function corresponding to the continuous random variable $X$ whose probability density function is

$$f_X(x) = \frac{2e^{-2x}}{(1 + e^{-2x})^2} \text{ for } -\infty < x < \infty.$$

**8.1.7** Suppose $X$ is a continuous random variable with cumulative distribution function

$$F(x) = \begin{cases} 0 & \text{if } x \leq 3, \\ 1 - (3/x)^2 & \text{if } x > 3. \end{cases}$$

Find the probability density function for $X$.

**8.1.8** Show that for all $\lambda > 0$ the function $f(x) = \lambda e^{-\lambda x}$ for $x \geq 0$ (and zero otherwise) is a probability density function (known as the exponential distribution). Find the cumulative distribution function as well.

**8.1.9** Suppose the probability density function for continuous random variable $X$ is $f_X(x) = e^{-x}$ for $x \geq 0$ and zero otherwise. Find the probability density function for $Y = e^X$.

**8.1.10** Suppose the probability density function for continuous random variable $X$ is given by

$$f_X(x) = \begin{cases} \frac{3}{4}(1 - x^2) & \text{if } -1 \leq x \leq 1, \\ 0 & \text{otherwise.} \end{cases}$$

Find the probability density function for $Y = X^2$.

**8.1.11** Suppose that Alice and Bob agree to meet in the library between 10:00AM and 11:00AM to study calculus. Because neither of them is very reliable, either may be late to the meeting or forget to come at all. They have an understanding that neither will wait more than 15 minutes for the other to arrive. If they arrive at independent times between 10:00AM and 11:00AM, what is the probability they will meet?

## 8.2    Expected Value of Continuous Random Variables

Now that continuous random variables and their probability distributions have been introduced formulae for determining their means and variances can be introduced. Unless otherwise stated, the following results hold for all continuous random variables and their distributions.

By definition the **expected value** or **mean** of a continuous random variable $X$ with probability density function $f_X(x)$ is

$$\mathbb{E}(X) = \int_{-\infty}^{\infty} x f_X(x) \, dx. \tag{8.5}$$

This equation is analogous to Eq. (2.11) for the case of a discrete random variable. In the continuous case the product of the value of the random variable and its probability density function is summed (this time through the use of a definite integral) over all values of the random variable. The expected value is only meaningful in cases in which the improper integral in Eq. (8.5) converges.

**Example 8.2.** Find the expected value of $X$ if $X$ is a continuously uniformly distributed random variable on the interval $[-50, 75]$.

**Solution.** Using Eq. (8.5),

$$\mathbb{E}\left(X\right) = \int_{-\infty}^{\infty} \frac{x}{75 - (-50)}\, dx = \left[\frac{1}{250}x^2\right]_{x=-50}^{x=75} = \frac{25}{2}.$$

The results which follow in this section will be similar to the results that were presented for discrete random variables. In most cases the notion of a discrete sum need only be replaced by an integral to justify these new results. To keep the exposition as brief as possible, many of the proofs of results in this section will be left to the reader. The expected value of a function $g$ of a continuously distributed random variable $X$ which has probability distribution function $f_X$ is defined as

$$\mathbb{E}\left(g(X)\right) = \int_{-\infty}^{\infty} g(x) f_X(x)\, dx, \tag{8.6}$$

provided the improper integral converges absolutely, *i.e.*, if and only if

$$\int_{-\infty}^{\infty} |g(x)| f_X(x)\, dx < \infty.$$

Joint probability distributions for continuous random variables are defined similarly to those for discrete random variables. A joint probability distribution for a pair of random variables, $X$ and $Y$, is a non-negative function $f_{X,Y}(x, y)$ for which

$$\int_{-\infty}^{\infty} \int_{-\infty}^{\infty} f_{X,Y}(x, y)\, dx\, dy = 1.$$

Consequently the expected value of a function $g : \mathbb{R}^2 \to \mathbb{R}$ of the two random variables $X$ and $Y$ is defined as

$$\mathbb{E}\left(g(X, Y)\right) = \int_{-\infty}^{\infty} \int_{-\infty}^{\infty} g(x, y) f_{X,Y}(x, y)\, dx\, dy,$$

again, provided the integral is absolutely convergent.

Before turning to sums and products of continuous random variables the notion of a marginal distribution must be defined. If $X$ and $Y$ are continuous random variables with joint distribution $f_{X,Y}(x,y)$ then the **marginal probability distribution** for $X$ is defined as the function

$$f_X(x) = \int_{-\infty}^{\infty} f_{X,Y}(x,y)\,dy.$$

The marginal distribution $f_Y(y)$ for the random variable $Y$ is defined similarly. Recall that jointly distributed discrete random variables $A$ and $B$ are independent if and only if $\mathbb{P}(A \cap B) = \mathbb{P}(A)\mathbb{P}(B)$. For continuous random variables a similar definition holds. Two continuous random variables are **independent** if and only if the joint probability distribution function factors as the product of the marginal distributions of $X$ and $Y$. In other words $X$ and $Y$ are independent if and only if

$$f_{X,Y}(x,y) = f_X(x)f_Y(y),$$

for all real numbers $x$ and $y$.

**Example 8.3.** Consider the jointly distributed random variables $(X,Y)$ whose distribution is the function $f_{X,Y}(x,y) = \frac{2}{3x^3}$ for $(1 \le x < \infty)$ and $-1 \le y \le 2$. Find the mean of $X + Y$.

**Solution.** The mean of $X + Y$ is determined as follows.

$$\mathbb{E}(X+Y) = \int_{-\infty}^{\infty}\int_{-\infty}^{\infty} (x+y)\frac{2}{3x^3}\,dx\,dy = \int_{-1}^{2}\int_{1}^{\infty}(x+y)\frac{2}{3x^3}\,dx\,dy$$

$$= \int_{-1}^{2}\int_{1}^{\infty}\frac{2}{3x^2}\,dx\,dy + \int_{-1}^{2}\int_{1}^{\infty}\frac{2y}{3x^3}\,dx\,dy$$

$$= \int_{1}^{\infty}\int_{-1}^{2}\frac{2}{3x^2}\,dy\,dx + \int_{-1}^{2}\int_{1}^{\infty}\frac{2y}{3x^3}\,dx\,dy$$

$$= 3\int_{1}^{\infty}\frac{2}{3x^2}\,dx + \int_{-1}^{2}\frac{y}{3}\,dy = 2 + \frac{1}{2} = \frac{5}{2}.$$

The improper integral above converges and hence the mean of $X + Y$ is $5/2$.

The reader should verify that $\mathbb{E}(X+Y) = \mathbb{E}(X) + \mathbb{E}(Y)$. This is true in general, not just for the previous example. The additivity property of the expected value of continuous random variables is stated in the following theorem.

**Theorem 8.2.** *If $X_1, X_2, \ldots, X_k$ are continuous random variables with joint probability distribution $f_{X_1,\ldots,X_k}(x_1, x_2, \ldots, x_k)$ then $\mathbb{E}(X_1 + X_2 + \cdots X_k) = \mathbb{E}(X_1) + \mathbb{E}(X_2) + \cdots + \mathbb{E}(X_k)$.*

*Proof.* See Exercise 8.2.5. □

The reader should again note that, just as was the case in Chapter 2, the previous theorem is true for random variables which are dependent or independent.

**Theorem 8.3.** *Let $X_1, X_2, \ldots, X_k$ be pairwise independent random variables with joint distribution $f_{X_1,\ldots,X_k}(x_1, x_2, \ldots, x_k)$, then*

$$\mathbb{E}(X_1 X_2 \cdots X_k) = \mathbb{E}(X_1)\mathbb{E}(X_2) \cdots \mathbb{E}(X_k).$$

*Proof.* See Exercise 8.2.6. □

If the continuous random variable $X$ is conditional on another continuous random variable $Y$ having value $Y = y$, the conditional probability density will be denoted $f_{X|Y}(x|y)$. The conditional density of $X$ given $Y = y$ can be calculated as

$$f_{X|Y}(x|y) = \frac{f_{X,Y}(x,y)}{f_Y(y)}. \tag{8.7}$$

To put this notion to use, assume the ordered pairs $(X, Y)$ are uniformly distributed over the triangle $T = \{(x,y) \mid 0 \le y \le x \le 1\}$. Since the distribution is uniform and the area of $T$ is $1/2$, the joint probability distribution can be defined as $f_{X,Y}(x,y) = 2$ on $T$ and zero elsewhere. The conditional probability distribution of $X$ given $Y$ is

$$f_{X|Y}(x|y) = \frac{f_{X,Y}(x,y)}{\int_y^1 f_{X,Y}(x,y)\, dx} = \frac{2}{\int_y^1 2\, dx} = \frac{1}{1-y},$$

provided $0 < y < 1$.

The conditional expected value for continuously distributed random variables is found by using the conditional probability density.

$$\mathbb{E}(X|Y = y) = \int_{-\infty}^{\infty} x f_{X|Y}(x|y)\, dx. \tag{8.8}$$

The reader will be asked to show in the exercises that conditional expected value has the linearity property of the familiar expected value.

**Example 8.4.** The joint probability distribution of ordered pairs $(X, Y)$ on the triangle $T = \{(x,y) \mid -3 \le y \le x \le 3\}$ is given by $f_{X,Y}(x,y) = x^2 y^2/162$. Find the conditional expected value $\mathbb{E}(X|Y = y)$.

**Solution.** The conditional expected value $\mathbb{E}\left(X|Y=y\right)$ can be found in two steps. First the conditional probability density function $f_{X|Y}(x|y)$ must be determined for $-3 < y < 3$.

$$f_{X|Y}(x|y) = \frac{f_{X,Y}(x,y)}{\int_y^3 f_{X,Y}(x,y)\,dx} = \frac{\frac{x^2 y^2}{162}}{\int_y^3 \frac{x^2 y^2}{162}\,dx} = \frac{x^2}{\int_y^3 x^2\,dx} = \frac{3x^2}{27 - y^3}.$$

Second the formula in Eq. (8.8) for conditional expectation yields

$$\mathbb{E}\left(X|Y=y\right) = \int_y^3 x\left(\frac{3x^2}{27 - y^3}\right)dx = \left[\frac{3x^4}{108 - 4y^3}\right]_{x=y}^{x=3} = \frac{3(y+3)(y^2+9)}{4(y^2+3y+9)},$$

which holds for $-3 < y < 3$. Note once again that the conditional expected value can be thought of as another random variable depending on $Y$.

Just as was the case for discrete random variables in Chapter 2, the expected value of a continuous random variable can be thought of as the average value of the outcome of an infinite number of experiments. The variance and standard deviation of a continuous random variable again mirror the concept earlier defined for discrete random variables.

### Exercises

**8.2.1** A random variable $X$ has a continuous Cauchy distribution with probability density function

$$f_X(x) = \frac{1}{\pi(1 + x^2)}.$$

Show that the mean of this random variable does not exist.

**8.2.2** If $X$ is a continuous random variable with probability density function $f(x)$, show that $\mathbb{E}\left(aX + b\right) = a\,\mathbb{E}\left(X\right) + b$ where $a, b \in \mathbb{R}$.

**8.2.3** Let $\lambda > 0$ and suppose $X$ is a continuous random variable with probability density function

$$f_X(x) = \begin{cases} \lambda e^{-\lambda x} & \text{if } x > 0, \\ 0 & \text{otherwise.} \end{cases}$$

Find $\mathbb{E}\left(X\right)$.

**8.2.4** Suppose $X$ is a continuous random variable with probability density function

$$f_X(x) = \begin{cases} 1/(2x) & \text{if } 1 \leq x \leq e^2, \\ 0 & \text{otherwise.} \end{cases}$$

Find $\mathbb{E}\left(\ln X\right)$.

**8.2.5** Prove Thm. 8.2.

**8.2.6** Prove Thm. 8.3.

**8.2.7** Referring to the joint probability distribution for $(X, Y)$ given in Example 8.4 find $\mathbb{E}(Y|X = x)$.

**8.2.8** Suppose that $X$, $Y$ and $Z$ are jointly distributed continuous random variables.

(a) Show that $\mathbb{E}(X + Y|Z = z) = \mathbb{E}(X|Z = z) + \mathbb{E}(Y|Z = z)$.

(b) Suppose $c$ is a constant and show that $\mathbb{E}(cX|Y = y) = c\,\mathbb{E}(X|Y = y)$.

**8.2.9** Suppose $X$ and $Y$ are continuously distributed random variables whose joint probability distribution function is

$$f_{X,Y}(x, y) = \begin{cases} \frac{3}{8}xy^3 & \text{if } 0 \leq y \leq x \leq 2, \\ 0 & \text{otherwise.} \end{cases}$$

(a) Find $\mathbb{E}(XY)$.

(b) Find $\mathbb{E}(X^2Y)$.

(c) Find $\mathbb{E}\left(X\sqrt{Y}\right)$.

**8.2.10** Suppose that random variables $X$ and $Y$ are jointly distributed as

$$f_{X,Y}(x, y) = \begin{cases} x/500000 & \text{if } 0 \leq x \leq 100 \text{ and } x \leq y \leq x + 100, \\ 0 & \text{otherwise.} \end{cases}$$

Find $\mathbb{E}(Y|X = x)$.

## 8.3 Variance and Standard Deviation

The variance of a continuous random variable is a measure of the spread of values of the random variable about the expected value of the random variable. The variance is defined as

$$\mathbb{Var}(X) = \mathbb{E}\left((X - \mu)^2\right) = \int_{-\infty}^{\infty} (x - \mu)^2 f_X(x)\,dx, \tag{8.9}$$

where $\mu = \mathbb{E}(X)$ and $f_X(x)$ is the probability distribution function of $X$. The variance may be interpreted as the squared deviation of a random variable from its expected value. An alternative formula for the variance is given in the following theorem.

**Theorem 8.4.** *Let $X$ be a random variable with probability density function $f_X$ and mean $\mu$, then the variance of $X$ is $\mathbb{E}(X^2) - \mu^2$.*

**Example 8.5.** Find the variance of the continuous random variable whose probability distribution is given by

$$f_X(x) = \begin{cases} 2x \text{ if } 0 \le x \le 1, \\ 0 \quad \text{otherwise.} \end{cases}$$

**Solution.** The reader may readily check that $\mu = \mathbb{E}(X) = 2/3$. Now by use of Thm. 8.4.

$$\text{Var}(X) = \int_{-\infty}^{\infty} x^2 f_X(x)\, dx - \left(\frac{2}{3}\right)^2 = \int_0^1 2x^3\, dx - \frac{4}{9}$$
$$= \frac{1}{2} - \frac{4}{9} = \frac{1}{18}.$$

In the exercises the reader will be asked to prove the following two theorems which extend to continuous random variables results already established for discrete random variables.

**Theorem 8.5.** *Let $X$ be a continuous random variable with probability density function $f_X(x)$ and let $a, b \in \mathbb{R}$, then*

$$\text{Var}(aX + b) = a^2 \text{Var}(X).$$

**Theorem 8.6.** *Let $X_1, X_2, \ldots, X_k$ be pairwise independent continuous random variables with joint probability distribution $f_{X_1,\ldots,X_k}(x_1, x_2, \ldots, x_k)$, then*

$$\text{Var}(X_1 + X_2 + \cdots + X_k) = \text{Var}(X_1) + \text{Var}(X_2) + \cdots + \text{Var}(X_k).$$

By definition the standard deviation of a continuous random variable is the square root of its variance. The standard deviation of $X$ is sometimes denoted by $\sigma_X$. On occasion the variance of a random variable is written as $\sigma_X^2$.

The covariance and correlation of continuous random variables are denoted as in Sec. 2.9 keeping in mind that expectation for continuous random variables is calculated by evaluating an integral.

**Example 8.6.** Suppose random variables $X$ and $Y$ have the joint probability distribution given by

$$f_{X,Y}(x, y) = \begin{cases} 2 \text{ if } 0 \le x \le y \text{ and } 0 \le y \le 1, \\ 0 \text{ otherwise.} \end{cases}$$

Find the covariance and correlation of $X$ and $Y$.

**Solution.** First find the marginal distribution of each random variable so that the expected value and variance of $X$ and $Y$ can be determined.

$$f_X(x) = \int_x^1 2 \, dy = 2 - 2x$$

$$f_Y(y) = \int_0^y 2 \, dx = 2y$$

$$\mathbb{E}(X) = \int_0^1 x(2 - 2x) \, dx = \frac{1}{3}$$

$$\mathbb{E}(Y) = \int_0^1 y(2y) \, dy = \frac{2}{3}$$

$$\mathbb{Var}(X) = \int_0^1 x^2(2 - 2x) \, dx - \left(\frac{1}{3}\right)^2 = \frac{1}{6} - \frac{1}{9} = \frac{1}{18}$$

$$\mathbb{Var}(Y) = \int_0^1 y^2(2y) \, dy - \left(\frac{2}{3}\right)^2 = \frac{1}{2} - \frac{4}{9} = \frac{1}{18}.$$

Now use the joint distribution to determine

$$\mathbb{E}(XY) = \int_0^1 \int_0^y 2xy \, dx \, dy = \int_0^1 y^3 \, dy = \frac{1}{4}.$$

Using the definition of covariance,

$$\mathbb{Cov}(X,Y) = \mathbb{E}(XY) - \mathbb{E}(X)\mathbb{E}(Y) = \frac{1}{4} - \left(\frac{1}{3}\right)\left(\frac{2}{3}\right) = \frac{1}{36}.$$

Therefore the correlation is determined to be

$$\mathbb{Cor}(X,Y) = \frac{\mathbb{Cov}(X,Y)}{\sqrt{\mathbb{Var}(X)\mathbb{Var}(Y)}} = \frac{1/36}{1/18} = \frac{1}{2}.$$

### *Exercises*

**8.3.1** Find the expected value and variance of the continuous random variable $X$ whose probability density function is given by

$$f_X(x) = \begin{cases} \frac{2}{5}|x| & \text{if } -1 \le x \le 2, \\ 0 & \text{otherwise.} \end{cases}$$

**8.3.2** Suppose random variable $X$ is uniformly distributed on the interval $[a,b]$ with $a < b$. Find the variance of $X$.

**8.3.3** Prove Thm. 8.4.

**8.3.4** Let $\lambda > 0$ and suppose $X$ is a continuous random variable with probability density function

$$f_X(x) = \begin{cases} \lambda e^{-\lambda x} & \text{if } x > 0, \\ 0 & \text{otherwise.} \end{cases}$$

Find $\mathbb{Var}(X)$.

**8.3.5** Prove Theorem 8.5.

**8.3.6** Suppose the cumulative distribution function for continuous random variable $X$ is given by

$$F(x) = \begin{cases} 1 - 1/x^3 & \text{if } x \geq 1, \\ 0 & \text{otherwise.} \end{cases}$$

Find $\text{Var}(X)$.

**8.3.7** Suppose random variable $X$ is uniformly distributed on $[0, 1]$. Find $\text{Var}(X^2)$.

**8.3.8** Prove Thm. 8.6.

**8.3.9** Let $X$ be a continuously distributed random variable with $\mathbb{E}(X) = \mu$ and $\text{Var}(X) = \sigma^2$.

(a) Find the $\mathbb{E}(X(X + 1))$.
(b) Find the $\mathbb{E}((X - C)^2)$ where $C$ is a constant.

**8.3.10** Suppose continuous random variable $X$ has the following probability density function.

$$f_X(x) = \begin{cases} \frac{15}{16}x^2(1 - x)^2 & \text{if } -1 \leq x \leq 1, \\ 0 & \text{otherwise.} \end{cases}$$

Find $\text{Var}(X)$.

**8.3.11** Suppose continuous random variable $X$ has the following probability density function.

$$f_X(x) = \begin{cases} 30x(1 - x)^4 & \text{if } 0 \leq x \leq 1, \\ 0 & \text{otherwise.} \end{cases}$$

Find $\text{Var}(X/(1 - X))$.

**8.3.12** Suppose $X$ and $Y$ are continuous random variables with probability distribution function

$$f_{X,Y}(x, y) = \begin{cases} e^x & \text{if } 0 \leq y \leq x \leq 1, \\ 0 & \text{otherwise.} \end{cases}$$

Find the $\text{Cor}(X, Y)$.

**8.3.13** Let $X$ and $Y$ be independent uniformly distributed continuous random variables on $[0, 1]$. Suppose that $Z = X + Y^2$. Find $\text{Cor}(Y, Z)$.

## 8.4 Normal Random Variables

One of the most important and useful continuous random variables is the **normally distributed random variable**. A continuous random variable obeying a normal probability distribution is frequently said to have a

PDF which resembles a "bell curve". Many measurable quantities found in nature seem to have normal distributions, for example adult heights and weights. Statisticians, mathematicians, and physical scientists frequently assume any quantity subject to a large number of small independently acting forces (regardless of their distributions) is normally distributed. The normal probability distribution even finds its way into the financial arena via the assumption that movements in the price of an asset are subject to a large number of incompletely understood political, economic, and social forces, thus justifying the assumption that these changes in value are related to a normal random variable (this assumption will be explored later).

The continuous normal random variable can also be thought of as the limiting behavior of the discrete binomial random variable introduced in Chapter 2. Recall the probability mass function for a binomial random variable with parameters $n$ and $p$ given in Eq. (2.10) and reproduced below for convenience.

$$\mathbb{P}\left(X = x\right) = \frac{n!}{x!(n-x)!}p^x(1-p)^{n-x} \text{ for } x = 0, 1, \ldots, n.$$

Under suitable assumptions the probability density function for a continuous normally distributed random variable is the limiting case of the probability mass function of a binomial random variable. The remainder of this section is dedicated to this derivation and to some elementary properties of normal random variables. This derivation is a generalization of the process outlined in [Bleecker and Csordas (1996), p. 139].

Suppose that a particle sits at the origin of the $x$-axis and may move to the left or to the right a distance $\Delta x > 0$ in each of $n$ independent identically distributed trials. After $n$ trials (steps) the particle will be located at one of the locations in the set $\{-n\Delta x, -(n-1)\Delta x, \ldots, n\Delta x\}$. Figure 8.2 illustrates a typical realization of this **random walk**. Define the constant quantity $\sigma^2 t = n(\Delta x)^2$. This particular form of constant is chosen to simplify the ultimate result. It is convenient to think of the time necessary to conduct the $n$ trials as $t$ and hence the time required by one trial is $\Delta t = t/n$. To mathematically model this random walk define the discrete random variables $X_k$ for $k = 1, 2, \ldots, n$ with probability mass function

| $x$ | $\mathbb{P}\left(X_k = x\right)$ |
|-----|------|
| 1 | $p$ |
| $-1$ | $1 - p$ |

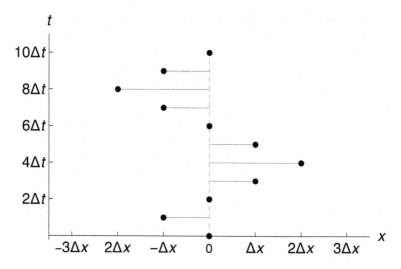

Fig. 8.2   A realization of a discrete random walk of a particle. The particle is initially at the origin and may take $n$ independent and identically distributed steps of magnitude $\Delta x$ left or right. In this realization the particle returns to the origin $x = 0$ three times in the first ten steps.

where naturally $0 < p < 1$. The random variable $W(t)$ where $t = n\Delta t$ is defined as

$$W(t) = \Delta x \sum_{k=1}^{n} X_k.$$

The expected value of $W(t)$ is calculated as

$$\mathbb{E}\left(W(t)\right) = \mathbb{E}\left(\Delta x \sum_{k=1}^{n} X_k\right) = \Delta x \sum_{k=1}^{n} \mathbb{E}\left(X_k\right)$$
$$= n\Delta x(2p - 1) = (2p - 1)\sigma\sqrt{t}.$$

Using the independence of the steps, the variance can be expressed as

$$\mathbb{V}\mathrm{ar}\left(W(t)\right) = \mathbb{V}\mathrm{ar}\left(\Delta x \sum_{k=1}^{n} X_k\right) = (\Delta x)^2 \sum_{k=1}^{n} \mathbb{V}\mathrm{ar}\left(X_k\right)$$
$$= 4n(\Delta x)^2 p(1 - p) = 4p(1 - p)\sigma^2 t.$$

Even more informative is the probability mass function for $W(t)$. If during the random walk $r$ of the $n$ total steps are to the right $(\Delta x)$ and $n - r$ steps are to the left $(-\Delta x)$ then $W(t) = (2r - n)\Delta x$. Let $(2r - n)\Delta x = w$, then

$$\mathbb{P}\left(W(t) = w\right) = \binom{n}{r}p^r(1 - p)^{n-r}, \tag{8.10}$$

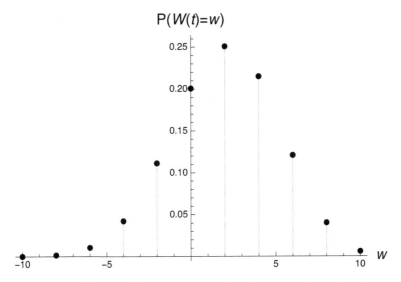

P(W(t)=w)

Fig. 8.3   A plot of the probability mass function for a discrete random walk with parameters $n = 10$, $p = 2/5$, and $\Delta x = 1$. Note that since $n$ is even the only final resting places of the $W(t)$ are even multiples of $\Delta x$.

for $r = 0, 1, \ldots, n$. A plot of the probability mass function for $W(t)$ with parameters $n = 10$, $p = 2/5$, and $\Delta x = 1$ is shown in Fig. 8.3. Consider the bijection,

$$w = (2r - n)\Delta x \iff r = \frac{1}{2}\left(n + \frac{w}{\Delta x}\right),$$

then

$$\mathbb{P}\left(W(t) = w\right) = \mathbb{P}\left(X = \frac{1}{2}\left(n + \frac{w}{\Delta x}\right)\right),$$

where random variable $X \sim \mathcal{B}(n, p)$. To determine the distribution of the random variable $W$ obtained in the limit as $n \to \infty$ or equivalently as $\Delta x \to 0$, follow the approach described in [Proschan (2008)]. The ratio of the values of two particular values of the probability mass function can be expressed as

$$\frac{\mathbb{P}\left(W(t) = w + 2\Delta x\right)}{\mathbb{P}\left(W(t) = w\right)} = \frac{\binom{n}{\frac{1}{2}(n+\frac{w}{\Delta x})+1} p^{\frac{1}{2}(n+\frac{w}{\Delta x})+1}(1-p)^{\frac{1}{2}(n-\frac{w}{\Delta x})-1}}{\binom{n}{\frac{1}{2}(n+\frac{w}{\Delta x})} p^{\frac{1}{2}(n+\frac{w}{\Delta x})}(1-p)^{\frac{1}{2}(n-\frac{w}{\Delta x})}}$$

$$= \frac{\left(n - \frac{w}{\Delta x}\right) p}{\left(n + \frac{w}{\Delta x} + 2\right)(1 - p)}.$$

The expected value of $W(t)/\Delta x$ is $(2p-1)n$ while its variance is $4np(1-p)$. Use the transformation $w = (2p-1)n\Delta x + 2z\sqrt{np(1-p)}\Delta x$ to rewrite the ratio of the probability mass functions as

$$\frac{\mathbb{P}\left(W(t) = w + 2\Delta x\right)}{\mathbb{P}\left(W(t) = w\right)} = \frac{(n - (2p-1)n - 2z\sqrt{np(1-p)})p}{(n + (2p-1)n + 2z\sqrt{np(1-p)} + 2)(1-p)}$$

$$\frac{\mathbb{P}\left(Z(t) = z + \Delta_n\right)}{\mathbb{P}\left(Z(t) = z\right)} = \frac{1 - pz\Delta_n}{1 + (1-p)z\Delta_n + (1-p)\Delta_n^2},$$

(8.11)

where $\Delta_n = 1/\sqrt{np(1-p)}$. The reader should note that $\Delta_n \to 0$ as $n \to \infty$. Suppose that $\mathbb{P}\left(Z(t) = z\right) \approx f_Z(z)\,dz$ where $f_Z$ is a probability density function for random variable $Z$ when $n$ is large. Equation (8.11) can be written in the form

$$\frac{f_Z(z + \Delta_n)\,dz}{f_Z(z)\,dz} = \frac{f_Z(z + \Delta_n)}{f_Z(z)} \approx \frac{1 - pz\Delta_n}{1 + (1-p)z\Delta_n + (1-p)\Delta_n^2}.$$

Taking the natural logarithm of both sides of this equation and dividing by $\Delta_n$ produce

$$\frac{\ln f_Z(z + \Delta_n) - \ln f_Z(z)}{\Delta_n}$$

$$\approx \frac{\ln(1 - pz\Delta_n)}{\Delta_n} - \frac{\ln(1 + (1-p)z\Delta_n + (1-p)\Delta_n^2)}{\Delta_n}.$$

Taking the limit of both sides of this approximation as $\Delta_n \to 0$ is equivalent to taking the limit as $n \to \infty$. The limit leads to

$$\frac{d}{dz}\left[\ln f_Z(z)\right] = -z.$$

(8.12)

The reader will be asked to provide the details of this derivation in the exercises. Integrating both sides of Eq. (8.12) yields

$$\ln f_Z(z) = \frac{-z^2}{2} + C,$$

$$f_Z(z) = e^C e^{-z^2/2}.$$

The function $f_Z(z)$ must satisfy the properties of a probability density function. For all real $z$ the expression is non-negative. Since for $|z| \geq 1$ it is true that $0 < e^{-z^2/2} \leq e^{-|z|/2}$, then the integral of $f_Z(z)$ over the entire real number line converges. Suppose that $S$ is finite and

$$S = \int_{-\infty}^{\infty} e^C e^{-z^2/2}\,dz.$$

The square of both sides of this equation can be written as

$$S^2 = \int_{-\infty}^{\infty} e^C e^{-z^2/2}\, dz \int_{-\infty}^{\infty} e^C e^{-y^2/2}\, dy$$

$$= e^{2C} \int_{-\infty}^{\infty} \int_{-\infty}^{\infty} e^{-(y^2+z^2)/2}\, dy\, dz.$$

Switching to polar coordinates by making the substitutions $y = \rho \cos\theta$, $z = \rho \sin\theta$, and $dy\, dz = \rho\, d\rho\, d\theta$ produces

$$S^2 = e^{2C} \int_0^{2\pi} \int_0^{\infty} \rho e^{-\rho^2/2}\, d\rho\, d\theta = 2\pi e^{2C} \int_0^{\infty} \rho e^{-\rho^2/2}\, d\rho$$

$$= 2\pi e^{2C} \int_0^{\infty} e^{-u}\, du = 2\pi e^{2C}.$$

To be a valid probability density function

$$2\pi e^{2C} = 1 \implies e^C = \frac{1}{\sqrt{2\pi}} \implies f_Z(z) = \frac{1}{\sqrt{2\pi}} e^{-z^2/2}.$$

Hence the function given above satisfies the non-negativity condition and unit area condition of a probability density function for a continuous random variable. The random variable $Z$ is a called the **standard normal random variable**. Since it occupies a rather special place in mathematics and statistics its probability density function is often denoted as

$$\varphi(z) = \frac{1}{\sqrt{2\pi}} e^{-z^2/2}. \tag{8.13}$$

The probability density function has the familiar bell shape as seen in Fig. 8.4. The mean and variance of random variable $Z$ are respectively $\mathbb{E}(Z) = 0$ and $\mathbb{Var}(Z) = 1$ (see the exercises at the end of this section). The cumulative distribution function for random variable $Z$ will be denoted

$$\Phi(z) = \int_{-\infty}^{z} \frac{1}{\sqrt{2\pi}} e^{-t^2/2}\, dt. \tag{8.14}$$

Now suppose $Y = aZ + b$ where $a$ and $b$ are constants and $a \neq 0$ then $Y$ is said to be a **normal random variable** with mean $b$ and variance $a^2$. According to Eq. (8.4) the probability density function for $Y$ is

$$f_Y(y) = \frac{1}{\sqrt{2\pi a^2}} e^{-(y-b)^2/(2a^2)}. \tag{8.15}$$

Random variable $Y$ is said to be normally distributed with parameters $b$ and $a$ denoted concisely as $Y \sim \mathcal{N}(b, a^2)$.

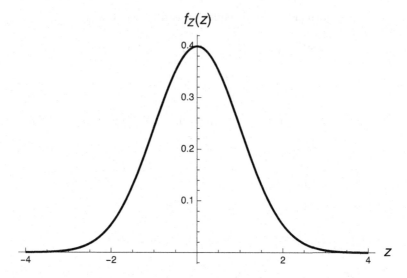

Fig. 8.4　The graph of the probability density function for the standard normal random variable is often described as a bell-shaped curve.

Since $w = (2p-1)\sigma\sqrt{t} + 2\sigma z\sqrt{p(1-p)t}$ then using Eq. (8.4) the probability density function for $W(t)$ the random walk, is

$$f_W(w) = \frac{1}{\sqrt{4p(1-p)\sigma^2 t}}\varphi\left(\frac{w - (2p-1)\sigma\sqrt{t}}{\sqrt{4p(1-p)\sigma^2 t}}\right). \tag{8.16}$$

Hence for a fixed $p$ in the limit as $\Delta x \to 0$ the random variable $W(t) \sim \mathcal{N}((2p-1)\sigma\sqrt{t}, 4p(1-p)\sigma^2 t)$. As $t$ increases the mean changes and the variance of $W(t)$ increases as well. The mean of $W(t)$ increases with increasing $t$ when $p > 1/2$ and decreases with increasing $t$ when $p < 1/2$. This makes sense because $p > 1/2$ implies that particle is more likely to take a rightward step than a leftward step while $p < 1/2$ reverses the probabilities. The surface plotted in Fig. 8.5 illustrates the evolving distribution of $W(t)$ for $t > 0$.

**Example 8.7.** Suppose the random walk $W(t) \sim \mathcal{N}(\frac{\sqrt{t}}{5}, \frac{24t}{25})$ for all $t > 0$. Find $\mathbb{P}(-1 < W(4) \le 1)$.

**Solution.** The random variable $W(4) \sim \mathcal{N}(2/5, 96/25)$, so

$$\mathbb{P}(-1 < W(4) \le 1) = \mathbb{P}\left(\frac{-1 - 2/5}{\sqrt{96/25}} < \frac{W(4) - 2/5}{\sqrt{96/25}} \le \frac{1 - 2/5}{\sqrt{96/25}}\right)$$

$$= \mathbb{P}(-0.7144 < Z \le 0.3062),$$

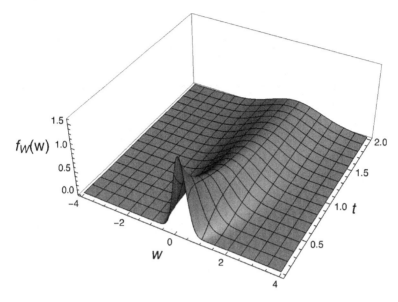

Fig. 8.5 The surface shown is the evolution as $t$ increases of the probability density function of the random walk $W(t)$. To create this surface parameter values were chosen to be $p = 4/5$ and $\sigma = 1$. For each $t > 0$ the $\mathbb{E}\,(W(t)) = 0.6\sqrt{t}$. The bell-shaped curve flattens for increasing $t$ since the variance is an increasing function of $t$, namely $\mathbb{V}\mathrm{ar}\,(W(t)) = 0.64t$.

where $Z \sim \mathcal{N}(0,1)$. The last probability can be estimated numerically from the definite integral

$$\int_{-0.7144}^{0.3062} \frac{1}{\sqrt{2\pi}} e^{-t^2/2}\, dt,$$

or evaluated on a calculator with a built-in function for the cumulative distribution function of the standard normal random variable. However the reader chooses to evaluate the quantity,

$$\mathbb{P}\,(-1 < W(4) \le 1) = \mathbb{P}\,(-0.7144 < Z \le 0.3062)$$
$$= \Phi\,(0.3062) - \Phi\,(-0.7144) = 0.3828.$$

### Exercises

**8.4.1** Show that $\displaystyle\lim_{\Delta_n \to 0} \frac{\ln(1 - pz\Delta_n)}{\Delta_n} = -p\,z$.

**8.4.2** Show that $\displaystyle\lim_{\Delta_n \to 0} \frac{\ln(1 + (1-p)z\Delta_n + (1-p)\Delta_n^2)}{\Delta_n} = (1-p)z$.

**8.4.3** Suppose $Z$ is the standard normal random variable with distribution $\mathcal{N}(0,1)$. Show $\mathbb{E}(Z) = 0$.

**8.4.4** Suppose $Z$ is the standard normal random variable with distribution $\mathcal{N}(0,1)$. Show $\mathbb{V}\text{ar}(Z) = 1$.

**8.4.5** Suppose $X \sim \mathcal{N}(\mu, \sigma^2)$ and show that $Y = aX + b$ is distributed as $\mathcal{N}(a\mu + b, a^2\sigma^2)$. Assume $a \neq 0$.

**8.4.6** Let $X$ be a normally distributed random variable for which $\mathbb{P}(X \le 205) = 0.5$ and $\mathbb{P}(X > 325) = 0.05$. Find the variance of $X$.

**8.4.7** Suppose $X \sim \mathcal{N}(3, 16)$ and find $\mathbb{P}(|X - 2| \ge 1)$.

**8.4.8** Suppose $X \sim \mathcal{N}(5, 9)$ and find $\mathbb{P}(X^2 - 3X \le -2)$.

**8.4.9** Suppose $X \sim \mathcal{N}(\mu, 25)$ and $\mathbb{P}(X \le 100) = 0.15$. Find $\mu$.

**8.4.10** Random variable $X$ is normally distributed with a mean of $\mu$ and variance of $\mu^3$ with $\mathbb{P}(X > 0) = 0.90$. Find $\mu$.

**8.4.11** Show that
$$\frac{\sigma}{\sqrt{2\pi}} \int_{(K-\mu)/\sigma}^{\infty} t e^{-t^2/2}\, dt = \frac{\sigma}{\sqrt{2\pi}} e^{-(K-\mu)^2/2\sigma^2}.$$

**8.4.12** Evaluate the following indefinite integral.
$$\int \frac{x}{2\sqrt{k\pi t}} e^{-\frac{x^2}{4kt}}\, dx.$$

**8.4.13** Use the technique of integration by parts to evaluate
$$\int \frac{x^2}{2\sqrt{k\pi t}} e^{-\frac{x^2}{4kt}}\, dx.$$

**8.4.14** Evaluate the following limit with $k > 0$ and $t > 0$.
$$\lim_{M \to \infty} 2ktM e^{-M^2/4kt}.$$

**8.4.15** Using a computer algebra system, graphing calculator, or some numerical method evaluate the following probabilities for a standard normal random variable:

(a) $\mathbb{P}(-1 < X < 1)$.
(b) $\mathbb{P}(-2 < X < 2)$.
(c) $\mathbb{P}(-3 < X < 3)$.
(d) $\mathbb{P}(1 < X < 3)$.

**8.4.16** Suppose that $X \sim \mathcal{N}(0,1)$. Show that $\Phi(-x) = 1 - \Phi(x)$ for all real numbers $x$.

**8.4.17** The annual rainfall amount in a geographical area is normally distributed with a mean of 14 inches and a standard deviation of 3.2 inches. What is the probability that the sum of the annual rainfalls in two consecutive years will exceed 30 inches? Assume that the amounts of rainfall received in the two different years are independent.

**8.4.18** Suppose $X \sim \mathcal{N}(\mu, \sigma^2)$. The random variable $Y = X^2$ is said to have the Chi-squared distribution. Find the probability density function for $Y$.

## 8.5 Lognormal Random Variables

While normal random variables are central to the discussion of probability and the upcoming Black–Scholes option pricing formula, of equal importance will be continuous random variables which are distributed in a lognormal fashion. A random variable $X$ is a **lognormal random variable** with parameters $\mu$ and $\sigma^2$ if $\ln X$ is a normally distributed random variable with mean $\mu$ and variance $\sigma^2$. This can be concisely denoted as $X \sim \mathcal{LN}(\mu, \sigma^2)$ and is equivalent to stating that $\ln X \sim \mathcal{N}(\mu, \sigma^2)$. When referring to a lognormal random variable, the parameters $\mu$ and $\sigma$ are often called the **drift** and **volatility** respectively. Suppose $X \sim \mathcal{LN}(\mu, \sigma^2)$ and $x$ is a real number. If $x \leq 0$ then $\mathbb{P}(X \leq 0) = 0$ since $\ln x$ is not a real number. If $x > 0$ then let $Y = \ln X$ and

$$\mathbb{P}(X \leq x) = \mathbb{P}(Y \leq \ln x) = \int_{-\infty}^{\ln x} \frac{1}{\sqrt{2\pi\sigma^2}} e^{(t-\mu)^2/(2\sigma^2)}\, dt$$

$$f_X(x) = \frac{1}{x\sqrt{2\pi\sigma^2}} e^{-(\ln x - \mu)^2/(2\sigma^2)}. \tag{8.17}$$

Thus a lognormal random variable is a continuous random variable which has a positive probability only for $x$ in the interval $(0, \infty)$. See Fig. 8.6 for a graph of a lognormal probability density function. The introduction of lognormal random variables does not complicate matters much since if $X \sim \mathcal{LN}(\mu, \sigma^2)$ and $x > 0$ then the probability that $X \leq x$ is

$$\mathbb{P}(X \leq x) = \Phi\left(\frac{\ln x - \mu}{\sigma}\right), \tag{8.18}$$

where $\Phi(z)$ is the cumulative probability distribution function for a normal random variable with mean zero and variance one. Similarly the $\mathbb{P}(X > x) = 1 - \Phi((\ln x - \mu)/\sigma)$.

**Example 8.8.** Suppose that random variable $S(t) = 10e^{\mu t + \sigma\sqrt{t}Z}$ where $\mu$ and $\sigma$ are constants and $Z \sim \mathcal{N}(0,1)$. Show that $S(t)$ is lognormally distributed for all $t > 0$ and its parameters.

**Solution.** Let $Y(t) = \ln S(t) = \ln 10 + \mu t + \sigma\sqrt{t}Z$, then

$$\mathbb{E}(Y(t)) = \mathbb{E}\left(\ln 10 + \mu t + \sigma\sqrt{t}Z\right) = \ln 10 + \mu t + \sigma\sqrt{t}\,\mathbb{E}(Z) = \ln 10 + \mu t,$$

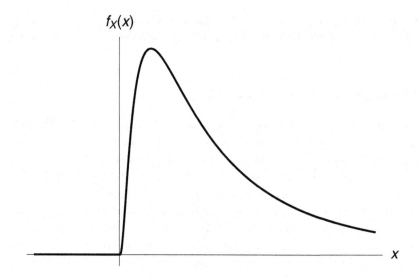

Fig. 8.6 The graph of the lognormal probability distribution function.

and
$$\mathsf{Var}\,(Y(t)) = \mathsf{Var}\left(\ln 10 + \mu t + \sigma\sqrt{t}Z\right) = \sigma^2 t\,\mathsf{Var}\,(Z) = \sigma^2 t.$$
Since $Z$ is normally distributed then $Y(t) \sim \mathcal{N}(\ln 10 + \mu t, \sigma^2 t)$ which implies $S(t) \sim \mathcal{LN}(\ln 10 + \mu t, \sigma^2 t)$.

The mean and variance of a lognormal random variable are readily found using the probability density function in Eq. (8.17).

**Lemma 8.1.** *If $X \sim \mathcal{LN}(\mu, \sigma^2)$ then*
$$\mathbb{E}\,(X) = e^{\mu + \sigma^2/2}, \tag{8.19}$$
$$\mathsf{Var}\,(X) = e^{2\mu + \sigma^2}\left(e^{\sigma^2} - 1\right). \tag{8.20}$$

*Proof.* According to the definition of the expected value of a continuous random variable and using the probability density function found in Eq. (8.17)
$$\begin{aligned}
\mathbb{E}\,(X) &= \int_0^\infty x\,\frac{1}{x\sqrt{2\pi\sigma^2}}e^{-(\ln x - \mu)^2/(2\sigma^2)}\,dx \\
&= \int_{-\infty}^\infty \frac{1}{\sqrt{2\pi\sigma^2}}e^t e^{-(t-\mu)^2/(2\sigma^2)}\,dt \\
&= e^{\mu + \sigma^2/2}\int_{-\infty}^\infty \frac{1}{\sqrt{2\pi\sigma^2}}e^{-(t-(\mu+\sigma^2))^2/(2\sigma^2)}\,dt \\
&= e^{\mu + \sigma^2/2}.
\end{aligned}$$

The second line of the equation makes use of the substitution $t = \ln x$. The last equality is true since the integral represents the area under the probability density curve for a normal random variable with mean $\mu + \sigma^2$ and variance $\sigma^2$.

Likewise by Eq. (8.9)

$$\begin{aligned}
\text{Var}\,(X) &= \mathbb{E}\left(X^2\right) - (\mathbb{E}\,(X))^2 \\
&= \int_0^\infty x^2 \frac{1}{x\sqrt{2\pi\sigma^2}} e^{-(\ln x - \mu)^2/(2\sigma^2)}\,dx - \left(e^{\mu+\sigma^2/2}\right)^2 \\
&= \int_{-\infty}^\infty \frac{1}{\sqrt{2\pi\sigma^2}} e^{2t} e^{-(t-\mu)^2/(2\sigma^2)}\,dt - e^{2\mu+\sigma^2} \\
&= e^{2(\mu+\sigma^2)} \int_{-\infty}^\infty \frac{1}{\sqrt{2\pi\sigma^2}} e^{-(t-(\mu+2\sigma))^2/(2\sigma^2)}\,dt - e^{2\mu+\sigma^2} \\
&= e^{2\mu+\sigma^2}\left(e^{\sigma^2} - 1\right).
\end{aligned}$$

Between the second and third lines of the equation the substitution $t = \ln x$ was used again. $\square$

**Example 8.9.** Suppose that random variable $S(t) = S(0)e^{\mu t + \sigma\sqrt{t}Z}$ where $S(0)$, $\mu$, and $\sigma$ are constants and $Z \sim \mathcal{N}(0,1)$. Find the mean and variance of $S(t)$ for all $t > 0$.

**Solution.** Following the same line of reasoning as in Example 8.8 $S(t)$ is lognormally distributed with $S(t) \sim \mathcal{LN}(\ln S(0) + \mu t, \sigma^2 t)$. Thus by Lemma 8.1,

$$\begin{aligned}
\mathbb{E}\,(S(t)) &= e^{\ln S(0)+\mu t+(\sigma\sqrt{t})^2/2} = S(0)e^{(\mu+\sigma^2/2)t} \\
\text{Var}\,(S(t)) &= e^{(\ln S(0)+\mu t+(\sigma\sqrt{t})^2/2)^2}\left(e^{(\sigma\sqrt{t})^2} - 1\right) \\
&= (S(0))^2 e^{(2\mu+\sigma^2)t}\left(e^{\sigma^2 t} - 1\right).
\end{aligned}$$

The next chapter will use the concepts of the continuous normal and lognormal distributions to develop a mathematical model of the movements of asset prices in a market.

## Exercises

**8.5.1** Show that the variance of a lognormally distributed random variable $X$ with parameters $\mu$ and $\sigma$ can be written as

$$\text{Var}\,(X) = (\mathbb{E}\,(X))^2 \left(e^{\sigma^2} - 1\right).$$

**8.5.2** Suppose $X \sim \mathcal{LN}(2,1)$ and determine $\mathbb{P}(X \le e)$.

**8.5.3** Suppose that $X$ is lognormally distributed and $k$ is a natural number. Show that $\mathbb{E}\left(X^k\right) = e^{k\mu + k^2\sigma^2/2}$ where $\mu$ and $\sigma^2$ are the parameters associated with the lognormal distribution of $X$.

**8.5.4** Suppose that random variable $X$ is lognormally distributed and $k$ is a non-zero integer. Show that $X^k$ is lognormally distributed.

**8.5.5** Suppose random variable $X \sim \mathcal{N}(0.1, 0.5)$ and $Y = e^X$. Determine the $\mathbb{P}(0.15 < Y \le 1.75)$.

**8.5.6** Suppose random variable $X \sim \mathcal{N}(0.25, 0.75)$ and $Y = e^{-X}$. Determine the $\mathbb{P}(Y \le \mathbb{E}(Y))$.

**8.5.7** The **median** of a random variable is the value of the random variable separating the lower 50% of the values of the variable from the upper 50% of values. Suppose $X \sim \mathcal{LN}(\mu, \sigma^2)$ and find the median of $X$.

**8.5.8** The ratio of prices of a security on consecutive days is a lognormally distributed random variable with parameters $\mu = 0.01$ and $\sigma = 0.05$. Price ratios recorded on different days are independent random variables.

(a) What is the probability of a 1-day increase in the security price?
(b) What is the probability of a 1-day decrease in the security price?
(c) What is the probability of the security price after 4 days is lower than the price on the first day?

## 8.6  Application: Portfolio Selection with Utility

The concept of an investor's utility function was introduced in Sec. 4.5. In this section the utility function will be coupled to the problem of deciding how an investor should optimally allocate funds between several potential investments. This is often called the **portfolio selection problem**. This provides an opportunity to apply the properties of normal and lognormal random variables.

Consider the following simple scenario. Suppose an investor has a total of $x$ amount of capital to invest. Assuming that the investor may use any proportion of this capital, let $\alpha \in [0, 1]$ be the proportion they invest. The investment is structured such that an allocation of $\alpha x$ will earn $\alpha x$ with probability $p$ and lose $\alpha x$ with probability $1 - p$. Thus for an investment of $\alpha x$ the random variable representing the investor's return after investment is

$$X = \begin{cases} x(1 + \alpha) & \text{with probability } p, \\ x(1 - \alpha) & \text{with probability } 1 - p. \end{cases}$$

The allocation proportion $\alpha$ will be optimal when the expected value of the investor's utility function is maximized.

$$\mathbb{E}\left(u(X)\right) = pu(x(1+\alpha)) + (1-p)u(x(1-\alpha))$$
$$\frac{d}{d\alpha}\left[\mathbb{E}\left(u(X)\right)\right] = pxu'(x(1+\alpha)) - x(1-p)u'(x(1-\alpha))$$
$$0 = pu'(x(1+\alpha)) - (1-p)u'(x(1-\alpha)).$$

The critical value of $\alpha$ which solves the equation above will correspond to a maximum for the expected value of the utility as long as the utility function is concave.

**Example 8.10.** If $u(x) = \ln x$, find the value of $\alpha$ which maximizes the investor's utility.

**Solution.** The derivative of the utility function is $u'(x) = 1/x$ and thus the critical value of $\alpha$ is the solution to the following equation.

$$0 = \frac{p}{x(1+\alpha)} - \frac{1-p}{x(1-\alpha)}$$
$$\alpha = 2p - 1.$$

Notice that by the choice of utility function, the total amount of capital to invest, $x$, becomes irrelevant when maximizing utility. For the sake of clarity assume $x = 1$ and then multiply by the appropriate scaling factor for other total amounts. If $p > 1/2$ the investor should allocate $100(2p - 1)\%$ of their capital. If $0 \le p \le 1/2$, then $\mathbb{E}\left(u(X)\right)$ is maximized when $\alpha = 0$, *i.e.*, when no investment is made. A curve depicting the expected value of the investor's utility as a function of $p$ is presented in Fig. 8.7.

With this simple example understood, now consider a more general situation. Suppose that an infinitely divisible unit amount of capital may be allocated among $n$ different investments. A weight $x_i$ will be allocated to security $i$ for $i = 1, 2, \ldots, n$. The vector notation $\mathbf{x} = \langle x_1, x_2, \ldots, x_n \rangle$ will be used to represent the portfolio of proportions of investments. The rate of return from investment $i$ will be denoted $R_i$ for $i = 1, 2, \ldots, n$. The portfolio selection problem is then defined to be that of determining $x_i$ for $i = 1, 2, \ldots, n$ such that

(1) $0 \le x_i \le 1$ for $i = 1, 2, \ldots, n$, and
(2) $\sum_{i=1}^{n} x_i = 1$,

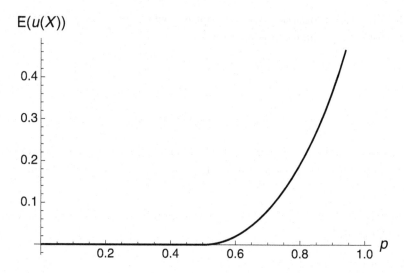

Fig. 8.7   The expected value of utility for a risk-averse investor allocating $100(2p-1)\%$ of their capital.

and which maximizes the investor's total expected utility $\mathbb{E}\left(u(W)\right)$, where

$$W = \sum_{i=1}^{n} x_i(1 + R_i).$$

Throughout this section it will be assumed that $n$ is large and that the returns $R_i$ are uncorrelated. Under these assumptions $W$ is a normally distributed random variable.

The allocation of investment funds can be driven by the objective of maximizing the expected value of the investor's utility function. Suppose an investor's utility function is given by $u(x) = 1 - e^{-bx}$ where $b > 0$. The reader was asked in Exercise 4.5.7 to show that this utility function is concave and monotonically increasing. Assuming that $W$ is a normally distributed random variable then $-bW$ is also normally distributed with

$$\mathbb{E}\left(-bW\right) = -b\,\mathbb{E}\left(W\right) \text{ and } \mathbb{V}\text{ar}\left(-bW\right) = b^2\,\mathbb{V}\text{ar}\left(W\right).$$

Consequently,

$$\mathbb{E}\left(u(W)\right) = \mathbb{E}\left(1 - e^{-bW}\right) = 1 - \mathbb{E}\left(e^{-bW}\right).$$

Making the assignment $Y = e^{-bW}$, then $Y$ is a lognormal random variable (see Sec. 8.5). According to Lemma 8.1,

$$\mathbb{E}\left(Y\right) = \mathbb{E}\left(e^{-bW}\right) = e^{-b\,\mathbb{E}(W)+b^2\,\mathbb{V}\text{ar}(W)/2},$$

which implies that

$$\mathbb{E}\left(u(W)\right) = 1 - e^{-b(\mathbb{E}(W) - b\,\mathbb{V}\mathrm{or}(W)/2)}.$$

Thus the expected utility is maximized when $\mathbb{E}\left(W\right) - b\,\mathbb{V}\mathrm{or}\left(W\right)/2$ is maximized. If two different portfolios, $\langle x_1, x_2, \ldots, x_n \rangle$ and $\langle y_1, y_2, \ldots, y_n \rangle$, give rise respectively to returns $X$ and $Y$ then the portfolio represented by $\langle x_1, x_2, \ldots, x_n \rangle$ will be preferable if

$$\mathbb{E}\left(X\right) \geq \mathbb{E}\left(Y\right) \text{ and } \mathbb{V}\mathrm{or}\left(X\right) \leq \mathbb{V}\mathrm{or}\left(Y\right).$$

To extend this discussion and make it more concrete numerically, suppose $b = 0.005$ and that an investor wishes to choose among an infinite number of different portfolios which can be created by investing in two securities denoted $A$ and $B$. The investor has $100 to invest and will allocate $y$ dollars to security $A$ and $100 - y$ dollars to $B$. The following table summarizes what is known about the rates of return on the hypothetical securities $A$ and $B$.

| Security | A | B |
|---|---|---|
| Expected rate of return | 0.16 | 0.18 |
| Volatility of return | 0.20 | 0.24 |

Assume that the correlation between the rates of return is $\rho = -0.35$. $W$ represents the wealth returned.

$$\mathbb{E}\left(W\right) = 100 + 0.16y + 0.18(100 - y) = 118 - 0.02y$$
$$\mathbb{V}\mathrm{or}\left(W\right) = y^2(0.20)^2 + (100 - y)^2(0.24)^2$$
$$+ 2y(100 - y)(0.20)(0.24)(-0.35)$$
$$= 0.04y^2 + 0.0576(100 - y)^2 - 0.0336y(100 - y)$$

Since the optimal portfolio is the one which maximizes

$$\mathbb{E}\left(W\right) - \frac{b}{2}\,\mathbb{V}\mathrm{or}\left(W\right)$$

$$= 118 - 0.02y - 0.0025(0.04y^2 + 0.0576(100 - y)^2 - 0.0336y(100 - y))$$
$$= -0.000328y^2 + 0.0172y + 116.56.$$

This occurs when $y \approx 26.2195$. For this choice of $y$

$$\mathbb{E}\left(W\right) \approx 117.476,$$
$$\mathbb{V}\mathrm{or}\left(W\right) \approx 276.049,$$
$$\mathbb{E}\left(u(W)\right) \approx 0.442296.$$

## Exercises

**8.6.1** Suppose an investor must choose between two portfolios, $A$ and $B$. The expected value of $A$ is \$15,000 with a standard deviation of \$10,000. The expected value of $B$ is \$15,000 with a standard deviation of \$12,500. Which portfolio is preferable?

**8.6.2** Consider the family of negative exponential utility functions $u(x) = 1 - e^{-bx}$. Find an expression for the marginal utility.

**8.6.3** Suppose an investment offers returns of 20% or $-5\%$ with equal probability. If an investor has utility function $u(x) = 1 - e^{-bx}$ where $x$ is the return of an investment. Find the certainty equivalent for the investment.

**8.6.4** Using the result of Exercise 8.6.3 describe the relationship between the certainty equivalent and the parameter $b$.

**8.6.5** Suppose the return on an investment is a random variable $X \sim \mathcal{N}(\mu, \sigma^2)$. If an investor has utility function $u(x) = 1 - e^{-bx}$ where $x$ represents the return of an investment, find the expected value of the utility.

## 8.7   Partial Expectation

In Chapter 6 the reader encountered frequent references to the positive part of the difference of two quantities. For instance, if an investor may purchase an asset worth $X$ for strike price $K$, then the excess profit (if any) of the transaction is positive part of the difference $(X-K)^+$. This section presents a theorem and several corollaries which enable the reader to calculate the expected value of $(X - K)^+$ when $X$ is a continuous random variable and $K$ is a constant. The expected value of $X$, when $X$ exceeds some constant $k$ called the **threshold** is known as the **partial expectation** of $X$. The partial expectation is defined as

$$g(K) = \int_K^\infty x \, f_X(x) \, dx. \qquad (8.21)$$

The partial expectation is also related to the conditional expectation of $X$.

**Theorem 8.7.** *Let $X$ be a continuous random variable with probability density function $f_X(x)$. The partial expectation of $X$ with respect to threshold $K$ is*

$$g(K) = \mathbb{E}\left(X|X > K\right) \mathbb{P}\left(X > K\right). \qquad (8.22)$$

*Proof.* Define the **indicator function** $\mathbb{1}_{(X>K)}$ to have value 0 when $X \leq K$ and value 1 when $X > K$.

$$\mathbb{E}\left(X|X>K\right) = \frac{\mathbb{E}\left(X\mathbb{1}_{(X>K)}\right)}{\mathbb{P}\left(X>K\right)} = \frac{\int_{-\infty}^{\infty} x\,\mathbb{1}_{(x>K)}\,f_X(x)\,dx}{\mathbb{P}\left(X>K\right)}$$

$$= \frac{\int_{K}^{\infty} x f_X(x)\,dx}{\mathbb{P}\left(X>K\right)} = \frac{g(K)}{\mathbb{P}\left(X>K\right)}. \qquad \square$$

Now suppose $X \sim \mathcal{LN}(\mu, \sigma^2)$ and suppose $K > 0$.

$$\mathbb{E}\left(X\mathbb{1}_{(X>K)}\right)$$

$$= \int_{0}^{\infty} x\,\mathbb{1}_{(x>K)} \frac{1}{x\sqrt{2\pi\sigma^2}} e^{-(\ln x - \mu)^2/(2\sigma^2)}\,dx$$

$$= \int_{K}^{\infty} \frac{1}{\sqrt{2\pi\sigma^2}} e^{-(\ln x - \mu)^2/(2\sigma^2)}\,dx$$

$$= \int_{0}^{\infty} \frac{1}{\sqrt{2\pi\sigma^2}} e^{-(\ln x - \mu)^2/(2\sigma^2)}\,dx - \int_{0}^{K} \frac{1}{\sqrt{2\pi\sigma^2}} e^{-(\ln x - \mu)^2/(2\sigma^2)}\,dx.$$

The first integral evaluates to $\mathbb{E}\left(X\right) = e^{\mu + \sigma^2/2}$ according to Eq. (8.19). Therefore,

$$\mathbb{E}\left(X\mathbb{1}_{(X>K)}\right) = \mathbb{E}\left(X\right)\left(1 - \int_{0}^{K} e^{-\mu - \sigma^2/2} \frac{1}{\sqrt{2\pi\sigma^2}} e^{-(\ln x - \mu)^2/(2\sigma^2)}\,dx\right).$$

Use the substitution

$$u = \frac{\ln x - \mu}{\sigma} \iff x = e^{\sigma u + \mu}$$

$$du = \frac{1}{\sigma x}\,dx \iff dx = \sigma e^{\sigma u + \mu}\,du,$$

and rewrite the definite integral as

$$\mathbb{E}\left(X\mathbb{1}_{(X>K)}\right)$$

$$= \mathbb{E}\left(X\right)\left(1 - \int_{-\infty}^{(\ln K - \mu)/\sigma} \frac{1}{\sqrt{2\pi}} e^{-(u-\sigma)^2/2}\,du\right)$$

$$= \mathbb{E}\left(X\right)\left(1 - \Phi\left(\frac{\ln K - \mu - \sigma^2}{\sigma}\right)\right) = \mathbb{E}\left(X\right)\Phi\left(\frac{\mu + \sigma^2 - \ln K}{\sigma}\right).$$

For the sake of conciseness let

$$v = \frac{\mu + \sigma^2 - \ln K}{\sigma}.$$

In Exercise 8.7.1 the reader will be asked to show that

$$\mathbb{E}\left(X\mathbb{1}_{(X>K)}\right) = \mathbb{E}\left(X\right)\Phi\left(v\right). \qquad (8.23)$$

In a similar fashion the reader can derive (see Exercise 8.7.4)

$$\mathbb{E}\left(X \, \mathbb{1}_{(X<K)}\right) = \mathbb{E}\left(X\right) \Phi\left(-v\right). \tag{8.24}$$

If Thm. 8.7 is applied to a lognormally distributed random variable the following corollary holds.

**Corollary 8.1.** *Suppose random variable $X$ is lognormally distributed with parameters $\mu$ and $\sigma$. If $K > 0$ then*

$$g(K) = \mathbb{E}\left(X\right) \Phi\left(v\right). \tag{8.25}$$

*Proof.* By the definition of partial expectation,

$$g(K) = \int_K^\infty x \, \frac{1}{x\sqrt{2\pi\sigma^2}} e^{-(\ln x - \mu)^2/(2\sigma^2)} \, dx.$$

Make the substitutions $x = e^{\mu+\sigma w}$ and $dx = \sigma e^{\mu+\sigma w} \, dw$ and the expression for $g(K)$ becomes

$$g(K) = \int_{(\ln K - \mu)/\sigma}^\infty \frac{1}{\sqrt{2\pi}} e^{\mu+\sigma w} e^{-w^2/2} \, dw$$

$$= e^{\mu+\sigma^2/2} \int_{(\ln K - \mu)/\sigma}^\infty \frac{1}{\sqrt{2\pi}} e^{-(w-\sigma)^2/2} \, dw.$$

After another substitution of $t = w - \sigma$ and $dt = dw$,

$$g(K) = \mathbb{E}\left(X\right) \int_{(\ln K - \mu - \sigma^2)/\sigma}^\infty \frac{1}{\sqrt{2\pi}} e^{-t^2/2} \, dt = \mathbb{E}\left(X\right) \Phi\left(\frac{\mu + \sigma^2 - \ln K}{\sigma}\right)$$

$$= \mathbb{E}\left(X\right) \Phi\left(v\right).$$

$\square$

Combining the results of Eqs. (8.22) and (8.25) produces the following equation for the conditional expectation of lognormally distributed random variables.

$$\mathbb{E}\left(X | X > K\right) = \mathbb{E}\left(X\right) \frac{\Phi\left(v\right)}{\mathbb{P}\left(X > K\right)} = \mathbb{E}\left(X\right) \frac{\Phi\left(v\right)}{\Phi\left(v - \sigma\right)}. \tag{8.26}$$

Development of the option pricing formulas in later chapters requires evaluation of the expected value of the positive part of $(X - K)$ when $X$ is a lognormally distributed continuous random variable. The following corollary is a consequence of Eq. (8.23).

**Corollary 8.2.** *If $X$ is a lognormally distributed random variable with parameters $\mu$ and $\sigma^2$ and $K > 0$ is a constant then*

$$\mathbb{E}\left((X - K)^+\right) = \mathbb{E}\left(X\right) \Phi\left(v\right) - K\Phi\left(v - \sigma\right). \tag{8.27}$$

*Proof.* Note that $(X - K)^+ = \max\{K, X\} - K$ and thus

$$\mathbb{E}\left((X - K)^+\right) = \mathbb{E}\left(\max\{K, X\}\right) - K = \int_0^\infty \max\{K, x\} f_X(x)\, dx - K$$

$$= \int_0^K K f_X(x)\, dx - K + \int_K^\infty x f_X(x)\, dx$$

$$= K\,\mathbb{P}\left(X \leq K\right) - K + \mathbb{E}\left(X \mathbb{1}_{(X>K)}\right)$$

$$= -K\,\mathbb{P}\left(X > K\right) + \mathbb{E}\left(X\right)\Phi\left(v\right).$$

The reader will complete the proof in Exercise 8.7.3. □

### Exercises

**8.7.1** Suppose that $X \sim \mathcal{LN}(\mu, \sigma^2)$ and $K > 0$ is a constant. Show that
$$v = \frac{\sigma^2 + \mu - \ln K}{\sigma} = \frac{\sigma^2/2 + \ln(\mathbb{E}\left(X\right)/K)}{\sigma}.$$

**8.7.2** Let $X$ be a continuous random variable with probability density function $f_X(x)$. Show that $\mathbb{P}\left(X > K\right) = \mathbb{E}\left(\mathbb{1}_{(X>K)}\right)$.

**8.7.3** Suppose that $X \sim \mathcal{LN}(\mu, \sigma^2)$ and $K > 0$ is a constant. Show that
$$\mathbb{P}\left(X > K\right) = \Phi\left(\frac{-\sigma^2/2 + \ln(\mathbb{E}\left(X\right)/K)}{\sigma}\right) = \Phi\left(v - \sigma\right).$$

**8.7.4** Suppose that $X \sim \mathcal{LN}(\mu, \sigma^2)$ and $K > 0$ is a constant. Show that
$$\mathbb{E}\left(X \mathbb{1}_{(X<K)}\right) = \mathbb{E}\left(X\right)\Phi\left(\frac{-\sigma^2/2 - \ln(\mathbb{E}\left(X\right)/K)}{\sigma}\right) = \mathbb{E}\left(X\right)\Phi\left(-v\right).$$

**8.7.5** Let $X$ be a uniformly distributed continuous random variable on the interval $[a, b]$. If $K$ is a constant, find an expression for $\mathbb{E}\left((X - K)^+\right)$.

**8.7.6** Suppose $X \sim \mathcal{N}(\mu, \sigma^2)$ and $K$ is a constant, then show that
$$\mathbb{E}\left((X - K)^+\right) = \sigma\varphi\left(\frac{\mu - K}{\sigma}\right) + (\mu - K)\Phi\left(\frac{\mu - K}{\sigma}\right).$$

**8.7.7** Suppose $X \sim \mathcal{LN}(\mu, \sigma^2)$ and $K$ is a constant. Show that
$$\mathbb{E}\left(X | X \leq K\right) = \mathbb{E}\left(X\right)\frac{\Phi\left(-v\right)}{\Phi\left(\sigma - v\right)}.$$

**8.7.8** Let $X \sim \mathcal{LN}(1, 0.5)$ and determine $\mathbb{E}\left(X | X > 0.25\right)$.

**8.7.9** Let $X \sim \mathcal{LN}(0.45, 1)$ and determine $\mathbb{E}\left(X | X \leq 0.5\right)$.

**8.7.10** Suppose $X \sim \mathcal{LN}(\mu, \sigma^2)$ and $0 < K_1 < K_2$ are constants. Show that
$$\mathbb{E}\left(X \mathbb{1}_{(K_1<X\leq K_2)}\right)$$
$$= \mathbb{E}\left(X\right)\left(\Phi\left(\frac{\ln K_2 - \mu - \sigma^2}{\sigma}\right) - \Phi\left(\frac{\ln K_1 - \mu - \sigma^2}{\sigma}\right)\right).$$

## 8.8 The Binormal Distribution

Suppose random variables $Z_1 \sim \mathcal{N}(0,1)$ and $Z_2 \sim \mathcal{N}(0,1)$ are independent random variables and define the joint probability density function,

$$f_{Z_1,Z_2}(z_1,z_2) = f_{Z_1}(z_1)f_{Z_2}(z_2) = \frac{1}{2\pi}e^{-(z_1^2+z_2^2)/2}. \tag{8.28}$$

The density in Eq. (8.28) is called the **standard bivariate normal distribution**  Let $\mu_X$ and $\mu_Y$ be real numbers, $-1 \leq \rho \leq 1$, $\sigma_X$ and $\sigma_Y$ be positive real numbers, and define the random variables

$$X = \mu_X + \sigma_X Z_1 \tag{8.29}$$

$$Y = \mu_Y + \sigma_Y \left(\rho Z_1 + \sqrt{1-\rho^2}Z_2\right). \tag{8.30}$$

Note that $X \sim \mathcal{N}(\mu_X, \sigma_X^2)$ and the reader will be asked to show in Exercise 8.8.1 that $Y$ has mean $\mu_Y$ and variance $\sigma_Y^2$. Random variable $Y$ is also normally distributed. Let $y = \mu_Y + \sigma_Y(\rho z_1 + \sqrt{1-\rho^2}z_2)$ then

$$\mathbb{P}\left(Y \leq y\right) = \int_{-\infty}^{\infty} \int_{-\infty}^{(y-\mu_Y-\rho\sigma_Y z_1)/(\sqrt{1-\rho^2}\sigma_Y)} f_{Z_2}(z_2)f_{Z_1}(z_1)\,dz_2\,dz_1.$$

Differentiating with respect to $y$ produces

$$f_Y(y) = \frac{1}{\sqrt{1-\rho^2}\sigma_Y} \int_{-\infty}^{\infty} f_{Z_2}\left(\frac{y-\mu_Y-\rho\sigma_Y z_1}{\sqrt{1-\rho^2}\sigma_Y}\right) f_{Z_1}(z_1)dz_1.$$

Using the probability density function for the standard normal random variable then

$$f_Y(y) = \frac{1}{\sqrt{2\pi(1-\rho^2)}\sigma_Y} \int_{-\infty}^{\infty} e^{-\left(\frac{y-\mu_Y-\rho\sigma_Y z_1}{\sqrt{1-\rho^2}\sigma_Y}\right)^2/2} \frac{1}{\sqrt{2\pi}}e^{-z_1^2/2}\,dz_1$$

$$= \frac{1}{\sqrt{2\pi(1-\rho^2)}\sigma_Y}e^{-\frac{(y-\mu_Y)^2}{2\sigma_Y^2}} \int_{-\infty}^{\infty} \frac{1}{\sqrt{2\pi}}e^{-\left(z_1-\frac{\rho(y-\mu_Y)}{\sigma_y}\right)^2/(2(1-\rho^2))}\,dz_1,$$

after completing the square in variable $z_1$. The last integral involves the probability density function for a normally distributed random variable. After a suitable substitution to evaluate the integral,

$$f_Y(y) = \frac{1}{\sqrt{2\pi}\sigma_Y}e^{-(y-\mu_Y)^2/(2\sigma_Y^2)},$$

which implies $Y \sim \mathcal{N}(\mu_Y, \sigma_Y^2)$.

The joint probability density function $f_{X,Y}(x,y)$ can be determined from the joint density function $f_{Z_1,Z_2}(z_1,z_2)$.

$$f_{X,Y}(x,y) = f_{Z_1,Z_2}(z_1,z_2) \left| \det \begin{bmatrix} \frac{\partial z_1}{\partial x} & \frac{\partial z_1}{\partial y} \\ \frac{\partial z_2}{\partial x} & \frac{\partial z_2}{\partial y} \end{bmatrix} \right|$$

$$= \frac{1}{2\pi} e^{-(z_1^2+z_2^2)/2} \left| \det \begin{bmatrix} \frac{1}{\sigma_X} & 0 \\ \frac{-\rho}{\sqrt{1-\rho^2}\sigma_X} & \frac{1}{\sqrt{1-\rho^2}\sigma_Y} \end{bmatrix} \right|.$$

Substituting expressions for $Z_1$ and $Z_2$ written in terms of $X$ and $Y$ (see Exercise 8.8.5) reveals

$$f_{X,Y}(x,y) = \frac{1}{2\pi\sqrt{1-\rho^2}\sigma_X\sigma_Y} e^{\frac{-1}{2}\left[\left(\frac{x-\mu_X}{\sigma_X}\right)^2 + \frac{1}{1-\rho^2}\left(\frac{y-\mu_Y}{\sigma_Y} - \frac{\rho(x-\mu_X)}{\sigma_X}\right)^2\right]}.$$

The standard format for writing this probability density function is

$$f_{X,Y}(x,y)$$

$$= \frac{1}{2\pi\sqrt{1-\rho^2}\sigma_X\sigma_Y} e^{\frac{-1}{2(1-\rho^2)}\left[\frac{(x-\mu_X)^2}{\sigma_X^2} - \frac{2\rho(x-\mu_X)(y-\mu_Y)}{\sigma_X\sigma_Y} + \frac{(y-\mu_Y)^2}{\sigma_Y^2}\right]}. \quad (8.31)$$

Two random variables $(X,Y)$ have a **binormal distribution** (or **bivariate normal distribution**), denoted $\mathcal{N}(\mu_X, \mu_Y, \sigma_X^2, \sigma_Y^2, \rho)$ if their joint probability distribution function takes the form shown in Eq. (8.31). Figure 8.8 shows the density plots of the probability density function for $(X,Y)$ where $(X,Y) \sim \mathcal{N}(0,0,1,2,\rho)$ for several different values of $\rho$. If $(X,Y) \sim \mathcal{N}(\mu_X, \mu_Y, \sigma_X, \sigma_Y, \rho)$ then the $\mathbb{P}(X \leq x, Y \leq y) = \Phi(x,y;\rho)$. This can be expressed as the improper integral,

$$\Phi(x,y;\rho) = \int_{-\infty}^{x} \int_{-\infty}^{y} f_{X,Y}(u,v)\,dv\,du.$$

This integral can be approximated numerically by functions included with many spreadsheets, programming languages, and computer algebra systems.

## Exercises

**8.8.1** Show that the random variable $Y$ defined earlier in this section has mean of $\mu_Y$ and variance $\sigma_Y^2$.

**8.8.2** Complete the square in variable $z_1$ to show that

$$-\frac{1}{2}\left(\frac{y-\mu_Y - \rho\sigma_Y z_1}{\sqrt{1-\rho^2}\sigma_Y}\right)^2 - \frac{1}{2}z_1^2$$

$$= -\frac{1}{2(1-\rho^2)}\left(z_1 - \frac{\rho(y-\mu_Y)}{\sigma_Y}\right)^2 - \frac{1}{2\sigma_Y^2}(y-\mu_Y)^2.$$

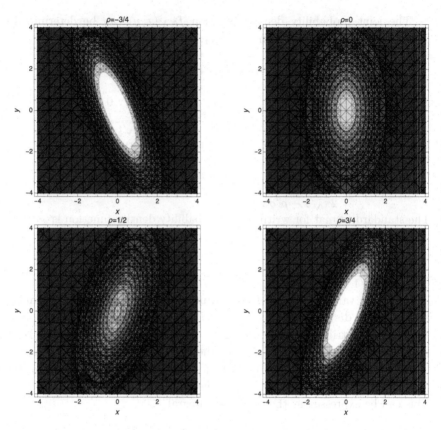

Fig. 8.8    Plots of the probability density function for $(X, Y)$ where $(X, Y) \sim \mathcal{N}(0, 0,$ $1, 2, \rho)$ for $\rho = -3/4, 0, 1/2, 3/4$.

**8.8.3** Use integration by substitution to show that

$$\int_{-\infty}^{\infty} \frac{1}{\sqrt{2\pi}} e^{-\left(z_1 - \frac{\rho(y - \mu_Y)}{\sigma_y}\right)^2 / (2(1 - \rho^2))} \, dz_1 = \sqrt{1 - \rho^2}.$$

**8.8.4** Find $\text{Cov}(X, Y)$.

**8.8.5** Solve Eqs. (8.29) and (8.30) for $Z_1$ and $Z_2$.

**8.8.6** Use Exercise 8.8.5 to show that if $(X, Y) \sim \mathcal{N}(\mu_1, \mu_2, \sigma_1^2, \sigma_2^2, \rho)$ then

$$\left( \frac{X - \mu_X}{\sigma_X}, \frac{1}{\sqrt{1 - \rho^2}} \frac{Y - \mu_Y}{\sigma_Y} - \frac{\rho(X - \mu_X)}{\sigma_X} \right) \sim \mathcal{N}(0, 0, 1, 1, 0).$$

**8.8.7** Find a expression for the conditional density $f_{X|Y}(x|y)$ of the bivariate normal distribution given in Eq. (8.31).

# Chapter 9

# Random Walks and Brownian Motion

This chapter will introduce and explain some of the concepts surrounding the stochastic models used to model the behavior of stock, security, option, and index prices. Limiting cases of random walks were used in Chapter 8 to introduce the notion of the continuous normal and lognormal random variables. This chapter will introduce Brownian motion and develop mathematical models of asset prices and returns on asset prices. Topics covered here could be expanded into an entire book of their own. This chapter explores just enough of the stochastic calculus to provide some justification for Itô's Lemma, the main result of this chapter.

## 9.1 Empirical Justification for the Mathematical Model

In order to assign rational, arbitrage-free prices to options on assets it is necessary to mathematically model the prices and price movements of assets. This section will examine some price data for a stock index and propose a model for the price. Mathematical models are approximations to reality, so in a sense mathematical models may fail to capture all the features of a phenomenon, but even simple mathematical models are preferable to guesswork or nothing at all.

The Wilshire 5000 Index (W5000) is a broadly based market index of approximately 3,500 publicly traded US-based companies. It is sometimes used as a surrogate for the market portfolio of stocks. The closing prices of the W5000 for the first half of 2020 are graphed in Fig. 9.1. Suppose the historical record of prices is $\{S(t_0), S(t_1), \ldots, S(t_n)\}$. The continuously

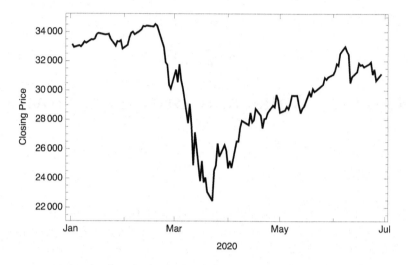

compounded rate of return on the index between time $t_i$ and time $t_j$ (with $t_j > t_i$) is denoted $R(t_i, t_j)$ where

$$R(t_i, t_j) = \ln \frac{S(t_j)}{S(t_i)}.$$

The prices reflected in Fig. 9.1 are the closing prices on different days, thus the histogram of continuously compounded daily returns on the index resembles that shown in Fig. 9.2. The histograms have an approximate bell shape, but two features are notable. First the peak of the histogram is higher that the peak of a normal probability density function. Second the histogram bars in the tails of the distribution are taller than the graph of the normal probability density function. These features are sometimes called **fat tails**. Collecting a larger sample of data may reveal a histogram with a more normal appearance, but there are other reasons why the histogram may deviate from the normal curve. The variance of the returns on the index in reality may be functions of time and not constant as assumed here. Also the prices of the index are not exactly continuous random variables, but are recorded only to the nearest $0.01. Nevertheless the model of stock and index prices will begin with the assumption that the continuously compounded returns are normally distributed.

A model for the price of an asset such as a stock should include parameters for the variance of the return on the asset, the dividends paid by the asset (which is assumed to be paid continuously at a rate proportional to the value of the asset), and the expected rate of return of the asset.

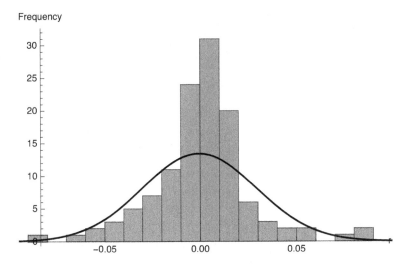

Fig. 9.2 A histogram of the continuously compounded daily returns for the Wilshire 5000 Total Market index. The graph of a normal probability density function with mean and variance equal to the sample mean and sample variance of the sample data is provided for comparison.

These parameters will be denoted $\alpha$ for the expected rate of return, $\delta$ for the continuously compounded dividend yield, and $\sigma^2$ for the variance of the return. Often $\sigma$ will be called the volatility of the asset. An important assumption of the model is that the continuously compounded return and the variance of the return are proportional to the length of time between observations of the price. Assuming that the price of the asset at time $t$ is $S(t)$ then for $h > 0$ the continuously compounded return on the asset over the interval $[t, t+h]$ is normally distributed.

$$\ln \frac{S(t+h)}{S(t)} \sim \mathcal{N}\left(\left(\alpha - \delta - \frac{\sigma^2}{2}\right)h, \sigma^2 h\right).$$

This implies that $S(t+h)/S(t)$ is lognormally distributed.

$$\frac{S(t+h)}{S(t)} \sim \mathcal{LN}\left(\left(\alpha - \delta - \frac{\sigma^2}{2}\right)h, \sigma^2 h\right).$$

The price of the asset at time $t+h$ can be expressed as

$$S(t+h) = S(t)e^{(\alpha - \delta - \sigma^2/2)h + \sigma\sqrt{h}Z}, \qquad (9.1)$$

where $Z \sim \mathcal{N}(0,1)$. Note that even if $S(t)$ is known, $S(t+h)$ is a (lognormally distributed) random variable since it is a function of the standard normal random variable $Z$. According to Lemma 8.1 the mean and variance of $S(t+h)$ are

$$\mathbb{E}\left(S(t+h)\right) = S(t)e^{(\alpha-\delta)h}, \tag{9.2}$$

$$\mathbb{Var}\left(S(t+h)\right) = (S(t))^2 e^{2(\alpha-\delta)h}\left(e^{\sigma^2 h} - 1\right). \tag{9.3}$$

Another important assumption of the model is that continuously compounded returns over non-overlapping time intervals are independent random variables. As a consequence, variances can be added.

**Example 9.1.** The historical price data collected for the Wilshire 5000 stock index used a time step $h$ equal to 1 day. For the sample the daily continuously compounded return had a mean of $-0.00050534$ and a standard deviation of $0.02973419$. Assume the index does not pay a dividend and let $S(0) = 33,142.20$. Find the annualized parameters for the index and determine the expected value of the index in 1 year.

**Solution.** Let $h = 1/365$ and $\delta = 0$, then

$$\left(\alpha - \delta - \frac{\sigma^2}{2}\right)\frac{1}{365} = -0.00050534$$

$$\sigma^2 \frac{1}{365} = (0.02973419)^2.$$

Thus $\sigma = 0.568071/\text{year}$ and $\alpha = -0.023096/\text{year}$. Using Eq. (9.2),

$$\mathbb{E}\left(S(1)\right) = 33,142.20 e^{(-0.023096-0)(1)} = 32,385.50.$$

The expected value of $S(t+h)$ given in Eq. (9.2) provided a point estimate of $S(t+h)$. A more informative piece of information may be the interval in which $S(t+h)$ is likely to be found. Since $S(t+h)$ is a function of the standard normal random variable $Z$, the range in which $Z$ is likely to be found determines an interval in which $S(t+h)$ is likely to be found. For example $\mathbb{P}\left(-1.64485 < Z \le 1.64485\right) = 0.90$ thus the **90% prediction interval** for $S(t+h)$ can be expressed as

$$S(t)e^{(\alpha-\delta-\sigma^2/2)h-1.64485\sigma\sqrt{h}} < S(t+h) < S(t)e^{(\alpha-\delta-\sigma^2/2)h+1.64485\sigma\sqrt{h}}.$$

The 90% prediction interval for the Wilshire 5000 index 1 year from when the historical pricing data was collected is approximately $(10826, 70158)$. This interval is so wide that it provides little help in forecasting the value of the index in 1 year.

One of the assumptions of this model of asset prices is that returns over non-overlapping intervals of time are independent. This plays an important role in determining the covariance in asset prices at two different times. Consider an asset with expected rate of return $\alpha$, dividend yield $\delta$, and volatility $\sigma$. Given $S(0)$ and times $0 < t_1 < t_2$ the $\mathbb{C}\text{ov}\,(S(t_1), S(t_2))$ can be readily found using the lognormal model of asset prices. By Eq. (2.15)

$$\mathbb{C}\text{ov}\,(S(t_1), S(t_2)) = \mathbb{E}\,(S(t_1)S(t_2)) - \mathbb{E}\,(S(t_1))\,\mathbb{E}\,(S(t_2)),$$

and the two factors following the minus sign are readily calculated from Eq. (9.2),

$$\mathbb{E}\,(S(t_1)) = S(0)e^{(\alpha-\delta)t_1}$$
$$\mathbb{E}\,(S(t_2)) = S(0)e^{(\alpha-\delta)t_2}.$$

Note that

$$\ln(S(t_1)S(t_2)) = 2\ln S(0) + R(0,t_1) + R(0,t_2).$$

However the intervals $[0,t_1]$ and $[0,t_2]$ overlap and hence random variables $R(0,t_1)$ and $R(0,t_2)$ are not independent. The variance of their sum can be calculated as follows.

$$\mathbb{V}\text{ar}\,(R(0,t_1) + R(0,t_2)) = \mathbb{V}\text{ar}\,(2R(0,t_1) + R(t_1,t_2))$$
$$= 4\,\mathbb{V}\text{ar}\,(R(0,t_1)) + \mathbb{V}\text{ar}\,(R(t_1,t_2))$$
$$= 4\sigma^2 t_1 + \sigma^2(t_2 - t_1) = \sigma^2(3t_1 + t_2).$$

The additivity of assets returns was used above to create random returns on non-overlapping intervals of time. Consequently $\ln(S(t_1)S(t_2))$ is lognormally distributed with parameters $2\ln S(0) + (\alpha - \delta - \sigma^2)(t_1 + t_2)$ and $\sigma^2(3t_1 + t_2)$. Applying the formula for the expected value of a lognormal random variable yields

$$\mathbb{E}\,(S(t_1)S(t_2)) = (S(0))^2 e^{(\alpha-\delta-\sigma^2/2)(t_1+t_2)+\sigma^2(3t_1+t_2)/2}$$
$$= (S(0))^2 e^{(\alpha-\delta)(t_1+t_2)}e^{\sigma^2 t_1}.$$

Therefore the

$$\mathbb{C}\text{ov}\,(S(t_1), S(t_2)) = (S(0))^2 e^{(\alpha-\delta)(t_1+t_2)}e^{\sigma^2 t_1} - S(0)e^{(\alpha-\delta)t_1}S(0)e^{(\alpha-\delta)t_2}$$
$$= (S(0))^2 e^{(\alpha-\delta)(t_1+t_2)}\left(e^{\sigma^2 t_1} - 1\right).$$

In Exercise 9.1.4 the reader will show that the covariance in the asset prices can be written succinctly as

$$\mathbb{C}\text{ov}\,(S(t_1), S(t_2)) = e^{(\alpha-\delta)(t_2-t_1)}\,\mathbb{V}\text{ar}\,(S(t_1)). \tag{9.4}$$

**Example 9.2.** Suppose the continuously compounded rate of return for a stock is 5% per annum, the volatility in the return is 40% per annum, and the stock pays dividends continuously at a rate of 1% per annum. If the initial price of the stock is \$75, find the expected value of the stock price at $t = 1$ and $t = 4$ and find the covariance between the stock prices at those times.

**Solution.** The stock price is lognormally distributed with

$$S(t) \sim \mathcal{LN}((0.05 - 0.01 - (0.40)^2/2)t, (0.40)^2 t)$$
$$\sim \mathcal{LN}(-0.04t, 0.16t).$$

Using Eq. (9.2) the expected values of the stock are

$$\mathbb{E}\left(S(1)\right) = 75e^{(0.05-0.01)(1)} = \$78.0608,$$
$$\mathbb{E}\left(S(4)\right) = 75e^{(0.05-0.01)(4)} = \$88.0133.$$

The covariance can be found using Eq. (9.4)

$$\mathbf{Cov}\left(S(1), S(4)\right) = e^{(0.05-0.01)(4-1)} \left(\mathbb{E}\left(S(1)\right)\right)^2 \left(e^{0.16(1)} - 1\right)$$
$$= 1192.0874.$$

Section 8.7 introduced the notion of partial expectation and derived several results for conditional expectation of lognormally distributed random variables such as the asset prices explored in this chapter. If $S(t)/S(0)$ is lognormally distributed as $\mathcal{LN}((\alpha - \delta - \sigma^2/2)t, \sigma^2 t)$ and $K > 0$ then

$$\mathbb{P}\left(S(t) > K\right) = \mathbb{P}\left(\ln \frac{S(t)}{S(0)} > \ln \frac{K}{S(0)}\right)$$
$$= \mathbb{P}\left(Z > \frac{\ln \frac{K}{S(0)} - (\alpha - \delta - \sigma^2/2)t}{\sigma\sqrt{t}}\right) = \Phi\left(\hat{d}_2\right),$$

where by definition

$$\hat{d}_2 = \frac{\ln \frac{S(0)}{K} + (\alpha - \delta - \sigma^2/2)t}{\sigma\sqrt{t}}. \tag{9.5}$$

Likewise define

$$\hat{d}_1 = \hat{d}_2 + \sigma\sqrt{t} = \frac{\ln \frac{S(0)}{K} + (\alpha - \delta + \sigma^2/2)t}{\sigma\sqrt{t}}. \tag{9.6}$$

Applying Eq. (8.26) to the situation of a stock price produces

$$\mathbb{E}\left(S(t)|S(t) > K\right) = \mathbb{E}\left(S(t)\right) \frac{\Phi\left(\hat{d}_1\right)}{\Phi\left(\hat{d}_2\right)}. \tag{9.7}$$

As a result of Exercise 8.7.7,

$$\mathbb{E}\left(S(t)|S(t) \le K\right) = \mathbb{E}\left(S(t)\right)\frac{\Phi\left(-\hat{d}_1\right)}{\Phi\left(-\hat{d}_2\right)}. \tag{9.8}$$

Equations (9.7) and (9.8) are integral in developing a formula for the arbitrage-free pricing of European put and call options explored in Chapter 10. The eager reader is free to jump ahead to these results. The remainder of this chapter will explore further some of the mathematical results surrounding mathematical models employing lognormal random variables.

## *Exercises*

**9.1.1** Given that $S(t) \sim \mathcal{LN}((\alpha - \delta - \sigma^2/2)t, \sigma^2 t)$ how is $R(t_1, t_2)$ distributed?

**9.1.2** Suppose that a stock $S(t)$ has the following distribution.

$$S(t) \sim \mathcal{N}(\ln 75 + 0.15t, 0.16t).$$

Find $\mathbb{E}\left(S(1)\right)$.

**9.1.3** Suppose that a stock $S(t)$ has the following distribution.

$$S(t) \sim \mathcal{N}(\ln 100 + 0.1t, 0.16t).$$

Find the 95% prediction interval for $S(1)$.

**9.1.4** Derive Eq. (9.4) for the covariance of $S(t_1)$ and $S(t_2)$.

**9.1.5** Suppose that $R(0, t) \sim \mathcal{N}((\alpha - \delta - \sigma^2/2)t, \sigma^2 t)$ and $0 < t_1 < t_2$. Assume that returns on non-overlapping intervals of time are independent random variables. Show that $\mathbb{C}\text{or}\left(R(0, t_1), R(0, t_2)\right) = \sqrt{t_1/t_2}$.

**9.1.6** Suppose the current price of a share of non-dividend-paying stock is $50. The continuously compounded rate of return for the stock is 8% per year and the volatility of the return is 20% per year. Find $\mathbb{P}\left(S(2) > 55\right)$.

**9.1.7** Suppose the current price of a share of non-dividend-paying stock is $80. The continuously compounded rate of return for the stock is 10% per year and the volatility of the return is 25% per year. Find $\mathbb{E}\left(S(4)|S(2) = 85\right)$.

**9.1.8** Suppose the current price of a share of non-dividend-paying stock is $95. The continuously compounded rate of return for the stock is 4% per year and the volatility of the return is 40% per year. Find $\mathbb{E}\left(S(1)|S(1) > 97\right)$.

**9.1.9** Suppose the current price of a share of non-dividend-paying stock is \$68. The continuously compounded rate of return for the stock is 10% per year and the volatility of the return is 30% per year. Find $\mathbb{E}\left(S(2)|S(2) \leq 65\right)$.

**9.1.10** Suppose the current price of a share of non-dividend-paying stock is \$60. The continuously compounded rate of return for the stock is 9% per year and the volatility of the return is 35% per year. Find the standard deviation of $(S(1) + S(2))/2$.

## 9.2 Advection/Diffusion Equation

The previous section made the case that returns of assets such as stocks or stock indices are normally distributed. This section will explore some of the properties of the dynamics of lognormal random variables. This will require the introduction of differential equations, but as stated before, no previous course or background in differential equations is required. All results will be developed from basic principles and elementary calculus.

During the development of the normal distribution as the limit of the binomial distribution, a general probability density function was derived in Eq. (8.16) and repeated here for convenience with the symbol used for the random variable changed to $X$.

$$f_X(x) = \frac{1}{\sqrt{4p(1-p)\sigma^2 t}} \varphi\left(\frac{x - (2p-1)\sigma\sqrt{t}}{\sqrt{4p(1-p)\sigma^2 t}}\right).$$

To make this expression more convenient to write let $\mu = (2p-1)\sigma$ and $k = 2p(1-p)\sigma^2$. Recall that $p$ is the probability a random walk takes a rightward step. For each $t > 0$, $X(t) \sim \mathcal{N}(\mu\sqrt{t}, 2kt)$. Thus the expected value of $X(t)$ is $\mu\sqrt{t} = (2p-1)\sigma\sqrt{t}$. The expected value, corresponding to the peak in the bell-shaped curve, shifts horizontally as time increases with a velocity given by

$$\frac{d}{dt}\left[(2p-1)\sigma\sqrt{t}\right] = \frac{(2p-1)\sigma}{2\sqrt{t}} = \frac{\mu}{2\sqrt{t}}.$$

As $t \to 0^+$ both the mean and the variance of $X(t)$ approach 0. Intuitively this implies the bell-shaped curve has its peak very close to the origin and the curve becomes very narrow with a high peak. Let $\epsilon > 0$ be an arbitrary, fixed positive number.

$$\mathbb{P}\left(|X - \mu\sqrt{t}| > \epsilon\right) = 2\,\mathbb{P}\left(X - \mu\sqrt{t} < -\epsilon\right) = 2\Phi\left(\frac{-\epsilon}{\sqrt{2kt}}\right).$$

Since the cumulative distribution function is continuous everywhere and approaches 0 as its argument approaches $-\infty$, there exists a $t > 0$ such that

$$\mathbb{P}\left(|X - \mu\sqrt{t}| > \epsilon\right) < \epsilon.$$

This can be interpreted as saying that the probability of $X$ being more than $\epsilon$ away from its mean is smaller than $\epsilon$ for all $t$ smaller than some positive value. Since this holds for all $\epsilon > 0$ the limit of this distribution can be thought as the **Dirac delta function** which will be denoted as $D(x)$. This function has many interesting properties, but the most commonly used ones are that $D(x) = 0$ for all $x \neq 0$ and

$$\int_{-\infty}^{\infty} D(x)\, dx = 1,$$

(which the reader will justify in Exercise 9.2.2). If $f(x)$ is a continuous function then

$$\int_{-\infty}^{\infty} D(x - x_0) f(x)\, dx = f(x_0),$$

(which the reader will prove in Exercise 9.2.3).

Since the probability density function in Eq. (8.16) depends on $x$ and $t$, define the function $U(x,t)$ to be this PDF. Taking partial derivatives of $U$ it is possible to see that $U(x,t)$ satisfies the following partial differential equation known as the **advection/diffusion equation**.

$$\frac{\partial U}{\partial t} + \frac{\mu}{2\sqrt{t}}\frac{\partial U}{\partial x} = k\frac{\partial^2 U}{\partial x^2}. \tag{9.9}$$

The constant $k$ is called the diffusion constant. In the special case that $p = 1/2$, Eq. (9.9) simplifies to another partial differential equation called the **diffusion equation**.

$$\frac{\partial U}{\partial t} = k\frac{\partial^2 U}{\partial x^2}. \tag{9.10}$$

Thus if the random walk is equally likely to take steps left and right, the probability density function solves the diffusion equation.

As $t \to 0^+$ the solution to the advection/diffusion equation converges to the Dirac delta function. In the study of the solutions to partial differential equations, the Dirac delta function would be called the initial condition of the initial value problem for which the partial differential equation is the advection/diffusion equation. Function $U(x,t)$ is not the only solution to the advection/diffusion equation, but it is the solution satisfying that

An Undergraduate Introduction to Financial Mathematics

particular initial condition. Suppose the initial condition is the bounded, piecewise continuous function $h(x)$, then the general solution to the initial value problem can be expressed as

$$u(x,t) = \begin{cases} \displaystyle\int_{-\infty}^{\infty} U(x-y,t)h(y)\,dy & \text{if } t > 0, \\ h(x) & \text{if } t = 0. \end{cases} \tag{9.11}$$

This claim can be justified along the lines used in [Buchanan and Shao (2018), Sec. 4.4].

**Example 9.3.** Find the solution to the advection diffusion equation satisfying the initial condition

$$h(x) = \begin{cases} 0 & \text{if } x < 0, \\ 1 & \text{if } x \geq 0. \end{cases}$$

**Solution.** If $t > 0$ then by Eq. (9.11) the solution $u(x,t)$ can be written as

$$u(x,t) = \int_{-\infty}^{\infty} \frac{1}{\sqrt{4\pi kt}} e^{-(x-y-\mu\sqrt{t})^2/(4kt)} h(y)\,dy$$

$$= \int_{0}^{\infty} \frac{1}{\sqrt{4\pi kt}} e^{-(x-y-\mu\sqrt{t})^2/(4kt)}\,dy.$$

Make the substitution $\sqrt{2kt}\,z = x - y - \mu\sqrt{t}$ and $-\sqrt{2kt}\,dz = dy$.

$$u(x,t) = \int_{(x-\mu\sqrt{t})/\sqrt{2kt}}^{-\infty} -\sqrt{2kt}\,\frac{1}{\sqrt{4\pi kt}} e^{-z^2/2}\,dz$$

$$= \int_{-\infty}^{(x-\mu\sqrt{t})/\sqrt{2kt}} \frac{1}{\sqrt{2\pi}} e^{-z^2/2}\,dz = \Phi\left(\frac{x-\mu\sqrt{t}}{\sqrt{2kt}}\right).$$

Figure 9.3 shows the graph of the solution when $p = 2/3$ and $\sigma = 1$ for several values of $t \geq 0$.

### Exercises

**9.2.1** An alternative means of deriving the Dirac delta function involves the family of functions $d_\epsilon(x)$ defined as follows.

$$d_\epsilon(x) = \begin{cases} \dfrac{1}{2\epsilon} & \text{if } -\epsilon < x < \epsilon, \\ 0 & \text{if } |x| \geq \epsilon, \end{cases}$$

where $\epsilon > 0$. Show that

$$\int_{-\infty}^{\infty} d_\epsilon(x)\,dx = 1.$$

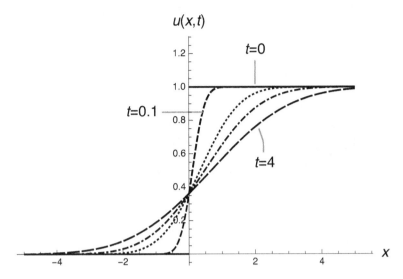

Fig. 9.3   A graph of the solution to the initial value problem associated with the advection/diffusion equation (with parameters $p = 2/3$ and $\sigma = 1$) for various values of $t$.

**9.2.2** Define the Dirac delta function $D(x)$ to be

$$D(x) = \lim_{\epsilon \to 0^+} d_\epsilon(x),$$

where $d_\epsilon(x)$ is defined in Exercise 9.2.1.

(a) Show that $D(x) = 0$ for all $x \neq 0$.
(b) Show that

$$\int_{-\infty}^{\infty} D(x)\, dx = 1.$$

**9.2.3** Show that if $f$ is continuous then

$$\int_{-\infty}^{\infty} D(x - x_0)\, f(x)\, dx = f(x_0).$$

Use the definition of the Dirac delta function from Exercise 9.2.2 and the Integral Mean Value Theorem [OpenStax (2018), Sec. 5.3].

For the remaining exercises in this section define the function $U(x,t)$ as

$$U(x,t) = \frac{1}{\sqrt{4\pi kt}} e^{-(x-\mu\sqrt{t})^2/(4kt)},$$

where $k = 2p(1-p)\sigma^2$ and $\mu = (2p-1)\sigma$.

**9.2.4** Show that $\dfrac{\partial U}{\partial x} = \dfrac{-x + \mu\sqrt{t}}{2kt} U(x,t)$.

**9.2.5** Show that $\dfrac{\partial^2 U}{\partial x^2} = \left( \dfrac{(x - \mu\sqrt{t})^2}{(2kt)^2} - \dfrac{1}{2kt} \right) U(x,t)$.

**9.2.6** Show that $\dfrac{\partial U}{\partial t} = \dfrac{1}{t} \left( \dfrac{x^2 - \mu\sqrt{t}x}{4kt} - \dfrac{1}{2} \right) U(x,t)$.

**9.2.7** Show that $U(x,t)$ satisfies Eq. (9.9).

**9.2.8** In the graph of the solution to the initial value problem for the advection/diffusion equation found in Example 9.3 the point of intersection of the solution on the vertical axis appears to be the same for all $t > 0$. Find this point of intersection.

**9.2.9** Find the solution to the advection/diffusion equation satisfying the initial condition $h(x) = e^{-x^2}$.

**9.2.10** Find the solution to the diffusion equation satisfying the initial condition $h(x) = e^{-|x|}$.

## 9.3  Continuous Random Walk

This section will introduce some of the properties of the continuous random walk. A fully rigorous treatment of this topic is beyond the undergraduate level, so this section will depend somewhat on intuition. Readers interested in more of the tails should consult any of the references on **stochastic calculus** such as Lawler (2006), Shreve (2004b), and Wilmott (2006).

The continuous random walk was developed as the limit of a discrete random walk in Sec. 8.4. The probability density function for the position of the moving particle is given in Eq. (8.16). The motion is considered continuous rather than discrete since the temporal and spatial steps were driven to zero in the limit in such a way that $(\Delta x)^2 \propto \Delta t$. However, the continuous random walk is not a smooth, differentiable function of $t$. A realization of a random walk is shown in Fig. 9.4. The mathematical model of continuous random motion is called **Brownian motion**, named after the Scottish botanist, Robert Brown, who first observed the random motion of particles of pollen suspended in a fluid. The function $W(t)$ is described as a **stochastic process** due to its random behavior. The mathematical description of Brownian motion was developed by the mathematician, Norbert Wiener. In recognition of this contribution, this stochastic process is often called a **Wiener process** and is denoted by $W(t)$. Even though $W(t)$ is called a function, keep in mind it is a family of continuous random variables indexed by $t$. In this section assume the mean of $W(t)$ is zero for all $t > 0$. For each $t > 0$ let $W(t) \sim \mathcal{N}(0, \sigma^2 t)$.

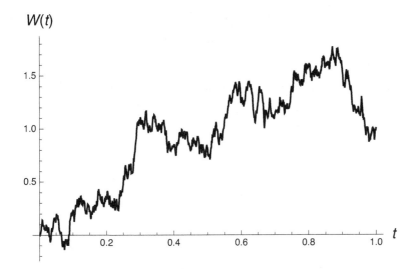

Fig. 9.4   As more frequent, but smaller, steps are taken by a random walk, the process limits on a continuous stochastic process called Brownian motion.

Standard Brownian motion possesses several important properties which are listed here.

(1) $W(0) = 0$ with probability one.
(2) $W(t)$ is a continuous function of $t$ with probability one.
(3) The increments of $W$ for non-overlapping intervals of time are independent random variables with means and variances proportional to the lengths of the time intervals.
(4) The increment $\Delta W_{[s,t]} = W(t) - W(s)$ is distributed as $\mathcal{N}(0, \sigma^2(t-s))$.

The condition of independent increments is made clearer by considering times $0 \le t_1 < t_2 \le t_3 < t_4$. The intervals $[t_1, t_2]$ and $[t_3, t_4]$ are non-overlapping. The increments are the random variables $\Delta W_{[t_1,t_2]}$ and $\Delta W_{[t_3,t_3]}$ and they are independent. Note that the distribution of $\Delta W_{[s,t]}$ depends on the length of the time interval $t - s$ and not on the beginning or ending times. This property is called **stationarity**.

In elementary calculus the definition of continuity at a point requires that if $f(x)$ is a function defined on an open set containing $x = a$ then

$$\lim_{h \to 0} |f(a+h) - f(a)| = 0.$$

Continuity of paths for Brownian motion must be interpreted in terms of probability. The probability of $W(t)$ making a jump of positive size in a short length of time must be shown to be arbitrarily small. Fix a value of $t > 0$ and consider

$$|W(t+h) - W(t)| = |\Delta W_{[t,t+h]}|,$$

which is the absolute value of an increment of Brownian motion. Since the increments are assumed to be normally distributed, then the probability density function for $|\Delta W_{[t,t+h]}|$ takes the form

$$f_X(x) = \frac{2}{\sqrt{2\pi\sigma^2 h}} e^{-x^2/(2\sigma^2 h)}.$$

Now if $\epsilon > 0$ then

$$\mathbb{P}\left(|\Delta W_{[t,t+h]}| > \epsilon\right) = 1 - \mathbb{P}\left(|\Delta W_{[t,t+h]}| \le \epsilon\right) = 2 - 2\Phi\left(\frac{\epsilon}{\sigma\sqrt{h}}\right).$$

As $h \to 0$ this probability approaches 0. Hence $|\Delta W_{[t,t+h]}|$ is almost surely less than $\epsilon$.

Besides the Brownian motion developed in previously, there are many related functions which also satisfy the definition of Brownian motion. Consider the function

$$X(t) = \begin{cases} 0 & \text{if } t = 0, \\ tW(1/t) & \text{if } t > 0. \end{cases}$$

Clearly $X(0) = 0$ with probability one. The random variable $X(t)$ is distributed as $\mathcal{N}(0, t^2\sigma^2/t)$ or simply as $\mathcal{N}(0, \sigma^2 t)$. If $0 < s < t$ then

$$\begin{aligned}
\mathbb{C}\text{ov}\left(X(s), X(t)\right) &= \mathbb{C}\text{ov}\left(sW(1/s), tW(1/t)\right) = st\,\mathbb{C}\text{ov}\left(W(1/s), W(1/t)\right) \\
&= st\,\mathbb{C}\text{ov}\left(W(1/s) - W(1/t) + W(1/t), W(1/t)\right) \\
&= st\,\mathbb{C}\text{ov}\left(W(1/s) - W(1/t), W(1/t)\right) \\
&\quad + st\,\mathbb{C}\text{ov}\left(W(1/t), W(1/t)\right) \\
&= st(0 + \sigma^2/t) = \sigma^2 s.
\end{aligned}$$

The increments of $X$ are independent since if $0 < t_1 < t_2 \le t_3 < t_4$ then

$$\begin{aligned}
&\mathbb{C}\text{ov}\left(X(t_2) - X(t_1), X(t_4) - X(t_3)\right) \\
&= \mathbb{C}\text{ov}\left(X(t_2), X(t_4)\right) - \mathbb{C}\text{ov}\left(X(t_2), X(t_3)\right) \\
&\quad - \mathbb{C}\text{ov}\left(X(t_1), X(t_4)\right) + \mathbb{C}\text{ov}\left(X(t_1), X(t_3)\right) \\
&= \sigma^2(t_2 - t_2 - t_1 + t_1) = 0.
\end{aligned}$$

Brownian motion $W(t)$ is continuous for $t > 0$ which implies $X(t)$ is continuous for $t > 0$. The continuity of $X(t)$ at $t = 0$ is confirmed using a probability argument. In essence as $t \to 0^+$, the probability of $X(t)$ being different from 0 vanishes. Let $\epsilon > 0$ then

$$\mathbb{P}\left(|X(t)| > \epsilon\right) = 1 - \mathbb{P}\left(|X(t)| \leq \epsilon\right) = 2 - 2\Phi\left(\frac{\epsilon}{\sigma\sqrt{t}}\right),$$

which approaches 0 as $t \to 0^+$. Therefore random process $X(t)$ is a standard Brownian motion.

Since a Brownian motion is continuous on the closed, bounded interval $[0, t]$ it must attain an absolute maximum and an absolute minimum on this interval [OpenStax (2018), Thm. 4.1]. Thus the following useful random functions, called respectively the minimum and maximum functions, are well defined.

$$\underline{M}(t) = \min_{0 \leq s \leq t} \{W(s)\},$$

$$\overline{M}(t) = \max_{0 \leq s \leq t} \{W(s)\}.$$

Define the **first-passage time** or **hitting time** as the first time the Brownian motion reaches a specified value.

$$T_a = \min\{t \mid W(t) = a\}.$$

The first-passage time can be infinite if $W(t)$ never reaches $a$. Like the Brownian motion itself, the first-passage time is a random variable.

**Theorem 9.1 (Reflection Principle).** *Suppose $W(t)$ is a standard Brownian motion and $a \geq 0$, then*

$$\mathbb{P}\left(\overline{M}(t) \geq a\right) = 2\,\mathbb{P}\left(W(t) \geq a\right) = 2 - 2\Phi\left(\frac{a}{\sigma\sqrt{t}}\right). \tag{9.12}$$

*Proof.* Fix $a \geq 0$ and consider

$$\mathbb{P}\left(W(t) \geq a\right) = \mathbb{P}\left((W(t) \geq a) \cap (\overline{M}(t) \geq a)\right) \;\; + \mathbb{P}\left((W(t) \geq a) \cap (\overline{M}(t) < a)\right).$$

The second probability on the right-hand side of the equation is 0 since $\overline{M}(t) \geq W(t)$ for all $t \geq 0$. Thus

$$\mathbb{P}\left(W(t) \geq a\right) = \mathbb{P}\left((W(t) \geq a) \cap (\overline{M}(t) \geq a)\right)$$
$$= \mathbb{P}\left(W(t) \geq a \mid \overline{M}(t) \geq a\right) \mathbb{P}\left(\overline{M}(t) \geq a\right)$$
$$= \mathbb{P}\left(W(t) \geq a \mid T_a \leq t\right) \mathbb{P}\left(\overline{M}(t) \geq a\right),$$

since $\overline{M}(t) \geq a$ if and only if $T_a \leq t$ as a result of the continuity of the Brownian motion. The random process $W(T_a + t) - a$ is also a Brownian motion for $t \geq 0$ and therefore is distributed as $\mathcal{N}(0, \sigma^2 t)$, the same as the distribution of $W(t)$. Using this

$$\begin{aligned}
\mathbb{P}\left(W(t) \geq a\right) &= \mathbb{P}\left(W(t) - a \geq 0 \,|\, T_a \leq t\right) \mathbb{P}\left(\overline{M}(t) \geq a\right) \\
&= \mathbb{P}\left(W(T_a + (t - T_a)) - a \geq 0 \,|\, t - T_a \geq 0\right) \mathbb{P}\left(\overline{M}(t) \geq a\right) \\
&= \frac{1}{2} \mathbb{P}\left(\overline{M}(t) \geq a\right).
\end{aligned}$$

Therefore

$$\mathbb{P}\left(\overline{M}(t) \geq a\right) = 2 \mathbb{P}\left(W(t) \geq a\right) = 2 - 2\Phi\left(\frac{a}{\sigma\sqrt{t}}\right).$$

$\square$

As a corollary to Thm. 9.1, $\mathbb{P}\left(\overline{M}(t) < a\right) = 2\Phi\left(a/(\sigma\sqrt{t})\right) - 1$.

The reader will show in Exercise 9.3.7 that $-W(t)$ is a Brownian motion process, thus

$$\min_{0 \leq s \leq t}\{-W(s)\} = -\max_{0 \leq s \leq t}\{W(s)\} \text{ and}$$

$$\max_{0 \leq s \leq t}\{-W(s)\} = -\min_{0 \leq s \leq t}\{W(s)\}.$$

Therefore the distribution of random variable $\overline{M}(t)$ is the same as the distribution of random variable $-\underline{M}(t)$. Furthermore Exercise 9.3.8 will establish that $aW(t/a^2)$ with $a \neq 0$ is a Brownian motion. In particular when $a > 0$

$$\max_{0 \leq s \leq t}\{aW(s/a^2)\} = a \max_{0 \leq s \leq t}\{W(s/a^2)\} = a \max_{0 \leq s \leq t/a^2}\{W(s)\} = a\overline{M}(t/a^2).$$

Note that

$$\begin{aligned}
\mathbb{P}\left(a\overline{M}(t/a^2) \leq a\right) &= \mathbb{P}\left(\overline{M}(t/a^2) \leq 1\right) \\
&= 2\Phi\left(\frac{1}{\sigma\sqrt{t/a^2}}\right) - 1 = 2\Phi\left(\frac{a}{\sigma\sqrt{t}}\right) - 1 \\
&= \mathbb{P}\left(\overline{M}(t) \leq a\right).
\end{aligned}$$

In other words $a\overline{M}(t/a^2)$ and $\overline{M}(t)$ have the same probability distribution for $a > 0$. The distribution of $\overline{M}(t)$ determines the distribution of first-passage times.

**Corollary 9.1.** *The first-passage time random variable $T_a$ for $a \geq 0$ has a probability density function*

$$f_{T_a}(t) = \frac{a}{\sqrt{2\pi\sigma^2 t^3}} e^{-a^2/(2\sigma^2 t)}, \tag{9.13}$$

*and $T_a < \infty$ with probability one.*

*Proof.* Using the maximum of the Brownian motion, the cumulative probability density function for $T_a$ is

$$\mathbb{P}\left(T_a \leq t\right) = \mathbb{P}\left(\overline{M}(t) \geq a\right) = 2 - 2\Phi\left(\frac{a}{\sigma\sqrt{t}}\right),$$

using Eq. (9.12). Differentiating this CDF with respect to $t$ yields

$$f_{T_a}(t) = \frac{d}{dt}\left[2 - 2\Phi\left(\frac{a}{\sigma\sqrt{t}}\right)\right] = \frac{a}{\sqrt{2\pi\sigma^2 t^3}}e^{-a^2/(2\sigma^2 t)},$$

for $t > 0$ and 0 otherwise. If $a = 0$ then $T_a = T_0 = 0 < \infty$, so suppose $a > 0$. $\mathbb{P}\left(T_a \leq t\right) \to 1$ as $t \to \infty$ and thus $T_a$ is almost surely finite. $\square$

While a Brownian motion $W(t)$ is continuous, $W(t)$ is not differentiable. Technically the probability that $dW/dt$ exists is zero. This property addresses the "roughness" of the path followed by Brownian motion as seen in Fig. 9.4. The claim can be proved rigorously, but consider an intuitive approach in order to avoid the complexities of the analysis. Recall the limit definition of the derivative from calculus,

$$f'(t) = \lim_{h \to 0} \frac{f(t+h) - f(t)}{h},$$

provided the limit exists. Suppose $f(t)$ is a Brownian motion $W(t)$. Fix a value of $t$, then the difference quotient $(W(t+h) - W(t))/h$ is not bounded as a function of $h$ in any open interval containing 0. Suppose to the contrary that it is bounded, then there exists a positive number $K$ such that

$$\frac{W(t+h) - W(t)}{h} < K \iff W(t+h) - W(t) < K h.$$

The reader can show that $X(h) = W(t+h) - W(t)$ is a standard Brownian motion and thus by Thm. 9.1

$$\mathbb{P}\left(\overline{M}(h) < K h\right) = 2\Phi\left(\frac{K h}{\sigma\sqrt{h}}\right) - 1 = 2\Phi\left(\frac{K\sqrt{h}}{\sigma}\right) - 1 \to 0,$$

as $h \to 0$. Hence there exists no bound for the difference quotient on any neighborhood of 0. Therefore the limit of the difference quotient and thus the derivative of Brownian motion do not exist.

The quadratic variation of a random process $X(t)$ is denoted $[X](t)$ and defined as

$$[X](t) = \lim_{\|P\| \to 0} \sum_{k=1}^{n} [X(t_k) - X(t_{k-1})]^2, \tag{9.14}$$

where $P = \{t_0, t_1, \ldots, t_n\}$ is any partition of the interval $[0, t]$ and $\|P\|$ denotes the norm of the partition. The quadratic variation of the process exists only if the limit exists and is the same for every partition where $\|P\| \to 0$. The quadratic variation property is important for the development of a **stochastic integral**. Theorem 9.2 demonstrates that Brownian motion $W(t)$ has quadratic variation proportional to $t$. The following theorem (adapted from [Seydel (2002), Chapter 1]), while technical in nature, will establish the necessary result.

**Theorem 9.2.** *Let $\{P^{(n)}\}$ for $n \in \mathbb{N}$ be a sequence of partitions of the interval $[0, t]$ such that*

$$0 = t_0^{(n)} < t_1^{(n)} < \cdots t_{i-1}^{(n)} < t_i^{(n)} < \cdots < t_n^{(n)} = t.$$

*For $i = 1, 2, \ldots, n$ let $\Delta t_i^{(n)} = t_i^{(n)} - t_{i-1}^{(n)}$ and $\Delta W_i^{(n)} = W(t_i^{(n)}) - W(t_{i-1}^{(n)})$ and let $\delta_n = \|P^{(n)}\| = \max_i\{\Delta t_i^{(n)}\}$. Then*

$$\mathbb{E}\left(\left(\sum_{i=1}^{n}\left((\Delta W_i^{(n)})^2 - \sigma^2 \Delta t_i^{(n)}\right)\right)^2\right) \to 0 \text{ as } \delta_n \to 0. \tag{9.15}$$

The convergence of the expected value to 0 is referred to as **convergence in probability**.

*Proof.* To simplify the notation drop the superscripts of the form $\cdot^{(n)}$. Define $X_i = (W(t_i) - W(t_{i-1}))^2 - \sigma^2(t_i - t_{i-1})$, then

$$X_i^2 = (\Delta W_i)^4 - 2\sigma^2(\Delta t_i)(\Delta W_i)^2 + \sigma^4(\Delta t_i)^2$$
$$\mathbb{E}\left(X_i^2\right) = \mathbb{E}\left((\Delta W_i)^4\right) - 2\sigma^2(\Delta t_i)\,\mathbb{E}\left((\Delta W_i)^2\right) + \sigma^4(\Delta t_i)^2$$
$$= 3\sigma^4(\Delta t_i)^2 - 2\sigma^4(\Delta t_i)^2 + \sigma^4(\Delta t_i)^2$$
$$= 2\sigma^4(\Delta t_i)^2.$$

The reader may use Exercises 9.3.5 and 9.3.6 to justify the result above. Now consider

$$\left(\sum_{i=1}^{n} X_i\right)^2 = \left(\sum_{i=1}^{n} X_i\right)\left(\sum_{j=1}^{n} X_j\right) = \sum_{i=1}^{n} X_i^2 + 2\sum_{i=1}^{n}\sum_{j=1}^{i-1}(X_i X_j)$$
$$\mathbb{E}\left(\left(\sum_{i=1}^{n} X_i\right)^2\right) = \sum_{i=1}^{n}\mathbb{E}\left(X_i^2\right) + 2\sum_{i=1}^{n}\sum_{j=1}^{i-1}\mathbb{E}\left(X_i X_j\right).$$

The expectation in the double summation can be re-written.

$$\mathbb{E}\left(X_i X_j\right) = \mathbb{E}\left((\Delta W_i)^2(\Delta W_j)^2\right) - \sigma^2(\Delta t_j)\,\mathbb{E}\left((\Delta W_i)^2\right)$$
$$- \sigma^2(\Delta t_i)\,\mathbb{E}\left((\Delta W_i)^2\right) + \sigma^4(\Delta t_i)(\Delta t_j).$$

Using the assumption of independence of the increments $\Delta W_i$ and $\Delta W_j$ and the result of Exercise 9.3.5 then $\mathbb{E}\left(X_i X_j\right) = 0$. Thus the double summation above vanishes. Therefore

$$\mathbb{E}\left(\left(\sum_{i=1}^n X_i\right)^2\right) = \sum_{i=1}^n \mathbb{E}\left(X_i^2\right) = 2\sigma^4 \sum_{i=1}^n (\Delta t_i)^2.$$

By assumption $\delta \geq \Delta t_i$ for all $i = 1, 2, \ldots, n$. Thus

$$\mathbb{E}\left(\left(\sum_{i=1}^n X_i\right)^2\right) \leq 2\sigma^4 \delta \sum_{i=1}^n \Delta t_i = \delta(2\sigma^4 t) \to 0,$$

as $\delta \to 0$. □

Later sections in this chapter will build upon and extend the results of this section for more general classes of random processes.

## Exercises

**9.3.1** Show that if $W(t) \sim \mathcal{N}(0, \sigma^2 t)$ then $\mathbb{E}\left(\Delta W_{[s,t]}\right) = 0$.

**9.3.2** Show that if $W(t) \sim \mathcal{N}(0, \sigma^2 t)$ and increments on non-overlapping time intervals are independent, then $\mathbb{Var}\left(\Delta W_{[s,t]}\right) = \sigma^2(t - s)$.

**9.3.3** Show that if $W(t) \sim \mathcal{N}(0, \sigma^2 t)$ then $\mathbb{P}\left(W(t) \leq 0\right) = 1/2$ for all $t > 0$.

**9.3.4** Let $W(t)$ be a standard Brownian motion, $a \geq 0$, and $T_a$ the first-passage time for $W(t)$. Define $X(t) = W(T_a + t) - a$ and show $X(t)$ is a standard Brownian motion.

**9.3.5** Suppose that $X \sim \mathcal{N}(\mu, \sigma^2)$. Show that $\mathbb{E}\left(X^2\right) = \mu^2 + \sigma^2$.

**9.3.6** Suppose that $X \sim \mathcal{N}(\mu, \sigma^2)$. Show that $\mathbb{E}\left(X^4\right) = \mu^4 + 6\mu^2\sigma^2 + 3\sigma^4$.

**9.3.7** Show that if $W(t)$ is a Brownian motion then $-W(t)$ is a Brownian motion.

**9.3.8** Suppose that $W(t)$ is a Brownian motion and $a \neq 0$ is a constant. Show that $aW(t/a^2)$ is a Brownian motion.

**9.3.9** The assumption that $W(t) \sim \mathcal{N}(0, \sigma^2 t)$ was very important in the proof of Thm. 9.2. Specifically this made it possible to state that $\mathbb{E}\left(X_i X_j\right) = 0$. Suppose $W(t) \sim \mathcal{N}(\mu t, \sigma^2 t)$ and show that $\mathbb{E}\left(X_i X_j\right) \neq 0$.

**9.3.10** Suppose the function $f(t)$ is continuous on $[0, T]$ and has a continuous derivative on the interval $(0, T)$. Show that the quadratic variation of $f$ is zero.

## 9.4   The Stochastic Integral

Of central importance in the remainder of the book will be the calculus of functions defined along a continuous random walk. In this section and those that follow assume $W(t) \sim \mathcal{N}(0, t)$, in other words set the parameter $\sigma = 1$. The derivative of a Brownian motion process does not exist, but what about the integral? If $P$ is a partition of $[0, t]$ with $0 = t_0 < t_1 < \cdots < t_n = t$ with $\|P\| = \max_{1 \le k \le n}\{t_k - t_{k-1}\}$ define the stochastic integral of a function $f(\tau)$ defined on $[0, t]$ informally as

$$\int_0^t f(\tau)\, dW(\tau) = \lim_{\|P\| \to 0} \sum_{k=1}^n f(t_{k-1})(W(t_k) - W(t_{k-1})). \qquad (9.16)$$

This integral is defined only if the limit exists and is the same for all partitions $P$. Note that the function $f$ is always evaluated at the left-hand endpoint of the subinterval $[t_{k-1}, t_k]$. Think of this as ensuring that the function being integrated cannot use any information about the future movements of the random walk as it is being integrated. To give a more concrete example, suppose a gambler places wagers defined by $f(\tau)$ continuously during the interval $[0, t]$ on the outcome of the continuous random walk $W(\tau)$. The stochastic integral given in Eq. (9.16) would represent the net winnings.

When convenient the stochastic integral can be written in a form reminiscent of the Fundamental Theorem of Calculus.

$$I_f(t) = \int_0^t f(\tau)\, dW(\tau). \qquad (9.17)$$

The letter "I" is chosen in honor of Kiyoshi Itô, the Japanese mathematician who developed much of the theory of the stochastic integral in the 1940's. For this reason the stochastic integral is sometimes called the Itô integral. The stochastic integral is sometimes written in a differential form obtained by informally differentiating Eq. (9.17).

$$dI_f(t) = f(t)\, dW(t). \qquad (9.18)$$

However, it is not possible to divide both sides of Eq. (9.18) by the differential $dt$ since $dW/dt$ is undefined. In general Eqs. (9.17) and (9.18) are considered to be synonymous.

The reader should keep in mind that $I_f(t)$ is a random variable, not the typical function encountered in elementary calculus. However, the stochastic integral does possess some familiar properties which hold with probability one.

**Theorem 9.3.** *If $f$ and $g$ are functions defined on $[0,t]$ with $0 \le s \le t$ such that their stochastic integrals exist and if $c$ is any constant, then*

$$\int_0^t f(\tau)\, dW(\tau) = \int_0^s f(\tau)\, dW(\tau) + \int_s^t f(\tau)\, dW(\tau)$$

$$\int_0^t (f(\tau) \pm g(\tau))\, dW(\tau) = \int_0^t f(\tau)\, dW(\tau) \pm \int_0^t g(\tau)\, dW(\tau)$$

$$\int_0^t c f(\tau)\, dW(\tau) = c \int_0^t f(\tau)\, dW(\tau).$$

The stochastic integral is a random variable. As the limit of a sum of normally distributed random variables, then $I_f(t)$ is normally distributed. The following theorem provides more detail on the distribution of the stochastic integral of non-random functions.

**Theorem 9.4.** *If $f$ is a deterministic (non-random) function defined on $[0,t]$ for which the stochastic integral exists and $0 \le s \le t$, then*

$$\mathbb{E}\left( \int_0^t f(\tau)\, dW(\tau) \right) = 0 \tag{9.19}$$

$$\mathbb{Var}\left( \int_0^t f(\tau)\, dW(\tau) \right) = \int_0^t (f(\tau))^2\, d\tau \tag{9.20}$$

$$\mathbb{Cov}\left( \int_0^s f(\tau)\, dW(\tau), \int_0^t f(\tau)\, dW(\tau) \right) = \int_0^s (f(\tau))^2\, d\tau. \tag{9.21}$$

*Proof.* Taking the expected value of both sides of Eq. (9.16) produces

$$\mathbb{E}\left( \int_0^t f(\tau)\, dW(\tau) \right) = \mathbb{E}\left( \lim_{\|P\| \to 0} \sum_{k=1}^n f(t_{k-1})(W(t_k) - W(t_{k-1})) \right)$$

$$= \lim_{\|P\| \to 0} \sum_{k=1}^n f(t_{k-1})\, \mathbb{E}\left( W(t_k) - W(t_{k-1}) \right),$$

assuming it is permissible to interchange the order of the operations of taking the limit and finding the expected value. Since $\mathbb{E}\left( W(t_k) - W(t_{k-1}) \right) = 0$, Eq. (9.19) is proved.

Proceeding in the same fashion by calculating the variance of both sides of Eq. (9.16) yields

$$\mathbb{V}\text{ar}\left(\int_0^t f(\tau)\, dW(\tau)\right) = \mathbb{V}\text{ar}\left(\lim_{\|P\|\to 0}\sum_{k=1}^n f(t_{k-1})(W(t_k)-W(t_{k-1}))\right)$$

$$= \lim_{\|P\|\to 0}\sum_{k=1}^n (f(t_{k-1}))^2\, \mathbb{V}\text{ar}\left(W(t_k)-W(t_{k-1})\right)$$

$$= \lim_{\|P\|\to 0}\sum_{k=1}^n (f(t_{k-1}))^2 (t_k - t_{k-1})$$

$$= \int_0^t (f(\tau))^2\, d\tau,$$

and Eq. (9.20) is established. The reader should note that the assumption that function $f$ is deterministic allows $f(t_{k-1})$ to be "pulled out" of the expected value and variance.

Turning to the covariance,

$$\mathbb{C}\text{ov}\left(I_f(s), I_f(t)\right) = \mathbb{C}\text{ov}\left(I_f(s), I_f(s) + \int_s^t f(\tau)\, dW(\tau)\right)$$

$$= \mathbb{C}\text{ov}\left(I_f(s), I_f(s)\right) + \mathbb{C}\text{ov}\left(I_f(s), \int_s^t f(\tau)\, dW(\tau)\right)$$

$$= \mathbb{V}\text{ar}\left(I_f(s)\right) + 0 = \int_0^s (f(\tau))^2\, d\tau.$$

This can be justified using the fact that the increments of $W(\tau)$ in the interval $[0, s]$ are independent of the increments of $W(\tau)$ in the interval $[s, t]$. □

In summary Thm. 9.4 establishes that

$$\int_0^t f(\tau)\, dW(\tau) \sim \mathcal{N}\left(0, \int_0^t (f(\tau))^2\, d\tau\right).$$

**Example 9.4.** Let $f(\tau) = \sin\tau$ and find the mean and variance of the stochastic integral of $f$ for $t \geq 0$.

According to Eq. (9.19)

$$\mathbb{E}\left(\int_0^t \sin\tau\, dW(\tau)\right) = 0,$$

while

$$\mathbb{V}\text{ar}\left(\int_0^t \sin\tau\, dW(\tau)\right) = \int_0^t \sin^2\tau\, d\tau = \frac{t}{2} - \frac{1}{4}\sin 2t.$$

Since Thm. 9.4 succinctly addressed the mean and variance of stochastic integrals of deterministic functions, attention now turns to an example in which the integrand is not deterministic. The following example and its analysis are taken from Example 4.3.2 of [Shreve (2004b)] with small modifications to the notation. Suppose the Itô integral of $W(t)$ exists then according to the definition of the integral

$$\int_0^t W(\tau)\,dW(\tau) = \lim_{\|P\|\to 0} \sum_{k=1}^n W(t_{k-1})\left[W(t_k) - W(t_{k-1})\right],$$

where $0 = t_0 < t_1 < \cdots < t_n = t$ is a partition of $[0, t]$. In Exercise 9.4.1 the reader is asked to verify that

$$\sum_{k=1}^n W(t_{k-1})\left[W(t_k) - W(t_{k-1})\right] = \frac{1}{2}(W(t))^2 - \frac{1}{2}\sum_{k=1}^n (W(t_k) - W(t_{k-1}))^2.$$

Using this allows the Itô integral to be expressed as

$$\int_0^t W(\tau)\,dW(\tau) = \frac{1}{2}(W(t))^2 - \frac{1}{2}\lim_{\|P\|\to 0}\sum_{k=1}^n (W(t_k) - W(t_{k-1}))^2$$

$$= \frac{1}{2}(W(t))^2 - \frac{1}{2}[W](t)$$

$$= \frac{1}{2}(W(t))^2 - \frac{t}{2}. \tag{9.22}$$

The second line of this equation follows from the definition of the quadratic variation of $W(t)$, see Eq. (9.14). The Itô integral $\int_0^t W(\tau)\,dW(\tau)$ is a random variable with expected value and variance as follows.

$$\mathbb{E}\left(\frac{1}{2}(W(t))^2 - \frac{t}{2}\right) = \frac{1}{2}\mathbb{E}\left((W(t))^2\right) - \frac{t}{2} = \frac{1}{2}(t) - \frac{t}{2} = 0$$

$$\mathbb{V}\mathrm{ar}\left(\frac{1}{2}(W(t))^2 - \frac{t}{2}\right) = \frac{1}{4}\mathbb{E}\left((W(t))^4\right) - \frac{t}{2}\mathbb{E}\left((W(t))^2\right) + \frac{t^2}{4}$$

$$= \frac{3t^2}{4} - \frac{t^2}{2} + \frac{t^2}{4} = \frac{t^2}{2}.$$

This example also illustrates a fundamental difference between the Itô integral and the Riemann integral of elementary calculus. If $W(t)$ is replaced with a continuously differentiable deterministic function $f(t)$ for which $f(0) = 0$ then

$$\int_0^t f(\tau)\,df(\tau) = \int_0^t f(\tau)f'(\tau)\,d\tau = \int_{f(0)}^{f(t)} u\,du = \frac{1}{2}(f(t))^2, \tag{9.23}$$

which lacks the final term of the Itô integral. The extra term of the form $-t/2$ is a result of the quadratic variation of the random process $W(t)$.

In Exercise 9.3.10 the reader will demonstrate that the quadratic variation of continuously differentiable functions is zero.

The final example of this introduction to the stochastic integral involves the Riemann integral of $W(t)$ itself.

**Example 9.5.** Use the technique of integration by parts to evaluate the integral,

$$\int_0^t W(\tau)\, d\tau.$$

**Solution.** Note that $d(\tau W(\tau)) = W(\tau)\, d\tau + \tau\, dW(\tau)$ and thus

$$\int_0^t d(\tau W(\tau)) = \int_0^t W(\tau)\, d\tau + \int_0^t \tau\, dW(\tau)$$

$$\int_0^t W(\tau)\, d\tau = tW(t) - \int_0^t \tau\, dW(\tau)$$

$$= \int_0^t t\, dW(\tau) - \int_0^t \tau\, dW(\tau)$$

$$= \int_0^t (t - \tau)\, dW(\tau).$$

Integration by parts has transformed the Riemann integral into a stochastic integral. The result is known as **integrated Brownian motion**. It is normally distributed with an expected value of 0 and a variance of

$$\mathrm{Var}\left(\int_0^t W(\tau)\, d\tau\right) = \int_0^t (t - \tau)^2\, d\tau = \frac{t^3}{3}.$$

*Exercises*

**9.4.1** Let $0 = t_0 < t_1 < \cdots < t_n = t$ be a partition of $[0, t]$ and let $W(s)$ be a Brownian motion process. Show that

$$\sum_{k=1}^n W(t_{k-1})\left[W(t_k) - W(t_{k-1})\right] = \frac{1}{2}(W(t))^2 - \frac{1}{2}\sum_{k=1}^n (W(t_k) - W(t_{k-1}))^2.$$

**9.4.2** Find the mean and variance of the stochastic integral

$$\int_0^2 t^3\, dW(t).$$

**9.4.3** Evaluate the following stochastic integral and find its mean and variance.

$$\int_0^t \tau\, dW(\tau).$$

**9.4.4** Let $c$ be a constant and evaluate the following stochastic integral and find its mean and variance.

$$\int_0^t c\,dW(\tau).$$

**9.4.5** Describe the distribution and find the mean and variance of the following stochastic integral.

$$\int_0^t e^{-\tau/2}\,dW(\tau).$$

## 9.5 Itô's Lemma

One of the most useful techniques of integration is integration by substitution which requires making a change of variable in the integral. Equation (9.23) argued that if the $f$ is non-random then a change of variable can be used to simplify the evaluation of an Riemann integral. However if a naïve change of variable is used in an Itô integral, for example in Eq. (9.22), the result may be incorrect. This section presents the proper method for performing a change of variable when Brownian motion $W(t)$ is involved.

Suppose function $f$ depends on two non-random variables $(x,t)$, then using the chain rule of multivariable calculus [OpenStax (2018), Thm. 4.8] the differential

$$df = \frac{\partial f}{\partial x}\,dx + \frac{\partial f}{\partial t}\,dt.$$

However, this does not hold if $x$ is replaced by $W(t)$. The proper procedure for finding the differential when stochastic variables are present can be derived using the multivariable version of Taylor's Theorem. A standard reference containing this result is [Taylor and Mann (1983)]. This section will give a brief overview of a two-variable version of Taylor's formula with remainder. Starting with the single-variable Taylor's formula, if $f(x)$ is an $(n+1)$-times differentiable function on an open interval containing $x_0$ then the function may be written as

$$f(x) = f(x_0) + f'(x_0)(x - x_0) + \frac{f''(x_0)}{2!}(x - x_0)^2$$
$$+ \cdots + \frac{f^{(n)}(x_0)}{n!}(x - x_0)^n + \frac{f^{(n+1)}(\theta)}{(n+1)!}(x - x_0)^{n+1}, \quad (9.24)$$

where $\theta$ lies between $x_0$ and $x$. The last term above is usually called the Taylor remainder formula and is denoted by $R_{n+1}$. The other terms form a polynomial in $x$ of degree at most $n$ and can be used as an approximation for $f(x)$ in a neighborhood of $x_0$. From Eq. (9.24) the two-variable form of Taylor's formula will be derived.

Suppose the function $F(y, z)$ has partial derivatives up to order three on an open disk containing the point with coordinates $(y_0, z_0)$. Define the function $f(x) = F(y_0 + xh, z_0 + xk)$ where $h$ and $k$ are chosen small enough that $(y_0 + h, z_0 + k)$ lies within the disk surrounding $(y_0, z_0)$. Since $f$ is a function of a single variable use the single-variable form of Taylor's formula in Eq. (9.24) with $x_0 = 0$ and $x = 1$ to write

$$f(1) = f(0) + f'(0) + \frac{1}{2}f''(0) + \frac{1}{3!}f^{(3)}(\theta). \tag{9.25}$$

Using the multivariable chain rule for derivatives, differentiate $f(x)$ and set $x = 0$,

$$f'(0) = hF_y(y_0, z_0) + kF_z(y_0, z_0) \tag{9.26}$$

$$f''(0) = h^2 F_{yy}(y_0, z_0) + 2hk F_{yz}(y_0, z_0) + k^2 F_{zz}(y_0, z_0). \tag{9.27}$$

The equality of the mixed second partial derivatives $F_{yz} = F_{zy}$ is made possible by the smoothness assumptions on $F(y, z)$. The remainder term contains only third order partial derivatives of $F$ evaluated somewhere on the line connecting the points $(y_0, z_0)$ and $(y_0 + h, z_0 + k)$. Substitute Eqs. (9.26) and (9.27) into (9.25) to obtain

$$\begin{aligned}
\Delta F &= f(1) - f(0) \\
&= F(y_0 + h, z_0 + k) - F(y_0, z_0) \\
&= hF_y(y_0, z_0) + kF_z(y_0, z_0) \\
&\quad + \frac{1}{2}\left(h^2 F_{yy}(y_0, z_0) + 2hk F_{yz}(y_0, z_0) + k^2 F_{zz}(y_0, z_0)\right) + R_3.
\end{aligned} \tag{9.28}$$

Equation (9.28) is used to derive Itô's formula, the proper method for performing a change of variable in an expression involving Brownian motion.

Suppose $Y = F(W(t), t)$ where $F$ is a non-random function, then using the Taylor series expansion for $Y$ detailed in Eq. (9.28),

$$\begin{aligned}
\Delta Y &= F_W(\Delta W) + F_t(\Delta t) + \frac{1}{2}F_{WW}(\Delta W)^2 + F_{Wt}(\Delta W)(\Delta t) \\
&\quad + \frac{1}{2}F_{tt}(\Delta t)^2 + R_3.
\end{aligned}$$

The Taylor remainder term $R_3$ contains terms of order $(\Delta t)^k$ where $k \geq 2$. As $\Delta t$ becomes small Itô's formula results,

$$dY = F_W \, dW(t) + \left(F_t + \frac{1}{2}F_{WW}\right)dt. \tag{9.29}$$

Note that all terms in the Taylor series involving $(\Delta t)^k$ with $k > 1$ have been ignored since they approach zero faster than $\Delta t$. Yet $(\Delta W)^2$ was retained since $\mathbb{E}\left((\Delta W)^2\right) = \Delta t$ and thus $(dW(t))^2 = dt$.

**Example 9.6.** Let $f(W(t), t) = (W(t))^2$ and find the differential $df$ and use it to evaluate the Itô integral,

$$\int_0^t W(s) \, dW(s).$$

Compare the result with the result obtained for this integral in Sec. 9.4.

**Solution.** Using Eq. (9.29),

$$d((W(t))^2) = 2W(t) \, dW(t) + (0) \, dt + \frac{1}{2}(2) \, dt = 2W(t) \, dW(t) + dt.$$

Rearranging terms in this equation produces,

$$W(t) \, dW(t) = \frac{1}{2} d((W(t))^2) - \frac{1}{2} dt$$

$$\int_0^t W(s) \, dW(s) = \frac{1}{2} \int_0^t d((W(s))^2) - \frac{1}{2} \int_0^t ds$$

$$= \frac{1}{2} (W(t))^2 - \frac{t}{2},$$

the same result found directly from the definition of the stochastic integral in Sec. 9.4.

The results of this informal introduction to stochastic differential equations can now be generalized. A familiar place to start generalizing is with the deterministic process of exponential growth and decay. Most introductions to calculus (see for example [Stewart (1999)]) contain the mathematical model which states that the rate of change of a non-negative quantity $P$, is proportional to $P$. Expressed as a differential equation this statement becomes

$$\frac{dP}{dt} = \mu P \tag{9.30}$$

where $\mu$ is the growth/decay proportionality constant and $t$ usually represents time. The proportionality constant is called a "growth rate" if $\mu > 0$ and a "decay rate" if $\mu < 0$. If the value of $P$ is known at a specified value of $t$ (usually at $t = 0$) then this differential equation has solution $P(t) = P(0)e^{\mu t}$ (see Exercise 9.5.4). This mathematical model is described as **deterministic** since there are no random variables or events in the model. Once the initial value of $P$ and the proportionality constant are set, the future evolution of $P(t)$ is completely determined.

Equation (9.30) can be re-written as

$$\frac{dP}{P} = \mu \, dt, \tag{9.31}$$

and making the change of variable $Z = \ln P$, this equation becomes

$$dZ = \mu \, dt. \tag{9.32}$$

Suppose a stochastic component is added introducing on the right-hand side of Eq. (9.32) a Brownian motion process with mean zero and standard deviation $\sigma \sqrt{dt}$. The mathematical model governing the time evolution of $Z$ becomes

$$dZ(t) = \mu \, dt + \sigma \, dW(t). \tag{9.33}$$

A stochastic differential equation of this form is called a **generalized Wiener process**. Notice it possesses a deterministic part and a stochastic part. The constant $\mu$ is called the **drift** and the constant $\sigma$ is called the **volatility**. The integral form of Eq. (9.33) is

$$Z(t) = Z(0) + \mu t + \int_0^t \sigma \, dW(\tau) = Z(0) + \mu t + \sigma W(t).$$

$Z(t)$ is a normally distributed random variable with a mean and variance determined as follows.

$$\mathbb{E}\left(Z(t)\right) = Z(0) + \mu t + \mathbb{E}\left(\int_0^t \sigma \, dW(\tau)\right) = Z(0) + \mu t, \tag{9.34}$$

by Thm. 9.4. Meanwhile the variance is calculated as

$$\mathbb{Var}\left(Z(t)\right) = \mathbb{Var}\left(Z(0) + \mu t + \int_0^t \sigma \, dW(\tau)\right)$$

$$= \mathbb{Var}\left(\int_0^t \sigma \, dW(\tau)\right) = \int_0^t \sigma^2 \, dt = \sigma^2 t.$$

Hence $Z(t) \sim \mathcal{N}(Z(0) + \mu t, \sigma^2 t)$.

Itô's formula can be extended to a more general setting. Let $X(t)$ be a random variable described by an Itô drift-diffusion process of the form

$$dX(t) = \mu(X(t), t) \, dt + \sigma(X(t), t) \, dW(t), \tag{9.35}$$

where $W(t)$ is the standard Brownian motion process and $\mu$ and $\sigma$ are functions of $X(t)$ and $t$. Let $Y(t) = F(X(t), t)$ be another random variable defined as a function of $X(t)$ and $t$. Given the equation describing the differential of $X(t)$ the equation which describes the differential of $Y(t)$

can be determined. Using the multivariable form of Taylor's theorem once again, produces

$$dY(t) = F_X \, dX(t) + F_t \, dt + \frac{1}{2} F_{XX}(dX(t))^2$$

$$= F_X \left( \mu(X(t), t) \, dt + \sigma(X(t), t) \, dW(t) \right) + F_t \, dt$$

$$+ \frac{1}{2} F_{XX} \left( \mu(X(t), t) \, dt + \sigma(X(t), t) \, dW(t) \right)^2$$

$$= \left( \mu(X(t), t) F_X + F_t + \frac{1}{2} (\sigma(X(t), t))^2 F_{XX} \right) dt$$

$$+ \sigma(X(t), t) F_X \, dW(t) + \cdots .$$

The terms not shown will involve powers of $dt$ greater than one and thus can be ignored. Therefore the result can be summarized as the following lemma.

**Lemma 9.1 (Itô's Lemma).** *Suppose that the random variable $X(t)$ is described by the Itô drift-diffusion process*

$$dX(t) = \mu(X(t), t) \, dt + \sigma(X(t), t) \, dW(t), \qquad (9.36)$$

*where $W(t)$ is the standard Brownian motion. Suppose the random variable $Y(t) = F(X(t), t)$ where $F(x, t) \in \mathcal{C}^2(\mathbb{R}, \mathbb{R})$. Then $Y(t)$ has the following differential.*

$$dY(t) = \left( \mu(X(t), t) F_X + F_t + \frac{1}{2} (\sigma(X(t), t))^2 F_{XX} \right) dt$$

$$+ \sigma(X(t), t) F_X \, dW(t). \qquad (9.37)$$

This is merely an outline of a proof of Itô's Lemma. The interested reader should consult [Neftci (2000), Chapter 10] for a more rigorous proof.

The next two examples will give the reader some idea of the utility of Itô's Lemma.

**Example 9.7.** Suppose $W(t)$ is the standard Brownian motion. Determine the stochastic process which $Y(t) = e^{W(t)}$ obeys.

**Solution.** The random variable $W(t)$ obeys the stochastic differential equation

$$dW(t) = (0) \, dt + (1) \, dW(t).$$

Applying Itô's Lemma with $X(t) = W(t)$, $\mu(X(t), t) = 0$, and $\sigma(X(t), t) = 1$ produces

$$dY(t) = \left[ (0)e^{W(t)} + 0 + \frac{1}{2} e^{W(t)} \right] dt + (1)e^{W(t)} \, dW(t)$$

$$= \frac{1}{2} Y(t) \, dt + Y(t) \, dW(t).$$

**Example 9.8.** The **Ornstein-Uhlenbeck equation** (sometimes called the **Langevin equation**) given below can be thought of as a stochastic extension to the deterministic exponential growth/decay ordinary differential equation.

$$dZ(t) = \mu Z(t)\, dt + \sigma\, dW(t).$$

Solve this equation assuming $Z(0) = Z_0$ and that $\mu$ and $\sigma$ are constants.

**Solution.** The key to solving this equation (and many other similar equations) is to perform the appropriate change of variables and then to use Itô's Lemma. Define $Y(t) = e^{-\mu t} Z(t)$. According to Itô's Lemma, $Y(t)$ has the following differential.

$$dY(t) = \left( \mu Z(t)e^{-\mu t} - \mu e^{-\mu t} Z(t) + \frac{1}{2}\sigma^2(0) \right) dt + \sigma e^{-\mu t}\, dW(t)$$

$$= \sigma e^{-\mu t}\, dW(t).$$

Using the fact that $Y(0) = Z_0$ and integrating both sides of the previous equation produces

$$Y(t) - Z_0 = \int_0^t \sigma e^{-\mu s}\, dW(s)$$

$$Z(t) = Z_0 e^{\mu t} + \int_0^t \sigma e^{\mu(t-s)}\, dW(s).$$

The reader should always keep in mind that the solution to a stochastic differential equation is a random variable. Using Thm. 9.4, $Z(t)$ is normally distributed with

$$\mathbb{E}\left( Z(t) \right) = Z_0 e^{\mu t}$$

$$\mathbb{V}\mathrm{ar}\left( Z(t) \right) = \int_0^t \sigma^2 e^{2\mu(t-s)}\, ds = \frac{\sigma^2}{2\mu}\left( e^{2\mu t} - 1 \right).$$

This chapter opened with a discussion of the distribution of the returns of assets such as stocks. Since $W(t) \sim \mathcal{N}(0,t)$ then then expression in Eq. (9.1) can be written as

$$S(t) = S(0)e^{(\alpha - \delta - \sigma^2/2)t + \sigma W(t)}. \tag{9.38}$$

Applying Itô's formula, the differential $dS(t)$ is found to be

$$dS(t) = \sigma S(t)\, dW(t) + \left( (\alpha - \delta - \frac{\sigma^2}{2})S(t) + \frac{\sigma^2}{2}S(t) \right)dt$$

$$= (\alpha - \delta)S(t)\, dt + \sigma S(t)\, dW(t). \tag{9.39}$$

Note that the coefficient of $dt$ does not contain the expression $-\sigma^2/2$ which would have been present had the chain rule of deterministic calculus been used instead of Itô's Lemma. The drift-diffusion Eq. (9.39) and Itô's Lemma play roles of central importance in the next chapter in which the Black–Scholes partial differential equation is derived.

### *Exercises*

**9.5.1** Evaluate the following stochastic integral and find its mean and variance.

$$\int_0^t (W(s))^2 \, dW(s).$$

*Hint:* $\mathbb{E}\left((W(t))^3 W(s)\right) = 3\,\mathbb{E}\left((W(t))^2\right)\mathbb{E}\left(W(t)W(s)\right).$

**9.5.2** Evaluate the following stochastic integral and find its mean and variance.

$$\int_0^t e^{W(s)-s/2} \, dW(s).$$

**9.5.3** Stochastic integrals have many properties similar to the properties of Riemann integrals. Using Itô's Lemma with $Y(t) = (W(t))^{n+1}$ to establish the reduction of order formula

$$\int_0^t (W(s))^n \, dW(s) = \frac{1}{n+1}(W(t))^{n+1} - \frac{n}{2}\int_0^t (W(s))^{n-1} \, ds.$$

**9.5.4** Assuming that $\mu$ and $P(0)$ are constants, verify by differentiation and substitution that $P(0)e^{\mu t}$ solves Eq. (9.30).

**9.5.5** Using the chain rule for derivatives from ordinary calculus (not the stochastic version governed by Itô's Lemma) verify the expressions found for $f'(0)$ and $f''(0)$ in Eqs. (9.26) and (9.27). Find an expression for the Taylor remainder $R_3$.

**9.5.6** Suppose that $P$ is governed by the stochastic process

$$dP = \mu P \, dt + \sigma P \, dW(t),$$

and that $Y = P^n$. Find the process which governs $Y$.

**9.5.7** Suppose that $P$ is governed by the stochastic process

$$dP = \mu P \, dt + \sigma P \, dW(t),$$

and that $Y = \ln P$. Find the stochastic process which governs $Y$.

**9.5.8** Suppose $W(t)$ is the standard Brownian motion and define $Y(t) = \dfrac{W(t)}{1+t^2}$. Determine the stochastic differential equation solved by $Y(t)$.

**9.5.9** The **mean reverting Ornstein-Uhlenbeck** stochastic differential equation [Øksendal (2003)] can be thought of as a stochastic version of Newton's Law of Cooling. It takes the form

$$dX(t) = k(\mu - X(t))\, dt + \sigma\, dW(t),$$

where $\mu$ represents the constant temperature of the environment and $k > 0$ and $\sigma > 0$ are constants.

(a) If $Z(t) = e^{kt}(\mu - X(t))$, find the stochastic differential equation solved by $Z(t)$.

(b) Solve the stochastic process found in Exercise for $Z(t)$.

(c) Find $\mathbb{E}\left(X(t)\right)$ and $\mathbb{Var}\left(X(t)\right)$.

**9.5.10** Suppose $f(t)$ is a continuous deterministic function and $X(t)$ obeys the stochastic process

$$dX(t) = \frac{1}{2}\left[f(t)\right]^2 dt + f(t)\, dW(t).$$

Define $Z(t) = e^{-X(t)}$ and find the stochastic differential equation solved by $Z(t)$.

**9.5.11** Suppose the stochastic processes $X_1(t)$ and $X_2(t)$ have the following respective stochastic differentials.

$$dX_1(t) = a_1(t)X_1(t)\, dt + b_1(t)X_1(t)\, dW(t),$$
$$dX_2(t) = a_2(t)X_2(t)\, dt + b_2(t)X_2(t)\, dW(t).$$

If $Y(t) = X_1(t)X_2(t)$, use the **product rule** for stochastic processes

$$d(X_1 X_2) = X_2\, dX_1 + X_1\, dX_2 + dX_1\, dX_2,$$

to find the stochastic differential for $Y(t)$.

**9.5.12** An extension of the logistic ordinary differential equation for bounded population growth to a stochastic differential equation has the form

$$dP(t) = rP(K - P)\, dt + \alpha P\, dW(t)$$
$$P(0) = P_0.$$

The constant $K$ is the carrying capacity of the environment. The quantity $rK$ is the *per capita* reproductive rate of the population at low population density. The constant $\alpha$ is related to the size of the perturbation in population size due to random events in the environment.

(a) Define $Y(t) = -1/P(t)$ and show the stochastic differential for $Y(t)$ [Gard (1988)] is

$$dY(t) = [(\alpha^2 - rK)Y(t) - r]\,dt - \alpha Y(t)\,dW(t).$$

(b) Consider the stochastic differential equation

$$dX(t) = (\alpha^2 - rK)X(t)\,dt - \alpha X(t)\,dW(t),$$

with initial condition $X(0) = 1$. Show the solution to this initial value problem is

$$X(t) = e^{(\alpha^2/2 - rK)t - \alpha W(t)}.$$

*Hint:* Let $Z(t) = \ln X(t)$.

(c) Define $Y_0(t) = 1/X(t) = e^{(rK - \alpha^2/2)t + \alpha W(t)}$ and show the stochastic differential for $Y_0$ is

$$dY_0(t) = rKY_0(t)\,dt + \alpha Y_0(t)\,dW(t).$$

(d) Notice that Exercise 9.5.12b solved a stochastic differential equation which is the same is the one in Exercise 9.5.12a save for one term on the right-hand side. Assume that $Y(t) = X(t)Z(t)$ where $Y(t)$ is the solution to Exercise 9.5.12a, $X(t)$ is the solution to Exercise 9.5.12b, and $Z(t)$ is an unknown stochastic process. Use the product rule for differentials of stochastic processes and the result of Exercise 9.5.12c to show that

$$dZ(t) = -rY_0(t)\,dt.$$

(e) Since $Y_0(t)$ has been found, write

$$Z(t) = Z(0) - r \int_0^t Y_0(s)\,ds.$$

Use this to show that the solution to the stochastic logistic equation can be written as

$$P(t) = \frac{P_0 e^{(rK - \alpha^2/2)t + \alpha W(t)}}{1 + rP_0 \int_0^t e^{(rK - \alpha^2/2)s + \alpha W(s)}\,ds}.$$

## 9.6 Maximum and Minimum of a Random Walk

The greatest and least values achieved by a random walk will be important when determining whether a random walk has passed through established barriers. This section will derive the distributions of the maximum and minimum of a random walk and the joint distributions of the walk and its extrema. The reader can safely skip this section until taking up the topic of barrier options in Sec. 14.8.

Consider a Brownian motion process with distribution $W(t) \sim \mathcal{N}(0, \sigma^2 t)$. Recall that $W(0) = 0$ with probability 1. Let $a > 0$ and define $T_a$ to be the first time $W(t) = a$. Formally this should be defined as $T_a = \min\{t : W(t) = a\}$ or $\infty$ if no such time exists. A capital letter $T_a$ is used since the first time the random walk hits level $a$ is a random variable. Since $W(t)$ is continuous, if $b \geq a$ and $\mathbb{P}(W(t) = b) = 1$ then $\mathbb{P}(W(\tau) = a) = 1$ for some $0 < \tau \leq t$. Suppose there is a real number $k$ such that $|W(n)| \leq k$ for all natural numbers $n = 1, 2, \ldots$. By assumption both $|W(n)| \leq k$ and $|W(n+1)| \leq k$ which implies $|W(n+1) - W(n)| \leq 2k$. The increment $W(n+1) - W(n) \sim \mathcal{N}(0, \sigma^2)$ so $\mathbb{P}(|W(n+1) - W(n)| \leq 2k) = p < 1$. By the assumption of independent increments for Brownian motion,

$$\mathbb{P}\left(\lim_{n \to \infty} |W(n)| \leq k\right)$$
$$= \mathbb{P}\left(\lim_{n \to \infty} |W(1) - W(0) + W(2) - W(1) + \cdots + W(n) - W(n-1)| \leq k\right)$$
$$\leq \lim_{n \to \infty} \prod_{i=1}^{n} \mathbb{P}(|W(i) - W(i-1)| \leq 2k) = \lim_{n \to \infty} p^n = 0.$$

Consequently $\mathbb{P}(\lim_{n \to \infty} |W(n)| = \infty) = 1$. As a result for any $a > 0$, with probability 1 there is a time $T_a < \infty$ such that $W(T_a) = a$.

The increment $W(t + T_a) - W(T_a) = W(t + T_a) - a \sim \mathcal{N}(0, \sigma^2 t)$ and is independent of $W(s)$ for $s \leq T_a$. The following theorem describes the distribution of $\overline{M}(t) = \max_{0 \leq \tau \leq t}\{W(\tau)\}$, the running maximum of $W(t)$.

**Theorem 9.5 (Reflection Principle).** *For all $a \geq 0$,*

$$\mathbb{P}\left(\overline{M}(t) \geq a\right) = 2\,\mathbb{P}(W(t) \geq a) = \frac{2}{\sqrt{2\pi\sigma^2 t}} \int_a^\infty e^{-\frac{x^2}{2\sigma^2 t}}\, dx. \tag{9.40}$$

*Proof.* Re-write $\mathbb{P}(W(t) \geq a)$ as

$$\mathbb{P}(W(t) \geq a) = \mathbb{P}\left((W(t) \geq a) \cap (\overline{M}(t) \leq a)\right)$$
$$+ \mathbb{P}\left((W(t) \geq a) \cap (\overline{M}(t) < a)\right)$$
$$= \mathbb{P}\left((W(t) \geq a) \cap (\overline{M}(t) \leq a)\right).$$

Using the conditional probability,

$$\begin{aligned}
\mathbb{P}\left(W(t) \geq a\right) &= \mathbb{P}\left(W(t) \geq a \,|\, \overline{M}(t) \geq a\right) \mathbb{P}\left(\overline{M}(t) \geq a\right) \\
&= \mathbb{P}\left(W(t) \geq a \,|\, T_a \leq t\right) \mathbb{P}\left(\overline{M}(t) \geq a\right) \\
&= \mathbb{P}\left(W((t - T_a) + T_a) - a \geq 0 \,|\, T_a \leq t\right) \mathbb{P}\left(\overline{M}(t) \geq a\right) \\
&= \frac{1}{2} \mathbb{P}\left(\overline{M}(t) \geq a\right).
\end{aligned}$$

Note the first probability equals $1/2$ since $W((t - T_a) + T_a) - a \sim \mathcal{N}(0, \sigma^2 t)$. This establishes Eq. (9.40). □

Since the event $\{\overline{M}(t) \geq a\}$ is identical to the event $\{T_a \leq t\}$ then the following corollary to the Reflection Principle holds.

**Corollary 9.2.** *For all* $a \geq 0$, $\mathbb{P}\left(T_a \leq t\right) = \mathbb{P}\left(\overline{M}(t) \geq a\right)$.

Suppose a modification is made to the path followed by $W(t)$. For $t > T_a$ replace $W(t)$ by $2a - W(t)$, the reflection of $W(t)$ across $a$. The relationship between the original and reflected paths is depicted in Fig. 9.5. Now let $0 \leq b < a$ and consider the joint distribution of $W(t)$ and $\overline{M}(t)$. If $T_a \leq t$ then by definition $W$ has reached $a > 0$ by time $t$. With equal probability

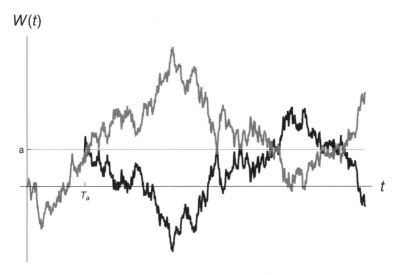

Fig. 9.5  At the earliest time $W(t)$ reaches $a \geq 0$ reflect the path followed by the Brownian motion across the horizontal line through $a$. Since the Brownian motion has no drift the lighter and darker paths are equally likely.

$W(t) \leq b$ or $W(t) \geq 2a - b$. Stated more formally as the reflection principle,

$$\mathbb{P}\left(W(t) \leq b \,|\, T_a \leq t\right) = \mathbb{P}\left(W(t) \geq 2a - b \,|\, T_a \leq t\right).$$

Multiplying both sides of this equation by $\mathbb{P}\left(T_a \leq t\right)$ yields,

$$\mathbb{P}\left((W(t) \leq b) \cap (T_a \leq t)\right) = \mathbb{P}\left((W(t) \geq 2a - b) \cap (T_a \leq t)\right).$$

Finally since the events $\{\overline{M}(t) \geq a\}$ and $\{T_a \leq t\}$ are identical,

$$\mathbb{P}\left((W(t) \leq b) \cap (\overline{M}(t) \geq a)\right) = \mathbb{P}\left((W(t) \geq 2a - b) \cap (\overline{M}(t) \geq a)\right).$$

Since $2a - b \geq a$ then

$$\mathbb{P}\left((W(t) \geq 2a - b) \cap (\overline{M}(t) \geq a)\right) = \mathbb{P}\left(W(t) \geq 2a - b\right)$$
$$= \frac{1}{\sqrt{2\pi\sigma^2 t}} \int_{2a-b}^{\infty} e^{-\frac{x^2}{2\sigma^2 t}} \, dx.$$

Hence the following theorem has been justified.

**Theorem 9.6.** *For all $a > b \geq 0$,*

$$\mathbb{P}\left((\overline{M}(t) > a) \cap (W(t) \leq b)\right) = \frac{1}{\sqrt{2\pi\sigma^2 t}} \int_{2a-b}^{\infty} e^{-\frac{x^2}{2\sigma^2 t}} \, dx. \qquad (9.41)$$

Let $f_{W(t),\overline{M}(t)}(w, m)$ be the joint probability distribution of $W(t)$, and $\overline{M}(t)$. By definition then

$$f_{W(t),\overline{M}(t)}(w, m) = \frac{\partial^2}{\partial w \partial m} \left[\mathbb{P}\left((W(t) \leq w) \cap (\overline{M}(t) \leq m)\right)\right]$$
$$= \frac{\partial^2}{\partial w \partial m} \left[\mathbb{P}\left(W(t) \leq w\right) - \mathbb{P}\left((W(t) \leq w) \cap (\overline{M}(t) > m)\right)\right]$$
$$= -\frac{\partial^2}{\partial w \partial m} \left[\mathbb{P}\left((W(t) \leq w) \cap (\overline{M}(t) > m)\right)\right]$$
$$= \frac{2(2m - w)}{\sigma^2 t \sqrt{2\pi\sigma^2 t}} e^{-\frac{(2m-w)^2}{2\sigma^2 t}},$$

for $m > 0$, $w \leq m$ and zero otherwise.

A Brownian motion process with drift $\mu$ is distributed as $\mathcal{N}(\mu t, \sigma^2 t)$. However, when $\mu \neq 0$ the reflection principle, key to the arguments used in the proofs above, no longer holds. The probability distributions can still be determined but require more sophisticated probability theoretic arguments which lie beyond the scope of this textbook. An *ad hoc* method of modifying the distribution is simply to shift the mean of the distribution of the Brownian motion from 0 to $\mu t$ while keeping the variance at $\sigma^2 t$. Multiplying the joint distribution of the un-drifted Brownian motion and its running maximum by $e^{\frac{\mu w}{\sigma^2} - \frac{\mu^2 t}{2\sigma^2}}$ will accomplish this. See Exercise 9.6.4 for

more details. Thus the following joint probability density function describes
a drifted Brownian motion and its running maximum.

$$\hat{f}_{W(t),\overline{M}(t)}(w,m) = f_{W(t),\overline{M}(t)}(w,m)e^{\frac{\mu w}{\sigma^2}-\frac{\mu^2 t}{2\sigma^2}}$$

$$= \frac{2(2m-w)}{\sigma^2 t\sqrt{2\pi\sigma^2 t}}e^{\frac{-(2m-w)^2+2\mu wt-\mu^2 t^2}{2\sigma^2 t}}, \qquad (9.42)$$

for $m > 0$, $w \le m$. From this joint distribution the probability distribution
of $\overline{M}(t)$ of a Brownian motion process with drift can be calculated as a
marginal distribution. The reader will be asked to provide the details in
Exercise 9.6.5. When $m > 0$,

$$\hat{f}_{\overline{M}(t)}(m) = 2e^{\frac{2\mu m}{\sigma^2}}\left[\frac{1}{\sqrt{2\pi\sigma^2 t}}e^{-\frac{(m+\mu t)^2}{2\sigma^2 t}} - \frac{\mu}{\sigma^2}\Phi\left(\frac{-m-\mu t}{\sigma\sqrt{t}}\right)\right] \qquad (9.43)$$

$$\hat{F}_{\overline{M}(t)}(m) = \Phi\left(\frac{m-\mu t}{\sigma\sqrt{t}}\right) - e^{\frac{2\mu m}{\sigma^2}}\Phi\left(\frac{-m-\mu t}{\sigma\sqrt{t}}\right). \qquad (9.44)$$

In a similar fashion, the running minimum of the drifted Brownian
motion is defined as $\underline{M}(t) = \min_{0\le\tau\le t}\{W(\tau)\}$. The joint distribution of
the drifted Brownian motion and its running minimum for $m < 0$ and
$w \ge m$ is

$$\hat{f}_{W(t),\underline{M}(t)}(w,m) = \frac{2(w-2m)}{\sigma^2 t\sqrt{2\pi\sigma^2 t}}e^{\frac{-(w-2m)^2+2\mu wt-\mu^2 t^2}{2\sigma^2 t}}. \qquad (9.45)$$

The probability density function and cumulative density function for the
running minimum of a drifted random walk are respectively,

$$\hat{f}_{\underline{M}(t)}(m) = 2e^{\frac{2\mu m}{\sigma^2}}\left[\frac{1}{\sqrt{2\pi\sigma^2 t}}e^{-\frac{(m+\mu t)^2}{2\sigma^2 t}} + \frac{\mu}{\sigma^2}\Phi\left(\frac{m+\mu t}{\sigma\sqrt{t}}\right)\right] \qquad (9.46)$$

$$\hat{F}_{\underline{M}(t)}(m) = \Phi\left(\frac{m-\mu t}{\sigma\sqrt{t}}\right) + e^{\frac{2\mu m}{\sigma^2}}\Phi\left(\frac{m+\mu t}{\sigma\sqrt{t}}\right), \qquad (9.47)$$

when $m < 0$ and zero otherwise.

### Exercises

**9.6.1** Find the probability density function for the maximum $\overline{M}(t)$ of
Brownian motion without drift.

**9.6.2** For a Brownian motion process without drift, find a formula for the
cumulative probability distribution $\mathbb{P}(T_a \le t)$ and the corresponding probability density function.

**9.6.3** Suppose $\mu < 0$ and show that as $t \to \infty$ the running maximum of
drifted Brownian motion is exponentially distributed with $\lambda = 2|\mu|/\sigma^2$.
(See Exercise 8.1.8.)

**9.6.4** Let $f_X(x) = \frac{1}{\sqrt{2\pi\sigma^2}} e^{-(x-\mu)^2/(2\sigma^2)}$ be the probability density function for a normally distributed random variable with mean $\mu$ and variance $\sigma^2$. Show that $f_X(y) e^{\frac{(\hat{\mu}-\mu)y}{\sigma^2} - \frac{\hat{\mu}^2-\mu^2}{2\sigma^2}}$ is the probability density function for a normally distributed random variable with mean $\hat{\mu}$ and variance $\sigma^2$.

**9.6.5** Starting from Eq. (9.42) find the probability density function and cumulative probability density function for the running maximum $\overline{M}(t)$.

**9.6.6** Let $W(t)$ be Brownian motion with drift and let $\overline{M}(t)$ be its running maximum. Show that for $a > 0$ and $b \le a$

$$\mathbb{P}\left((\overline{M}(t) > a) \cap (W(t) \le b)\right) = e^{\frac{2\mu a}{\sigma^2}} \Phi\left(\frac{b - 2a - \mu t}{\sigma\sqrt{t}}\right).$$

**9.6.7** Let $W(t)$ be Brownian motion with drift and let $\overline{M}(t)$ be its running maximum. Show that for $a > 0$ and $b \le a$

$$\mathbb{P}\left((\overline{M}(t) \le a) \cap (W(t) \le b)\right) = \Phi\left(\frac{b - \mu t}{\sigma\sqrt{t}}\right) - e^{\frac{2\mu a}{\sigma^2}} \Phi\left(\frac{b - 2a - \mu t}{\sigma\sqrt{t}}\right),$$

and is zero otherwise.

**9.6.8** Let $W(t)$ be Brownian motion with drift and let $\underline{M}(t)$ be its running minimum. Find an expression for $\mathbb{P}\left(\underline{M}(t) \le a, W(t) \le b\right)$ for $a < 0$ and $b \ge a$. Show that for $a < 0$ and $b \ge a$

$$\mathbb{P}\left((\underline{M}(t) \le a) \cap (W(t) \le b)\right)$$
$$= e^{\frac{2\mu a}{\sigma^2}} \left( \Phi\left(\frac{b - 2a - \mu t}{\sigma\sqrt{t}}\right) - \Phi\left(\frac{-a - \mu t}{\sigma\sqrt{t}}\right)\right) + \Phi\left(\frac{a - \mu t}{\sigma\sqrt{t}}\right),$$

and is zero otherwise.

**9.6.9** Let $W(t)$ be Brownian motion with drift and let $\overline{M}(t)$ be its running maximum. Show that for $a > 0$ and $b \le a$

$$\mathbb{P}\left((\overline{M}(t) \le a) \cap (W(t) > b)\right) = \Phi\left(\frac{a - \mu t}{\sigma\sqrt{t}}\right) - \Phi\left(\frac{b - \mu t}{\sigma\sqrt{t}}\right)$$
$$+ e^{\frac{2\mu a}{\sigma^2}} \left( \Phi\left(\frac{b - 2a - \mu t}{\sigma\sqrt{t}}\right) - \Phi\left(\frac{-a - \mu t}{\sigma\sqrt{t}}\right)\right),$$

and is zero otherwise.

**9.6.10** Let $W(t)$ be Brownian motion with drift and let $\underline{M}(t)$ be its running minimum. Show that for $a < 0$ and $b \ge a$

$$\mathbb{P}\left((\underline{M}(t) \le a) \cap (W(t) > b)\right) = e^{\frac{2\mu a}{\sigma^2}} \Phi\left(\frac{2a - b + \mu t}{\sigma\sqrt{t}}\right),$$

and is zero otherwise.

Chapter 10

# Black–Scholes Equation and Option Formulas

This chapter derives the Black–Scholes partial differential equation, composes an initial, boundary-value problem for European-style options, and uses the fundamental solution to the advection/diffusion equation to find a closed-form pricing formula for European call and put options. Many European-style options can be handled in this framework. The differences in the European option pricing formulas are due to different boundary and payoff conditions. Even though the discussion revolves around partial differential equations, the mathematics is kept at the multivariable calculus level. The reader need not have studied PDEs extensively to follow the derivation presented in this chapter. Several approaches can be taken to derive the European option pricing formulas. Some textbooks use the continuous limit of a discrete time binomial model of option prices (see [Ross (1999)]). Some use a risk-neutral expectation argument. This text emphasizes the mathematical modeling approach of [Wilmott *et al.* (1995)]. The Black–Scholes PDE is transformed through a sequence of changes of variables to the diffusion equation which is solved by the technique outlined Sec. 9.2, particularly Eq. (9.11).

If the boundary and final conditions of the Black–Scholes PDE are changed from those of the European calls and puts (commonly called the vanilla options) to more esoteric payoff conditions (the exotic options), more sophisticated solution techniques or even numerical approximations of the solution may be required. Some types of exotic options will be priced in Chapter 14.

## 10.1  Black–Scholes Partial Differential Equation

This section will derive the fundamental equation governing the pricing of options, the famous **Black–Scholes partial differential equation**. The ideas of arbitrage, stochastic processes, and present value converge at this point in the study of financial mathematics.

Suppose $S(t)$ is the time $t$ value of a security such as a stock and that it obeys a stochastic process of the form

$$dS(t) = (\alpha - \delta)S(t)\,dt + \sigma S(t)\,dW(t). \tag{10.1}$$

The constants $\alpha$, $\delta$, and $\sigma$ are respectively the expected growth rate, continuous dividend yield, and volatility of the stock. $W(t)$ is the standard Brownian motion. If $F(S,t)$ is the value of an option (more generally called a financial derivative), then according to Itô's Lemma (Lemma 9.1), $F$ obeys the following stochastic process.

$$dF = \left( (\alpha - \delta)SF_S + \frac{1}{2}\sigma^2 S^2 F_{SS} + F_t \right) dt + \sigma S F_S \,dW(t). \tag{10.2}$$

Suppose a portfolio $P$ is created by selling the option, buying $\Delta$ units of the security, and lending or borrowing $B$ at the continuously compounded, risk-free interest rate $r$ so that the net cost of the portfolio is $P = 0$. Assume shares are infinitely divisible and an investor can purchase or sell any necessary fraction of a share. The value of the portfolio can be written as $P = F - \Delta S + B$. The notation $\Delta$ for the number of shares of the stock purchased is standard in the derivation and analysis of the Black–Scholes equation. The reader should not let the notation $\Delta S$ suggest "change in $S$", it merely means $\Delta$ multiplied by the share price $S$.

Since the portfolio is a combination of the option, a bond, and the stock then the stochastic process governing the portfolio is

$$dP = d(F - \Delta S + B) = dF - \Delta(dS + \delta S\,dt) + rB\,dt.$$

The price of the stock obeys the stochastic process $dS$ given in Eq. (10.1) and also pays dividends in the amount of $\delta S\,dt$ in the infinitesimal time interval $dt$. The bond pays/earns interest in the amount of $rB\,dt$ during that same time interval.

$$dP = F_S\,dS + \left( F_t + \frac{1}{2}\sigma^2 S^2 F_{SS} \right) dt - \Delta(dS + \delta S\,dt) + rB\,dt$$

$$= (F_S - \Delta)\,dS + \left( F_t + \frac{1}{2}\sigma^2 S^2 F_{SS} - \delta \Delta S + rB \right) dt. \tag{10.3}$$

Notice the coefficient of the differential $dS$ in Eq. (10.3), is $F_S - \Delta$, thus Eq. (10.3) can be simplified if $\Delta = F_S$. This implies the amount of the bond is $B = SF_S - F$. The differential of the portfolio becomes,

$$dP = \left( F_t + \frac{1}{2}\sigma^2 S^2 F_{SS} + (r - \delta)SF_S - rF \right)dt.$$

By the choice of $B$ the portfolio has an initial cash flow of zero. The choice of $\Delta = F_S$ removes the randomness from the portfolio value. In the absence of arbitrage, this non-random, zero-cost portfolio must have zero rate of return. Therefore,

$$F_t + (r - \delta)SF_S + \frac{1}{2}\sigma^2 S^2 F_{SS} - r\,F = 0 \qquad (10.4)$$

which is the well-known Black–Scholes equation for pricing financial derivatives. Equation (10.4) is an example of a **partial differential equation or PDE**, for short. While the general theory of solving PDEs is beyond the scope of this book, brief remarks concerning a few concepts related to the Black–Scholes PDE are appropriate. More extensive background information can be found in [Buchanan and Shao (2018)]. PDEs are often described by their order, type, and linearity properties. The Black–Scholes PDE is a **second order** equation since the highest order derivative of the unknown function $F$ present in the equation is the second derivative. This PDE is of **parabolic type**. The best known example of the parabolic PDE is the **diffusion equation** which can be used for example to describe the distribution of temperature along a one-dimensional object. The diffusion equation was encountered in Eq. (9.10). Since the coefficients of $F_t$ and $F_{SS}$ have the same algebraic signs, the Black–Scholes PDE is sometimes referred to as a **backwards** parabolic equation. In [Wilmott *et al.* (1995)] the Black–Scholes PDE is solved by appropriate changes of variables until it becomes the diffusion equation. The diffusion equation can be solved by several independent techniques which all naturally, ultimately yield the same solution. Lastly, the Black–Scholes PDE is an example of a **linear** partial differential equation since if $F_1$ and $F_2$ are two solutions to the equation then $c_1 F_1 + c_2 F_2$ is also a solution where $c_1$ and $c_2$ are any constants, see Exercise 10.1.1.

Financial derivative products of many types obey the Black–Scholes equation. Different solutions correspond to different initial/final and boundary side conditions imposed while solving the equation. The next section will describe in more detail the initial and boundary conditions relevant to the Black–Scholes PDE. Section 10.4 will derive a solution to the Black–Scholes equation for the price of a European style call option.

## Exercises

**10.1.1** Suppose $F_1$ and $F_2$ are two solutions to Eq. (10.4) and $c_1$ and $c_2$ are any constants. Show that $c_1 F_1 + c_2 F_2$ is also a solution to Eq. (10.4).

**10.1.2** Show that the prepaid forward delivering a stock at time $T$ which pays a continuous dividend at rate $\delta$ is a solution to the Black–Scholes partial differential equation.

**10.1.3** Show that a bond whose present value is $Be^{-r(T-t)}$ where $T$ is the maturity date of the bond satisfies the Black–Scholes partial differential equation.

**10.1.4** Find a value for the exponent $a$ such that $F(S,t) = S^a$ solves the Black–Scholes partial differential equation.

**10.1.5** Find a value for the exponent $a$ such that $F(S,t) = e^t S^a$ solves the Black–Scholes partial differential equation.

**10.1.6** A type of option on a stock has a value at time $t$ of

$$F(S(t),t) = e^{-0.07(2-t)}\left(0.03 - 0.015t + \ln S(t)\right),$$

for $0 \le t \le 2$. The underlying stock has a volatility of 25% and pays a continuous dividend with yield $\delta$. The continuously compounded, risk-free interest rate is 7%. Find the dividend yield of the stock. *Hint*: Use Eq. (10.4).

## 10.2   Boundary and Initial Conditions

Without the imposition of side conditions in the form of initial or final conditions and boundary conditions, a partial differential equation can have many different solutions. The exercises from Sec. 10.1 provided a few examples of the variety of solutions to the Black–Scholes partial differential equation. Mathematicians refer to a problem with many possible solutions as "ill-posed". This section will discuss the appropriate extra conditions to impose in order to obtain a solution to Eq. (10.4) which describes the value of a European call option. The reader will be asked to develop conditions to describe other types of option solutions to Eq. (10.4) in the exercises.

The domain of the unknown function $F$ present in Eq. (10.4) is a region in $(S,t)$-space. Refer to this region as $\Omega \subset (S,t)$-space. A European call option has a value during the time interval $[0,T]$, where $T$ is the exercise or strike time of the option. During this time interval the price of the security underlying the option may be any non-negative value. Thus when modeling the price of a European call option on an underlying stock, the domain of the solution to Eq. (10.4) is

$$\Omega = \{(S,t) \, | \, 0 \le S < \infty \text{ and } 0 \le t \le T\}.$$

At time $t = T$ the value of the security will either exceed the strike price (in which case the call option is in-the-money and will be exercised generating an income flow of $S(T) - K > 0$) or the security will have a value less than or equal to the strike price (in which case the call option is out-of-the-money and expires unused with value 0). Thus the terminal value of a European call option is $(S(T) - K)^+$, where $S(T)$ is the value of the underlying security at the strike time and $K$ is the strike price of the option. Graphically the payoff of the portfolio is said to resemble a "hockey stick". See Fig. 6.1 in Chapter 6. Thus if $F$ represents a European call option, then $F(S, T) = (S(T) - K)^+$ is the **final condition** for the Black–Scholes PDE.

According to Eq. (10.1) if $S(t) = 0$, then $dS(t) = 0$, *i.e.*, $S(t)$ never changes and hence remains zero. The boundary at $S(t) = 0$ is said to be invariant. Thus on the portion of the boundary of $\Omega$ where $S(t) = 0$, the call option would never be exercised and hence must have zero value. Thus one boundary condition is $F(0, t) = 0$. Now suppose that $S(t)$ is approaching infinity. It becomes increasingly likely that the call option will be exercised, since as $S(t) \to \infty$, $S(t)$ will exceed any finite value of $K$. Likewise as $S(t) \to \infty$ a put option would never be exercised. Thus according to the Put-Call Parity formula in Eq. (6.8), as $S \to \infty$, $C \to Se^{-\delta(T-t)} - Ke^{-r(T-t)}$. Hence the second boundary condition, sometimes called the boundary condition at infinity, is $F(S, t) = Se^{-\delta(T-t)} - Ke^{-r(T-t)}$ as $S \to \infty$. In summary the Black–Scholes equation and its final and boundary conditions for a European call option consist of the following set of equations.

$$rF = F_t + (r - \delta)SF_S + \frac{1}{2}\sigma^2 S^2 F_{SS} \text{ for } (S,t) \text{ in } \Omega, \qquad (10.5)$$

$$F(S,T) = (S(T) - K)^+ \text{ for } S > 0, \qquad (10.6)$$

$$F(0,t) = 0 \text{ for } 0 \le t < T, \qquad (10.7)$$

$$F(S,t) = Se^{-\delta(T-t)} - Ke^{-r(T-t)} \text{ as } S \to \infty. \qquad (10.8)$$

In the next section, a method for determining the solution to this initial, boundary-value problem will be described.

## Exercises

**10.2.1** Show that the initial boundary-value problem for the price of a European put option with strike price $K$ and strike time $T$ on a stock is as follows.

$$rF = F_t + (r - \delta)SF_S + \frac{1}{2}\sigma^2 S^2 F_{SS} \text{ for } (S,t) \text{ in } \Omega,$$

$$F(S,T) = (K - S(T))^+ \text{ for } S > 0,$$

$$F(0,t) = Ke^{-r(T-t)} \text{ for } 0 \le t < T,$$

$$F(S,t) = 0 \text{ as } S \to \infty.$$

**10.2.2** Suppose a European-style option will pay the option holder \$1 if the price of the underlying stock exceeds $K > 0$ at time $T$ and pays nothing otherwise. State the initial, boundary-value problem for this option known as a **cash-or-nothing call option**.

**10.2.3** Suppose a European-style option will pay the option holder \$1 if the price of the underlying stock does not exceed $K > 0$ at time $T$ and pays nothing otherwise. State the initial, boundary-value problem for this option known as a **cash-or-nothing put option**.

**10.2.4** Find a Put-Call Parity formula for cash-or-nothing call and put options. *Hint*: What is the payoff to an investor who holds a long position in both a cash-or-nothing put and a cash-or-nothing call?

**10.2.5** Suppose a European-style option will pay the option holder $S(T)$ if the price of the underlying stock exceeds $K > 0$ at time $T$ and pays nothing otherwise. State the initial, boundary-value problem for this option known as an **asset-or-nothing call option**.

**10.2.6** Show that a European call option on an underlying stock is a linear combination of a cash-or-nothing call option and an asset-or-nothing call option.

**10.2.7** Suppose a European-style option will pay the option holder $S(T)$ if the price of the underlying stock does not exceed $K > 0$ at time $T$ and pays nothing otherwise. State the initial, boundary-value problem for this option known as an **asset-or-nothing put option**.

**10.2.8** Find a Put-Call Parity formula for asset-or-nothing call and put options. *Hint*: What is the payoff to an investor who holds a long position in both an asset-or-nothing put and an asset-or-nothing call?

## 10.3   Changing Variables in the Black–Scholes PDE

The condition given in Eq. (10.6) states that when the expiry date for the option arrives, the call option will be worth nothing if the value of the stock is less than the strike price. Otherwise the call option is worth the excess of the stock's value over the strike price. The boundary conditions are specified in Eqs. (10.7) and (10.8). The first condition implies that if the stock itself becomes worthless before maturity, the call option is also worthless. An investor could buy the stock for nothing and then let the call option expire unused. The second part of this condition follows from the European call option property discussed in Chapter 6, namely $C^e \geq Se^{-\delta(T-t)} - Ke^{-r(T-t)}$. By the Put-Call Parity formula in Eq. (6.8), as $S \to \infty$ a put option becomes worthless and the value of a call option approaches $Se^{-\delta(T-t)} - Ke^{-r(T-t)}$ asymptotically.

In this section a change of variables will be employed to convert the Black–Scholes equation and its side conditions into an equivalent form in which the partial differential equation is the diffusion equation Eq. (9.10). In the next section, the diffusion equation will be solved and the change of variables reversed. Suppose $F$, $S$, and $t$ are defined in terms of the new variables $v$, $x$, and $\tau$ as in the following equations.

$$S = Ke^x \iff x = \ln \frac{S}{K}, \tag{10.9}$$

$$t = T - \frac{2\tau}{\sigma^2} \iff \tau = \frac{\sigma^2}{2}(T - t), \tag{10.10}$$

$$F(S,t) = Ke^{-2\delta\tau/\sigma^2} v(x,\tau) \iff v(x,\tau) = e^{\delta(T-t)}\frac{F(S,t)}{K}. \tag{10.11}$$

How are the Black–Scholes equation and its side conditions altered by this change of variables? The multivariable form of the chain rule can be used to determine new expressions for the derivatives present in Eq. (10.5).

$$F_S = \frac{\partial}{\partial S}\left[Ke^{-2\delta\tau/\sigma^2} v(x,\tau)\right] = Ke^{-2\delta\tau/\sigma^2}(v_x x_S + v_\tau \tau_S)$$

$$= Ke^{-2\delta\tau/\sigma^2}\left(v_x \frac{1}{S} + 0\right) = e^{-2\delta\tau/\sigma^2 - x}v_x \tag{10.12}$$

The reader will be asked to verify that

$$F_t = \frac{K\sigma^2}{2}\left(\frac{2\delta}{\sigma^2} - v_\tau\right) \quad \text{and} \tag{10.13}$$

$$F_{SS} = \frac{1}{K}e^{-2\delta\tau/\sigma^2 - 2x}(v_{xx} - v_x), \tag{10.14}$$

in Exercises 10.3.1 and 10.3.2. Substituting the results in Eqs. (10.11)–(10.14) in the Black–Scholes Eq. (10.5) and simplifying produces the equation

$$v_\tau = v_{xx} + (k - 1)v_x - kv, \tag{10.15}$$

where $k = 2(r - \delta)/\sigma^2$. See Exercise 10.3.3.

The final condition for $F$ is converted by this change of variables into an initial condition since when $t = T$, $\tau = 0$. The initial condition $v(x,0)$ is then found to be

$$Kv(x,0) = F(S,T) = (S-K)^+ = K(e^x - 1)^+$$
$$v(x,0) = (e^x - 1)^+. \tag{10.16}$$

Since $\lim_{S \to 0^+} x = -\infty$, then

$$0 = \lim_{S \to 0^+} F(S,t) = \lim_{x \to -\infty} Kv(x,\tau) \implies \lim_{x \to -\infty} v(x,\tau) = 0.$$

Likewise since as $\lim_{S \to \infty} x = \infty$,

$$\lim_{S \to \infty} F(S,t) = S - Ke^{-r(T-t)} = \lim_{x \to \infty} Kv(x,\tau)$$
$$e^x - e^{-k\tau} = \lim_{x \to \infty} v(x,\tau).$$

These equations constitute a pair of boundary conditions for the partial differential equation for $v(x,\tau)$ in Eq. (10.15). So the original Black–Scholes initial, boundary value problem can be recast in the form of the following.

$$v_\tau = v_{xx} + (k-1)v_x - kv \text{ for } x \in (-\infty, \infty), \tau \in (0, \tfrac{T\sigma^2}{2}) \tag{10.17}$$
$$v(x,0) = (e^x - 1)^+ \text{ for } x \in (-\infty, \infty) \tag{10.18}$$
$$v(x,\tau) \to 0 \text{ as } x \to -\infty \text{ and} \tag{10.19}$$
$$v(x,\tau) \to e^x - e^{-k\tau} \text{ as } x \to \infty, \tau \in (0, \tfrac{T\sigma^2}{2}). \tag{10.20}$$

The independent variable $x$ can be thought of as corresponding to a spatial variable. The reader should note that where previously the price of the security $S$ was assumed to take on only non-negative values, now $x$ can be any real number.

If the last two terms on the right-hand side of Eq. (10.17) were absent, this would be the diffusion equation. In order to eliminate these terms another change of variables is needed. Let $\alpha$ and $\beta$ be constants and introduce the new dependent variable $u$.

$$v(x,\tau) = e^{\alpha x + \beta \tau} u(x,\tau) \tag{10.21}$$
$$v_x = e^{\alpha x + \beta \tau}(\alpha u(x,\tau) + u_x) \tag{10.22}$$
$$v_{xx} = e^{\alpha x + \beta \tau}(\alpha^2 u(x,\tau) + 2\alpha u_x + u_{xx}) \tag{10.23}$$
$$v_\tau = e^{\alpha x + \beta \tau}(\beta u(x,\tau) + u_\tau). \tag{10.24}$$

Substituting the expressions found in Eqs. (10.21)–(10.24) into Eq. (10.17) produces

$$u_\tau = (\alpha^2 + (k-1)\alpha - k - \beta)u + (2\alpha + k - 1)u_x + u_{xx}. \tag{10.25}$$

Since $\alpha$ and $\beta$ are arbitrary constants they can now be chosen appropriately to simplify Eq. (10.25). Ideally the coefficients of $u_x$ and $u$ would be zero. Solving the two equations:

$$0 = \alpha^2 + (k-1)\alpha - k - \beta$$
$$0 = 2\alpha + k - 1,$$

yields $\alpha = (1-k)/2$ and $\beta = -(k+1)^2/4$. The initial condition for $u$ can be derived from the initial condition given in Eq. (10.18).

$$v(x,0) = (e^x - 1)^+$$
$$u(x,0) = e^{(k-1)x/2}(e^x - 1)^+ = (e^{(k+1)x/2} - e^{(k-1)x/2})^+.$$

Likewise boundary conditions at $x = \pm\infty$ for $u$ can be derived from Eqs. (10.19) and (10.20). To summarize, the original Black–Scholes partial differential equation, initial, and boundary conditions have been converted to the following system of equations.

$$u_\tau = u_{xx} \text{ for } x \in (-\infty, \infty) \text{ and } \tau \in (0, T\sigma^2/2) \tag{10.26}$$

$$u(x,0) = (e^{(k+1)x/2} - e^{(k-1)x/2})^+ \text{ for } x \in (-\infty, \infty) \tag{10.27}$$

$$u(x,\tau) \to 0 \text{ as } x \to -\infty \text{ for } \tau \in (0, T\sigma^2/2) \tag{10.28}$$

$$u(x,\tau) \to e^{\frac{(k+1)}{2}(x+(k+1)\tau/2)} - e^{\frac{(k-1)}{2}(x+(k-1)\tau/2)}, \tag{10.29}$$

as $x \to \infty$ for $\tau \in (0, T\sigma^2/2)$. The PDE for $u(x,\tau)$ in Eq. (10.26) is the diffusion equation which was solved back in Sec. 9.2.

## Exercises

**10.3.1** Using the new variables described in Eqs. (10.9), (10.10), (10.11), and the multivariable chain rule, verify the identity shown in Eq. (10.13).

**10.3.2** Using the new variables described in Eqs. (10.9), (10.10), (10.11), the expression for $F_S$ shown in Eq. (10.12), and the multivariable chain rule, verify the identity shown in Eq. (10.14).

**10.3.3** Verify that the Black–Scholes partial differential Eq. (10.5) can be simplified to the form shown in Eq. (10.15) by using the variables described in Eqs. (10.9), (10.10), and (10.11) and the partial derivatives found in Eqs. (10.12), (10.13), and (10.14).

**10.3.4** Fill in some of the details in the derivation of Eqs. (10.26)–(10.29) by showing that

$$u(x,\tau) \to e^{\frac{(k+1)}{2}[x+(k+1)\tau/2]} - e^{\frac{(k-1)}{2}[x+(k-1)\tau/2]},$$

as $x \to \infty$.

**10.3.5** A judicious change of variable often simplifies an equation. Show that the ordinary differential equation

$$t^2 \frac{d^2y}{dt^2} + \alpha t \frac{dy}{dt} + \beta y(t), = 0,$$

simplifies to

$$\frac{d^2y}{dx^2} + (\alpha - 1)\frac{dy}{dx} + \beta y(x) = 0,$$

when $t = e^x$.

## 10.4   Solving the Black–Scholes Equation

The piecewise-defined solution to the advection/diffusion equation was outlined in Sec. 9.2. The fundamental solution to the diffusion equation takes on the form

$$U(x,\tau) = \frac{1}{\sqrt{4\pi\tau}} e^{-x^2/(4\tau)}.$$

Since the diffusion constant in Eq. (10.26) is one. Applying the solution technique to the transformed Black-Schole initial, boundary-value problem in Eq. (10.26) requires evaluating the following integral.

$$u(x,\tau) = \int_{-\infty}^{\infty} \frac{1}{\sqrt{4\pi\tau}} e^{-(x-y)^2/(4\tau)} \left( e^{(k+1)y/2} - e^{(k-1)y/2} \right)^+ dy$$

$$= e^{(k+1)x/2+(k+1)^2\tau/4} \int_{0}^{\infty} \frac{1}{\sqrt{4\pi\tau}} e^{-(y-x-(k+1)\tau)^2/(4\tau)} dy$$

$$- e^{(k-1)x/2+(k-1)^2\tau/4} \int_{0}^{\infty} \frac{1}{\sqrt{4\pi\tau}} e^{-(y-x-(k-1)\tau)^2/(4\tau)} dy.$$

Using integration by substitution on each of the remaining integrals produces

$$u(x,\tau) = e^{(k+1)x/2+(k+1)^2\tau/4} \Phi\left( \frac{x+(k+1)\tau}{\sqrt{2\tau}} \right)$$

$$- e^{(k-1)x/2+(k-1)^2\tau/4} \Phi\left( \frac{x+(k-1)\tau}{\sqrt{2\tau}} \right). \tag{10.30}$$

This is the solution to the initial, boundary-value problem in Eqs. (10.26)–(10.29). The reader should take a moment to check that the expression for $u(x,\tau)$ given in Eq. (10.30) satisfies the boundary conditions as $x \to \pm\infty$ specified in Eqs. (10.28) and (10.29).

Now begins the task of re-writing this solution in terms of the variables $S$ and $t$ used in the Black–Scholes initial, boundary-value problem stated in

Eqs. (10.5)–(10.8). Using the change of variables in Eq. (10.21) this solution can be re-written in terms of the function $v(x, \tau)$, where

$$v(x, \tau) = e^{-(k-1)x/2 - (k+1)^2 \tau/4} u(x, \tau)$$

$$= e^x \Phi \left( \frac{x + (k+1)\tau}{\sqrt{2\tau}} \right) - e^{-k\tau} \Phi \left( \frac{x + (k-1)\tau}{\sqrt{2\tau}} \right). \qquad (10.31)$$

Now using the change of variables described in Eqs. (10.9) and (10.10) the reader can show in Exercise 10.4.5 that

$$d_1 = \frac{x + (k+1)\tau}{\sqrt{2\tau}} = \frac{\ln \frac{S}{K} + \left( r - \delta + \frac{\sigma^2}{2} \right)(T - t)}{\sigma \sqrt{T - t}}, \qquad (10.32)$$

$$d_2 = \frac{x + (k-1)\tau}{\sqrt{2\tau}} = \frac{\ln \frac{S}{K} + \left( r - \delta - \frac{\sigma^2}{2} \right)(T - t)}{\sigma \sqrt{T - t}}. \qquad (10.33)$$

Substituting $d_1$ and $d_2$ and replacing $e^x$ with $S/K$ and $k\tau$ with $(r - \delta)(T - t)$ in Eq. (10.31) yield

$$v(x, \tau) = \frac{S}{K} \Phi(d_1) - e^{-(r-\delta)(T-t)} \Phi(d_2),$$

which upon using Eq. (10.11) produces the sought after **Black–Scholes European call option pricing formula**. For the sake of simplicity and meaningful notation, the value of the European call will be denoted $C^e$.

$$C^e(S, t) = S e^{-\delta(T-t)} \Phi(d_1) - K e^{-r(T-t)} \Phi(d_2). \qquad (10.34)$$

A surface plot of the value of the European call option as a function of $S$ and $t$ is shown in Fig. 10.1.

The value of a European put option could be found via a similar set of calculations or the already determined value of the European call could be used along with the Put-Call Parity formula in Eq. (6.8) to derive

$$P^e(S, t) = K e^{-r(T-t)} \Phi(-d_2) - S e^{-\delta(T-t)} \Phi(-d_1). \qquad (10.35)$$

A surface plot of the value of the European put option as a function of $S$ and $t$ is shown in Fig. 10.2.

**Example 10.1.** Suppose the current price of a non-dividend paying stock is \$62 per share. The continuously compounded interest rate is 10% per year. The volatility of the price of the security is 20% per year. Find the cost of a 5-month European call option with a strike price of \$60.

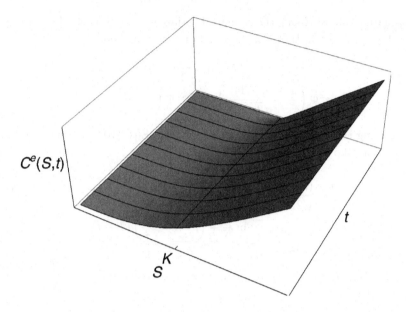

Fig. 10.1    A surface plot of $C^e$ as a function of $S$ and $t$.

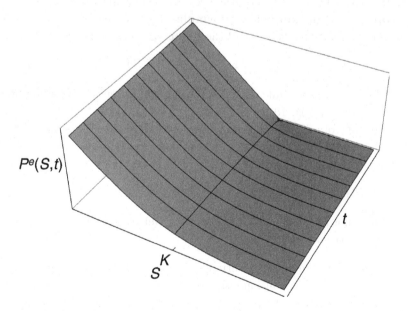

Fig. 10.2    A surface plot of $P^e$ as a function of $S$ and $t$.

**Solution.** The call premiums can be found using Eqs. (10.32), (10.33) and (10.34). Summarizing the known quantities in the notation of this section: $T = 5/12$, $t = 0$, $r = 0.10$, $\delta = 0$, $\sigma = 0.20$, $S = 62$, and $K = 60$. Thus,

$$d_1 = \frac{\ln\frac{62}{60} + \left(0.10 - 0 + \frac{(0.20)^2}{2}\right)\left(\frac{5}{12} - 0\right)}{0.20\sqrt{\frac{5}{12} - 0}} \approx 0.641287$$

$$d_2 = 0.641287 - 0.20\sqrt{\frac{5}{12} - 0} \approx 0.512188.$$

Substituting these values in Eq. (10.34) yields the price of the European call option.

$$C^e = 62e^{0(5/12-0)}\Phi(0.641287) - 60e^{-0.10(5/12-0)}\Phi(0.512188)$$
$$= (62)(0.739332) - (57.5514)(0.695740)$$
$$= \$5.7978.$$

**Example 10.2.** Suppose the current price of a stock is \$97 per share. The stock pays dividends continuously at a rate of 2% annually. The continuously compounded interest rate is 8% per year. The volatility of the price of the security is $\sigma = 45\%$ per year. Find the cost of a 3-month European put option with a strike price of \$95.

**Solution.** The premium for the put can be calculated as follows. With $T = 1/4$, $t = 0$, $r = 0.08$, $\delta = 0.02$, $\sigma = 0.45$, $S = 97$ and $K = 95$,

$$d_1 = \frac{\ln\frac{97}{90} + \left(0.08 - 0.02 + \frac{(0.45)^2}{2}\right)\left(\frac{3}{12} - 0\right)}{0.45\sqrt{\frac{3}{12} - 0}} \approx 0.271763$$

$$d_2 = 0.271763 - 0.45\sqrt{\frac{3}{12} - 0} \approx 0.046763.$$

Substituting these values in Eq. (10.35) yields the price of the European put option.

$$P^e = 95e^{-0.08(3/12-0)}\Phi(-0.046763) - 97e^{-0.02(3/12-0)}\Phi(-0.271763)$$
$$= (93.1189)(0.481351) - (96.5162)(0.392902)$$
$$= \$6.9014.$$

Before concluding this section the reader should make note of the similarity of the expressions for $d_1$ and $d_2$ given in Eqs. (10.32) and (10.33) to the expressions for $\hat{d}_1$ and $\hat{d}_2$ defined in Eqs. (9.6) and (9.5) respectively. The appearance of $t$ in the definitions of $\hat{d}_1$ and $\hat{d}_2$ where $T - t$ appears in $d_1$ and $d_2$ is just a superficial difference. In both set of formulas this can be interpreted as the time remaining until expiry. The significant difference is the replacement of $\alpha$ (the expected rate of growth) in $\hat{d}_1$ and $\hat{d}_2$ with $r$ the risk-free continuously compounded interest rate. Therefore $\Phi(d_1)$ and $\Phi(d_2)$ are risk-neutral probabilities rather than real-world probabilities. This makes possible a natural probabilistic interpretation of the Black–Scholes option pricing formulas for calls and puts. According to Eq. (9.7)

$$\mathbb{E}\left(S(T) \mid S(T) > K\right) \Phi\left(\hat{d}_2\right) = \mathbb{E}\left(S(T)\right) \Phi\left(\hat{d}_1\right)$$
$$= S(t)e^{(\alpha-\delta)(T-t)} \Phi\left(\hat{d}_1\right),$$

where $\mathbb{P}\left(S(T) > K\right) = \Phi\left(\hat{d}_2\right)$. In terms of risk-neutral probabilities and risk-neutral expectation

$$\mathbb{E}^*\left(S(T) \mid S(T) > K\right) \mathbb{P}^*\left(S(T) > K\right) = S(t)e^{(r-\delta)(T-t)} \Phi(d_1),$$

The risk-neutral price of a European call option is

$$C^e = S(t)e^{-\delta(T-t)} \Phi(d_1) - Ke^{-r(T-t)} \Phi(d_2)$$
$$= e^{-r(T-t)} \left(S(t)e^{(r-\delta)(T-t)} \Phi(d_1) - K\Phi(d_2)\right)$$
$$= e^{-r(T-t)} \left(\mathbb{E}^*\left(S(T) \mid S(T) > K\right) \mathbb{P}^*\left(S(T) > K\right)\right.$$
$$\left. - \mathbb{E}^*\left(K \mid S(T) > K\right) \mathbb{P}^*\left(S(T) > K\right)\right)$$
$$= e^{-r(T-t)} \mathbb{E}^*\left(S(T) - K \mid S(T) > K\right) \mathbb{P}^*\left(S(T) > K\right).$$

In other words, the price of a European call option is the time $t$ present value of the risk-neutral partial expectation of the amount that $S(T)$ exceeds $K$. A similar statement can be made for the put option pricing formula.

The formulas for the prices of European call and put options on stocks can also be interpreted in terms of asset-or-nothing and cash-or-nothing options discussed earlier. The European call option gives its owner the option to exchange cash for the asset, therefore the price of the asset-or-nothing call and cash-or-nothing call are respectively,

$$C^{an}(S(t), K, T) = S(t)e^{-\delta(T-t)} \Phi(d_1) \qquad (10.36)$$
$$C^{cn}(S(t), K, T) = Ke^{-r(T-t)} \Phi(d_2). \qquad (10.37)$$

It is assumed the cash-or-nothing call pays $K$ when $S(T) > K$. Therefore $C^e(S(t), K, T) = C^{an}(S(t), K, T) - C^{cn}(S(t), K, T)$. Similarly the European put is $P^e(S(t), K, T) = P^{cn}(S(t), K, T) - P^{an}(S(t), K, T)$, where the asset-or-nothing put and cash-or-nothing put are respectively,

$$P^{an}(S(t), K, T) = S(t)e^{-\delta(T-t)}\Phi(-d_1) \qquad (10.38)$$

$$P^{cn}(S(t), K, T) = Ke^{-r(T-t)}\Phi(-d_2). \qquad (10.39)$$

## Exercises

**10.4.1** Verify that the expression

$$\left(e^{(k+1)y/2} - e^{(k-1)y/2}\right)^+$$

is non-zero whenever $y > 0$.

**10.4.2** Show that

$$\frac{-(x-y)^2}{4\tau} + \frac{(k \pm 1)y}{2} = \frac{(k \pm 1)x}{2} + \frac{(k \pm 1)^2\tau}{4} - \frac{(y - x - (k \pm 1)\tau)^2}{4\tau}.$$

**10.4.3** Show that

$$\int_0^\infty \frac{1}{\sqrt{4\pi\tau}} e^{-(y-x-(k\pm 1)\tau)^2/(4\tau)}\, dy = \Phi\left(\frac{x + (k \pm 1)\tau}{\sqrt{2\tau}}\right).$$

**10.4.4** Verify that the expression for $u(x, \tau)$ given in Eq. (10.30) satisfies the initial condition as specified in Eq. (10.27).

**10.4.5** Verify using the change of variable formulas in Eqs. (10.9) and (10.10) and the fact that $k = 2(r - \delta)/\sigma^2$ that

$$x/\sqrt{2\tau} + \frac{1}{2}(k+1)\sqrt{2\tau} = \frac{\ln(S/K) + (r + \sigma^2/2)(T - t)}{\sigma\sqrt{T - t}}$$

$$x/\sqrt{2\tau} + \frac{1}{2}(k-1)\sqrt{2\tau} = \frac{\ln(S/K) + (r + \sigma^2/2)(T - t)}{\sigma\sqrt{T - t}} - \sigma\sqrt{T - t}.$$

Also show that $d_2 = d_1 - \sigma\sqrt{T - t}$.

**10.4.6** Find the value of a 6-month European call option on a stock currently worth \$85 per share which pays dividends at a continuous rate of 1.75% per year. The risk-free interest rate is 4.5% per annum. The volatility of the stock is 35% per year. The strike price for the option will be \$90.

**10.4.7** Find the value of a 4-month European put option on a stock currently worth \$55 per share which pays dividends at a continuous rate of

2.5% per year. The risk-free interest rate is 5.5% per annum. The volatility of the stock is 28% per year. The strike price for the option will be $58.

**10.4.8** What is the price of a European call option on a stock when the stock price is $52, the strike price is $50, the interest rate is 12%, the stock's volatility is 30%, the stock pays a continuous dividend proportional to the stock price at a rate of 3%, and the exercise time is 3 months?

**10.4.9** What is the price of a European put option on a stock when the stock price is $69, the strike price is $70, the risk-free interest rate is 5% compounded continuously, the stock's volatility is 35%, the stock pays a continuous dividend at a rate of 1%, and the exercise time is 6 months?

**10.4.10** Find the value of a 3-month European call option on a stock currently worth $110 per share which pays dividends at a continuous rate of 1% per year. The risk-free interest rate is 5.5% per annum. The volatility of the stock is 25% per year. The strike price for the option will be $120.

**10.4.11** Find the value of the European put option for the stock described in Exercise 10.4.10.

**10.4.12** A European call option on a non-dividend-paying stock has a market value of $2.50. The stock price is $15, the strike price is $13, the interest rate is 5%, and the exercise time is 3 months. What is the volatility of the stock?

**10.4.13** A non-dividend-paying stock has an expected growth rate compounded continuously of 10%, its volatility is 25%, and the mean of the stock price at 6 months is $65.87. The continuously compounded risk-free interest rate is 4%. Find the price of a 1-year, 70-strike European put option on the stock.

**10.4.14** A non-dividend-paying stock has a time $t$ price of $S(t)$ with $S(0) = 100$. The $\mathbb{V}\mathrm{ar}\,(\ln S(t)) = 0.36t$ and the continuously compounded, risk-free interest rate is 7%. Find the price of a straddle expiring in 1 year with a strike price of $100.

**10.4.15** Show that the price of a European put option is the time $t$ present value of the risk-neutral partial expectation of the amount $K$ exceeds $S(T)$.

**10.4.16** Consider options (American or European) on a non-dividend-paying stock which are identical except for the value of the underlying stock $S_1 < S_2$. Show that the following two inequalities hold.

$$C(S_2) - C(S_1) \le S_2 - S_1,$$
$$P(S_1) - P(S_2) \le S_2 - S_1.$$

**10.4.17** Consider American options written on an underlying stock whose value is $S$. The strike prices of the options are both $K$. Show that $C^a(S(t)) - (S(t) - K)^+$ and $P^a(S(t)) - (K - S(t))^+$ are maximized when $S(t) = K$.

## 10.5   Derivation Using the Binomial Model

In Sec. 7.2 an approximation formula for the price of a European call option based on a binomial lattice was derived. The current section will show that as the size of the time step approaches zero, the limit of this discrete model is the Black–Scholes formula for the European call option. The argument will rely on the previous development of the continuous random walk. This section can safely be skipped if the reader is satisfied with the derivation of the Black–Scholes formulas for European options already established.

If the current price of a stock is $S(t)$, the risk-free, continuously compounded interest rate is $r$, the stock pays dividends continuously at a rate $\delta$ proportional to $S(t)$, the stock's volatility is $\sigma$, then the price of a $K$-strike European call option expiring at time $T$ is approximately $e^{-r(T-t)}\, \mathbb{E}^* \left( (S(T) - K)^+ \right)$. The risk-neutral expectation is taken with respect to a binomial distribution $\mathcal{B}(p^*, n)$. If the time step used to construct the binomial lattice is $\Delta t$, then

$$p^* = \frac{1}{1 + e^{\sigma\sqrt{\Delta t}}},$$

with $u = e^{(r-\delta)\Delta t + \sigma\sqrt{\Delta t}}$ and $d = e^{(r-\delta)\Delta t - \sigma\sqrt{\Delta t}}$. The price of the stock at time $T$ is a random variable and $S(T) = S(t)u^X d^{n-X}$ where $X \sim \mathcal{B}(p^*, n)$.

$$\ln S(T) = \ln S(t) + X \ln u + (n - X) \ln d$$

$$\ln \frac{S(T)}{S(t)} = X((r - \delta)\Delta t + \sigma\sqrt{\Delta t}) + (n - X)((r - \delta)\Delta t - \sigma\sqrt{\Delta t})$$

$$\ln \frac{S(T)}{S(t)} = (r - \delta)(T - t) + (2X - n)\sigma\sqrt{\Delta t}.$$

Thus $\ln(S(T)/S(t)) - (r - \delta)(T - t)$ can be thought of as the final value of a discrete $n = (T - t)/\Delta t$ step random walk which starts at 0 and with probability $p^*$ takes $X$ rightward steps of size $\sigma\sqrt{\Delta t}$ and $n - X$ leftward steps of the same magnitude. The mean of this random variable is

$$\mathbb{E}\left( \ln \frac{S(T)}{S(t)} - (r - \delta)(T - t) \right) = \mathbb{E}\left( (2X - n)\sigma\sqrt{\Delta t} \right)$$

$$= (2\,\mathbb{E}\,(X) - n)\sigma\sqrt{\Delta t}$$

$$= (2p^* - 1)\frac{\sigma(T - t)}{\sqrt{\Delta t}}.$$

As $\Delta t \to 0^+$ the expected value approaches $-\sigma^2(T - t)/2$. The variance of the random variable is

$$\mathbb{V}\text{ar}\left(\ln\frac{S(T)}{S(t)} - (r - \delta)(T - t)\right) = \mathbb{V}\text{ar}\left((2X - n)\sigma\sqrt{\Delta t}\right)$$
$$= 4\sigma^2 \Delta t\, \mathbb{V}\text{ar}(X)$$
$$= 4\sigma^2(T - t)p^*(1 - p^*).$$

As $\Delta t \to 0^+$ the variance approaches $\sigma^2(T - t)$.

In summary using the risk-neutral probability, as $\Delta t \to 0^+$, random variable $\ln(S(T)/S(t))$ converges in distribution to a normally distributed random variable with mean $(r - \delta - \sigma^2/2)(T - t)$ and variance $\sigma^2(T - t)$, the same result as the continuous time stochastic model of $S(T)$ presented earlier. Hence the binomial approximation to an option price will converge to the Black–Scholes option price as the number of time steps in the binomial model is increased. A further investigation of binomial option pricing formulas can be found in [Chance (2008)]. Numerical confirmation of this convergence is demonstrated in the following example.

**Example 10.3.** What is the price of a \$120-strike European put option on a stock when the stock price is \$125, the stock pays dividends continuously at a rate of 2%, the risk-free interest rate is 10%, the stock's volatility is 45%, and the exercise time is 1 year? Use the binomial lattice model with an increasing number of time steps and observe the prices converge to the Black–Scholes formula price.

**Solution.** Using the Black–Scholes formula for the European put produces,

$$d_1 = \frac{\ln\frac{125}{120} + \left(0.10 - 0.02 + \frac{(0.45)^2}{2}\right)(1)}{0.45\sqrt{1}} = 0.493493$$

$$d_2 = 0.493493 - 0.45\sqrt{1} = 0.043493$$

$$P^e = 120e^{-0.10}\Phi(-0.043493) - 125e^{-0.02}\Phi(-0.493493) = \$14.3222.$$

The following table lists various approximations to the European put price employing different numbers of time steps.

| $\Delta t$ (years) | $u$ | $d$ | $p^*$ | $P^e$ |
|---|---|---|---|---|
| 1 | 1.69893 | 0.690734 | 0.389361 | \$18.5971 |
| 1/2 | 1.43075 | 0.757147 | 0.421115 | \$14.6579 |
| 1/12 | 1.14634 | 0.884054 | 0.467570 | \$14.7086 |
| 1/24 | 1.09987 | 0.915282 | 0.477052 | \$14.4812 |
| 1/120 | 1.04263 | 0.960393 | 0.489732 | \$14.3608 |

As $\Delta t$ approaches zero $p^* = 1/2$ and thus in the limit steps up and down are equally likely (Exercise 10.5.3). Similarly in the limit $d = 1/u$ or equivalently $\ln d = -\ln u$ so that upward and downward steps are of the same magnitude (Exercise 10.5.4). In conclusion, whether the value of European options is derived from the continuous model via the solution of the Black–Scholes partial differential equation or by developing a multi-time step binomial model and taking its limit as the time step becomes infinitesimally small, the result remains the same.

## Exercises

**10.5.1** Using the formula for the risk-neutral probability in the binomial model of stock price movements, show that

$$\lim_{\Delta t \to 0^+} (2p^* - 1)\frac{\sigma(T-t)}{\sqrt{\Delta t}} = \frac{-\sigma^2}{2}(T-t).$$

**10.5.2** Using the formula for the risk-neutral probability in the binomial model of stock price movements, show that

$$\lim_{\Delta t \to 0^+} 4\sigma^2(T-t)p^*(1-p^*) = \sigma^2(T-t).$$

**10.5.3** Show that as $\Delta t \to 0$, $p^* \to 1/2$.

**10.5.4** Show that as $\Delta t \to 0$, $u\,d = 1$.

**10.5.5** What is the price of a European call option on a non-dividend-paying stock when the stock price is \$43, the strike price is \$42, the risk-free interest rate is 11%, the stock's volatility is 25%, and the exercise time is 4 months? Use the binomial lattice model with a time step of 1 month to approximate the price of the option. Compare the result to the Black–Scholes formula price.

**10.5.6** What is the price of a European put option on a stock when the stock price is \$96, the stock pays dividends continuously at a rate of 1%, the strike price is \$100, the risk-free interest rate is 6%, the stock's volatility is 33%, and the exercise time is 3 months? Use the binomial lattice model with a time step of 1 month to approximate the price of the option. Compare the result to the Black–Scholes formula price.

**10.5.7** What is the price of an at-the-money European call option on a stock when the stock price is \$100, the stock pays dividends continuously at a rate of 2%, the risk-free interest rate is 8%, the stock's volatility is 40%, and the exercise time is 6 months? Use the binomial lattice model with a time step of 1 month to approximate the price of the option. Compare the result to the Black–Scholes formula price.

# Chapter 11

# Extensions of the Black–Scholes Model

The work done to develop the option relationships and the Black–Scholes option pricing formulas in Chapters 6–10 can be readily extended to options on underlying assets other than stocks. This chapter will develop variations on the Black–Scholes model to price options on foreign exchange rates, futures, and options on stocks which pay discrete dividends. Unless otherwise specified, all options discussed in this chapter will still have the European style of exercise.

## 11.1 Options on Exchange Rates

Often in cases of international business transactions, goods must be paid for in a foreign currency. The number of units of domestic currency which can be exchanged for a unit of a specific foreign currency is driven by a variety of market forces. A plot of the number US dollars (USD) which can be exchanged for one Euro during the first half of 2020 is presented in Fig. 11.1. Payers may manage the risk in movements of the foreign exchange rate by purchasing options on the exchange rate. This section will show that foreign exchange options can be priced with only minor changes to the Black–Scholes call and put pricing formulas. This section will also recast the Put-Call parity formula for foreign exchange options and derive an equivalence between foreign exchange calls and puts.

Denote by $X(t)$ the current (spot) exchange rate between a domestic currency and a foreign currency. That is, $X(t)$ units of the domestic currency may be exchanged for 1 unit of the foreign currency. Often the exchange rate is said to be "domestic currency-denominated" since it refers to the situation of exchanging the domestic currency for the foreign currency. Assume $X(t)$ is lognormally distributed and that the domestic

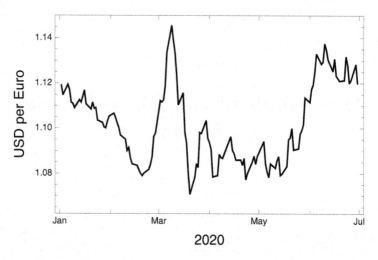

Fig. 11.1   A sample of the historical record of US dollar for Euro exchange rate for the first 6 months of 2020.

risk-free, continuously compounded interest rate is $r$. The foreign currency may also be invested at a foreign risk-free, continuously compounded rate denoted $r_f$. With these terms defined, the following Put-Call parity relationship holds.

**Theorem 11.1.** *If the current exchange rate is $X(t)$, the domestic risk-free interest rate is $r$ per annum compounded continuously, the foreign risk-free interest rate is $r_f$ per annum compounded continuously, $C^e(X(t), K, T)$ is a European call on the underlying exchange rate with a strike of $K$ (units of domestic currency for one unit of foreign currency), $P^e(X(t), K, T)$ is a European put on the exchange rate, and $T$ is the strike time of the options, then in the absence of arbitrage,*

$$C^e(X(t), K, T) + Ke^{-r(T-t)} = P^e(X(t), K, T) + X(t)e^{-r_f(T-t)}. \quad (11.1)$$

Notice that the exchange rate $X(t)$ replaces the spot price of the underlying stock $S(t)$ and the foreign risk-free rate $r_f$ replaces the dividend yield rate $\delta$ in Eq. (6.8) but otherwise the formula is unchanged. The quantity $X(t)e^{-r_f(T-t)}$ is the prepaid forward cost (in the domestic currency) of obtaining one unit of the foreign currency at time $T$.

*Proof.* If $C^e(X(t), K, T) + Ke^{-r(T-t)} < P^e(X(t), K, T) + X(t)e^{-r_f(T-t)}$ then an investor may purchase the call, sell the put, sell a prepaid forward $X(t)e^{-r_f(T-t)}$, and lend $Ke^{-r(T-t)}$. The cash flows at times $t$ and $T$ are counted in the following table.

| | CF time $t$ | CF time $T$ | |
| :---: | :---: | :---: | :---: |
| | | $X(T) < K$ | $X(T) \geq K$ |
| | $-C^e$ | 0 | $X(T) - K$ |
| | $P^e$ | $-(K - X(T))$ | 0 |
| | $X(t)e^{-r_f(T-t)}$ | $-X(T)$ | $-X(T)$ |
| | $-Ke^{-r(T-t)}$ | $K$ | $K$ |
| Aggregate | $> 0$ | 0 | 0 |

This investment strategy produced a positive cash flow at time $t$ with no risk of loss at time $T$, thus arbitrage is present. The reader will show in Exercise 11.1.1 that if the opposite inequality holds, arbitrage is present as well. Hence Eq. (11.1) holds. □

Mimicking the development of the initial, boundary-value problem for European options on stocks, replace the continuous dividend yield $\delta$ with the risk-free foreign interest rate $r_f$ and the underlying asset $S(t)$ with the current exchange rate $X(t)$ in Eqs. (10.5)–(10.8). This produces the following initial, boundary-value problem for a foreign exchange European call option.

$$rF = F_t + (r - r_f)XF_X + \frac{1}{2}\sigma^2 X^2 F_{XX} \text{ for } (X,t) \text{ in } \Omega, \qquad (11.2)$$

$$F(X,T) = (X(T) - K)^+ \text{ for } X > 0, \qquad (11.3)$$

$$F(0,t) = 0 \text{ for } 0 \leq t < T, \qquad (11.4)$$

$$F(X,t) = Xe^{-r_f(T-t)} - Ke^{-r(T-t)} \text{ as } X \to \infty. \qquad (11.5)$$

The domain of the solution is the set $\Omega = \{(X,t) \,|\, 0 \leq X < \infty, 0 \leq t < T\}$. Hence the prices of European calls and puts on foreign exchange denominated in the domestic currency are respectively,

$$C^e(X(t), K, T) = X(t)e^{-r_f(T-t)}\Phi(d_1) - Ke^{-r(T-t)}\Phi(d_2), \qquad (11.6)$$

$$P^e(X(t), K, T) = Ke^{-r(T-t)}\Phi(-d_2) - X(t)e^{-r_f(T-t)}\Phi(-d_1), \qquad (11.7)$$

where

$$d_1 = \frac{\ln\frac{X(t)}{K} + \left(r - r_f + \frac{\sigma^2}{2}\right)(T - t)}{\sigma\sqrt{T - t}}$$

$$d_2 = \frac{\ln\frac{X(t)}{K} + \left(r - r_f - \frac{\sigma^2}{2}\right)(T - t)}{\sigma\sqrt{T - t}}.$$

As before the symbol $\sigma$ represents volatility, in this situation of the exchange rate.

**Example 11.1.** Suppose the current exchange rate is \$1.10/€, the dollar-denominated continuously compounded, risk-free interest rate is 5%, the euro-denominated continuously compounded risk-free rate is 6%, and the volatility of the exchange rate is 25%. Find the dollar-denominated prices of 3-month, 1.15-strike European call and put options on the foreign exchange rate.

**Solution.** Using the formulas above,

$$d_1 = \frac{\ln\frac{1.10}{1.15} + \left(0.05 - 0.06 + \frac{(0.25)^2}{2}\right)(3/12 - 0)}{0.25\sqrt{3/12 - 0}} = -0.313114.$$

$$d_2 = d_1 - 0.25\sqrt{3/12 - 0} = -0.438144.$$

$$C^e = 1.10e^{-0.06(3/12)}\Phi(-0.313114) - 1.15e^{-0.05(3/12)}\Phi(-0.438144)$$
$$= \$0.0331.$$

The price of the European put can be determined from Eq. (11.7) or from Eq. (11.1) since the price of the call is already known.

$$P^e = 1.10e^{-0.06(3/12)} - 0.0331 - 1.15e^{-0.05(3/12)} = \$0.0852.$$

The call option on the foreign exchange is the option to buy one unit of the foreign currency for $K$ units of the domestic currency. This can also be thought of as the put option to sell $1/K$ units of the foreign currency for one unit of the domestic currency. Thus there is an equivalence between call and put options. Let $C_d^e(X(t), K, T)$ be the domestic currency-denominated price of the call option to buy one unit of the foreign currency for $K$ units of the domestic currency at time $T$ where $X(t)$ (units of domestic currency per unit of foreign currency) is the spot exchange rate. Let $P_f^e(1/X(t), 1/K, T)$ be the foreign currency-denominated price of the put to sell one unit of foreign currency for $1/K$ units of domestic currency. Thus calls and puts are related through the following equation.

$$C_d^e(X(t), K, T) = K\,X(t)\,P_f^e\left(\frac{1}{X(t)}, \frac{1}{K}, T\right). \qquad (11.8)$$

This equivalence is readily justified using the established formulas for the call and put. If $X(t)$ is the domestic currency-denominated exchange rate, then $1/X(t)$ is the foreign currency-denominated exchange rate. The premium for the foreign currency-denominated put is therefore

$$P_f^e\left(\frac{1}{X(t)}, \frac{1}{K}, T\right) = \frac{1}{K}e^{-r_f(T-t)}\Phi(-d_{2,f}) - \frac{1}{X(t)}e^{-r(T-t)}\Phi(-d_{1,,f}),$$

where $d_{1,f}$ and $d_{2,f}$ are the quantities calculated by reversing the roles of the domestic and foreign risk-free interest rates. Note that in this situation,

$$
\begin{aligned}
-d_{1,f} &= \frac{-\ln \frac{1/X(t)}{1/K} - \left(r_f - r + \frac{\sigma^2}{2}\right)(T-t)}{\sigma\sqrt{T-t}} \\[2mm]
&= \frac{\ln \frac{X(t)}{K} + \left(r - r_f - \frac{\sigma^2}{2}\right)(T-t)}{\sigma\sqrt{T-t}} \\[2mm]
&= d_2,
\end{aligned}
$$

where $d_2$ is the quantity calculated assuming the original roles of the domestic and foreign interest rates. Similarly the reader can show that $-d_{2,f} = d_1$. Therefore,

$$
\begin{aligned}
KX(t)P_f^e\left(\frac{1}{X(t)}, \frac{1}{K}, T\right) &= X(t)e^{-r_f(T-t)}\Phi(d_1) - Ke^{-r(T-t)}\Phi(d_2) \\[2mm]
&= C_d^e(X(t), K, T),
\end{aligned}
$$

and the equivalence is shown. More information on currency exchange options can be found in [Garman and Kohlhagan (1983)].

## Exercises

**11.1.1** Show that if $C^e(X(t), K, T) + Ke^{-r(T-t)} > P^e(X(t), K, T) + X(t)e^{-r_f(T-t)}$ under the assumptions of Thm. 11.1 then an investor may create an arbitrage opportunity.

**11.1.2** Suppose the current exchange rate is \$1.10/€, the dollar-denominated continuously compounded, risk-free interest rate is 8%, the euro-denominated continuously compounded risk-free rate is 3%, and the volatility of the exchange rate is 50%. Find the dollar-denominated price of a 6-month, at-the-money European put option on the foreign exchange rate.

**11.1.3** The price of a 4-month, 110-strike yen-denominated call on the euro is ¥4.05. The spot exchange rate is 117 ¥/€. The continuously compounded, risk-free interest rate for the yen is 1% and the continuously compounded , risk-free interest rate for the euro is 4%. What is the yen-denominated price of the corresponding put?

**11.1.4** The price of a 6-month, yen-denominated call on the euro is ¥11.25 while the otherwise identical put has price ¥15.67. The spot exchange rate is 120 ¥/€. What is the strike price of the two options?

312 *An Undergraduate Introduction to Financial Mathematics*

**11.1.5** The price of a 1.05-strike dollar-denominated call on the euro is $0.03. The spot exchange rate is $1.10/€. Find the strike of the corresponding euro-denominated dollar put option and its price.

**11.1.6** An important detail was glossed over in the justification of Eq. (11.8). Show that $\mathbb{V}\text{ar}\,(X(t)) = \mathbb{V}\text{ar}\,(1/X(t))$. *Hint*: $X(t)$ is lognormally distributed.

## 11.2 Options on Futures

Suppose the underlying asset is a futures contract whose current value is $F(t)$. The Put-Call parity relationships for futures can be expressed as

$$C^e(F(t), K, T) + Ke^{-r(T-t)} = P^e(F(t), K, T) + F(t)e^{-r(T-t)}. \quad (11.9)$$

Note that the prepaid forward price of the futures contract is simply the present value of the futures contract. As before $K$ is the strike price of the European call and put options and $r$ is the continuously compounded, risk-free interest rate. Assuming that the futures price is lognormally distributed then the initial, boundary-value problem for the European call given in Eqs. (10.5)–(10.8) simplifies to the set of equations below.

$$rV = V_t + \frac{1}{2}\sigma^2 F^2 V_{FF} \text{ for } (F, t) \text{ in } \Omega, \quad (11.10)$$

$$V(F, T) = (F(T) - K)^+ \text{ for } F > 0, \quad (11.11)$$

$$V(0, t) = 0 \text{ for } 0 \le t < T, \quad (11.12)$$

$$V(F, t) = (F - K)e^{-r(T-t)} \text{ as } F \to \infty. \quad (11.13)$$

The domain of the solution is the set $\Omega = \{(F, t)\,|\,0 \le F < \infty, 0 \le t \le T\}$. Equation (11.10) lacks one of the terms present in the Black–Scholes partial differential equation and thus Eq. (11.10) is often called the **Black equation**. Using the same reasoning to solve this initial, boundary-value problem as in Sec. 10.4 yields the formulas for the value of European calls and puts with underlying futures contracts. Make the assignments,

$$d_1 = \frac{\ln\frac{F(t)}{K} + \frac{\sigma^2}{2}(T - t)}{\sigma\sqrt{T - t}}$$

$$d_2 = \frac{\ln\frac{F(t)}{K} - \frac{\sigma^2}{2}(T - t)}{\sigma\sqrt{T - t}},$$

then the value of the European call and put are respectively

$$C^e(F(t), K, T) = F(t)e^{-r(T-t)}\Phi(d_1) - Ke^{-r(T-t)}\Phi(d_2) \quad (11.14)$$

$$P^e(F(t), K, T) = Ke^{-r(T-t)}\Phi(-d_2) - F(t)e^{-r(T-t)}\Phi(-d_1). \quad (11.15)$$

**Example 11.2.** At the time of writing, the 1-year futures price for an ounce of gold was \$2,005.50, the risk-free continuously compounded interest rate was 2.85% per annum, and the volatility in the price of the futures contract was 57%. Find the premiums for 1-year, 2000-strike European call and put options on gold futures.

**Solution.** Using the quantities $F(0) = 2005.50$, $K = 2000$, $r = 0.0285$, $\sigma = 0.57$, and $T = 1$ calculate,

$$d_1 = \frac{\ln \frac{2005.50}{2000} + \frac{(0.57)^2}{2}(1-0)}{0.57\sqrt{1-0}} = 0.289818.$$

$$d_2 = 0.289818 - 0.57\sqrt{1-0} = -0.280182.$$

$$C^e = e^{-0.0285(1)} \left(2005.50\Phi\left(0.289818\right) - 2000\Phi\left(-0.280182\right)\right) = \$439.381.$$

$$P^e = e^{-0.0285(1)} \left(2000\Phi\left(0.280182\right) - 2005.50\Phi\left(-0.289818\right)\right) = \$434.036.$$

Further examples and properties of options on futures contracts will be explored in the exercises.

### *Exercises*

**11.2.1** Show that at-the-money European calls and puts on the same underlying futures contract have the same price.

**11.2.2** Show that at-the-money European calls and puts on the same underlying futures contract have the price:

$$C^e = P^e = Ke^{-r(T-t)}(2\Phi\left(d_1\right) - 1).$$

**11.2.3** The 6-month futures price for an ounce of silver is \$27.55, the risk-free continuously compounded interest rate is 3.5% per annum, and the volatility in the price of the futures contract was 70%. Find the price of a 6-month, 25-strike European put option on silver futures.

**11.2.4** The 1-year futures price for a barrel of crude oil is \$44.63, the risk-free continuously compounded interest rate is 4.25% per annum, and the volatility in the price of the futures contract was 62%. Find the price of a 1-year, 45-strike European call option on crude oil futures.

**11.2.5** The current price of a stock is \$150, the stock pays a continuous dividend proportional to the value of the stock with a dividend yield of 2%. The continuously compounded, risk-free interest rate is 6% per annum. The volatility of the stock is 40%.

(a) Find the price of a 6-month, 148-strike put on the stock.
(b) Find the 6-month forward price of the stock.

(c) Find the price of a 6-month, 148-strike put on the futures contract.
(d) Why are the two put premiums the same?

## 11.3 Options on Stocks Paying Discrete Dividends

Developing the pricing formula for an option on a stock paying discrete dividends is not as straight forward as pricing an option on a stock paying continuous dividends, yet it is also not difficult if the absence of arbitrage assumption is kept in mind. In this section the case of a stock which pays a single discrete dividend during the life of the option will be considered. The case of multiple, discrete dividends is readily seen as an extension of the approach described here.

Recall that if a stock pays a dividend of $d_y S(t)$ at $t = t_d$ then in the absence of arbitrage the price of the stock immediately before and after the dividend payment must satisfy Eq. (6.6), repeated below for convenience.

$$S(t_d^+) = S(t_d^-)(1 - d_y).$$

The stock price follows a random walk, but any realization of the random variable must exhibit a jump discontinuity at $t = t_d$. The vertical distance of the jump must be the amount of the dividend paid. A typical realization of a random walk is illustrated in Fig. 11.2.

If $S(t)$ is the continuous random variable describing the value of the stock on which the single, discrete proportional dividend $d_y S(t_d^-)$ is paid then $S(t)$ obeys the following stochastic differential equation.

$$dS = (\alpha - d_y D(t - t_d))S\,dt + \sigma S\,dW(t). \tag{11.16}$$

As before $\alpha$ and $\sigma$ are the growth rate (or drift) and volatility associated with this stock. The expression $d_y D(t - t_d) S$ is the value of the dividend paid. The function $D(t - t_d)$ is the **Dirac delta function**. The Dirac delta function is frequently used in mathematical models to represent an instantaneous impulse or shock to a system. Properties of this function were explored in the exercises of Sec. 9.2. Equation (11.16) will be used to derive a relationship between $d_y$ and $\delta$ which was used in Eq. (6.6) to represent the fraction of the stock's value paid as a discrete dividend. Defining $Y = \ln S$ and applying Itô's Lemma (Lemma 9.1) produces the stochastic differential equation

$$dY = \left(\alpha - d_y D(t - t_d) - \frac{1}{2}\sigma^2\right)dt + \sigma\,dW(t). \tag{11.17}$$

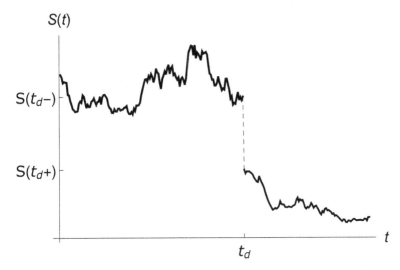

Fig. 11.2   The discontinuous jump in the price of a stock across a dividend date. In the absence of arbitrage the amount by which the stock price decreases must be the amount of the dividend per share paid to the stockholders.

Integrating yields the following piecewise-defined function.

$$Y(t) - Y(0) = \int_0^t \left( \alpha - d_y D\left(s - t_d\right) - \frac{1}{2}\sigma^2 \right) ds + \int_0^t \sigma \, dW(s)$$

$$= \left( \alpha - \frac{1}{2}\sigma^2 \right) t - d_y \int_0^t D\left(s - t_d\right) dt + \sigma W(t),$$

$$Y(t) = Y(0) + \begin{cases} \left( \alpha - \frac{1}{2}\sigma^2 \right) t + \sigma W(t) & \text{if } t < t_d, \\[2mm] \left( \alpha - \frac{1}{2}\sigma^2 \right) t - d_y + \sigma W(t) & \text{if } t \geq t_d. \end{cases}$$

Note that $Y(t)$ is a normally distributed random variable with mean $(\alpha - \sigma^2/2)t$ for $t < t_d$, with mean $(\alpha - \sigma^2/2)t - d_y$ for $t \geq t_d$ and with variance $\sigma^2 t$. Since $Y(t) = \ln S(t)$ then the stock price is a lognormal random variable with

$$S(t) = \begin{cases} S(0)e^{(\alpha - \sigma^2/2)t + \sigma W(t)} & \text{if } t < t_d, \\[2mm] S(0)e^{(\alpha - \sigma^2/2)t - d_y + \sigma W(t)} & \text{if } t \geq t_d. \end{cases} \tag{11.18}$$

Consider the one-sided limits at $t = t_d$.

$$\lim_{t \to t_d^-} S(t) - \lim_{t \to t_d^+} S(t)$$

$$= \lim_{t \to t_d^-} \left[ S(0)e^{(\alpha - \sigma^2/2)t + \sigma W(t)} \right] - \lim_{t \to t_d^+} \left[ S(0)e^{(\alpha - \sigma^2/2)t - d_y + \sigma W(t)} \right].$$

Since the expression $(\alpha - \sigma^2/2)t + \sigma W(t)$ is continuous the limits from the left and the right as $t$ approaches $t_d$ are the same and thus

$$S(t_d^-) - S(t_d^+) = S(0)e^{(\alpha-\sigma^2/2)t_d+\sigma W(t_d)} - S(0)e^{(\alpha-\sigma^2/2)t_d-d_y+\sigma W(t_d)}$$

$$= S(0)e^{(\alpha-\sigma^2/2)t_d+\sigma W(t_d)}\left[1 - e^{-d_y}\right].$$

Using the fact that $S(t_d^-) = S(0)e^{(\alpha-\sigma^2/2)t_d+\sigma W(t_d)}$ simplifies the last equation to the following.

$$S(t_d^+) = S(t_d^-)e^{-d_y}. \tag{11.19}$$

Thus when a discrete dividend of size $d_y S(t_d^-)$ is paid, the relative change in the value of the stock is $1 - e^{-d_y}$. Since $S(t)$ is a lognormal random variable, apply Lemma 8.1 to see that

$$\mathbb{E}\left(S(t)\right) = \begin{cases} S(0)e^{\alpha t} & \text{if } t < t_d, \\ S(0)e^{\alpha t - d_y} & \text{if } t \geq t_d. \end{cases}$$

Figure 11.3 illustrates the expected value of a stock for which $\alpha = 1$ and $d_y = 1/3$.

Naïvely one might expect the jump discontinuity in the price of the stock to induce a jump discontinuity in the value of a European option on the

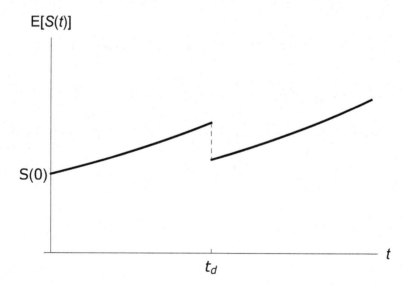

Fig. 11.3 The expected value of a stock with growth rate $\alpha = 1$ for which a single discrete dividend is paid at $t = t_d$. The dividend is one third the value of the stock at $t = t_d$.

stock. In fact the value of an option must be a continuous function of time across the dividend date or arbitrage results. Expressed mathematically

$$\lim_{t \to t_d^-} C^e(S(t), K, T) = \lim_{t \to t_d^+} C^e(S(t), K, T)$$

$$C^e(S(t_d^-), K, T) = C^e(S(t_d^+), K, T)$$

$$C^e(S(t_d^-), K, T) = C^e(S(t_d^-)e^{-d_y}, K, T), \qquad (11.20)$$

using Eq. (11.19). The reader should carefully consider Eq. (11.20) and its implications. This equation states that the value of the call option is continuous across the dividend date even though the price of the underlying stock has a jump discontinuity. The price of the option is made continuous by equating the value of the option just before the dividend is paid $C^e(S(t_d^-), K, T)$ with the value of the option just after the dividend is paid $C^e(S(t_d^-)e^{-d_y}, K, T)$ for which the value of the underlying stock has been decreased to $S(t_d^+) = S(t_d^-)e^{-d_y}$.

Now consider the mechanics of pricing an option on a stock for which a single discrete dividend will be paid during the life of the option. This will require solving the Black–Scholes equation twice. Once for the interval $(t_d, T]$ from immediately after the dividend is paid until expiry and a second time for the interval $[0, t_d)$, from the moment of the option is written until immediately prior to the dividend date. To cross the dividend date the two solutions of the Black–Scholes equation will be connected using Eq. (11.20).

On the interval $(t_d, T]$ no dividends are paid and thus the call option can be priced using the formula given in Eq. (10.34) with $\delta = 0$. For $t_d < t < T$ the value of the European call is

$$C^e(S(t), K, T) = S(t_d^-)e^{-d_y} \Phi(d_1) - Ke^{-r(T-t)} \Phi(d_2),$$

where

$$d_1 = \frac{\ln \frac{S(t_d^-)e^{-d_y}}{K} + \left(r + \frac{\sigma^2}{2}\right)(T - t)}{\sigma\sqrt{T - t}}$$

$$d_2 = d_1 - \sigma\sqrt{T - t}.$$

Note that the underlying stock price has been discounted by the factor $e^{-d_y}$ and the stock is treated as if it pays no continuous dividend. Immediately before the dividend date the value of the call option is according to Eq. (11.20)

$$C^e(S(t_d^-), K, T) = S(t_d^-)e^{-d_y} \Phi(d_1) - Ke^{-r(T-t_d^+)} \Phi(d_2), \qquad (11.21)$$

where $d_1$ and $d_2$ have been evaluated as $t \to t_d^+$. On the right-hand side of Eq. (11.21) the price of the underlying stock has been scaled by the factor $e^{-d_y}$. Consider the effect of scaling the value of the underlying stock by $e^{-d_y}$ on the Black–Scholes initial, boundary-value problem given in Eqs. (10.5)–(10.8). Making the change of variable $\hat{S} = Se^{-d_y}$ then the partial derivatives

$$\frac{\partial}{\partial S}\left[F(S,t)\right] = e^{-d_y}\frac{\partial}{\partial \hat{S}}\left[F(S,t)\right] \text{ and}$$

$$\frac{\partial^2}{\partial S^2}\left[F(S,t)\right] = e^{-2d_y}\frac{\partial^2}{\partial \hat{S}^2}\left[F(S,t)\right].$$

Therefore upon setting the continuous dividend yield to zero, Eq. (10.5) is transformed to

$$rF = F_t + r\hat{S}F_{\hat{S}} + \frac{1}{2}\sigma^2\hat{S}^2F_{\hat{S}\hat{S}}. \tag{11.22}$$

Consequently scaling the stock price leaves the Black–Scholes partial differential equation unchanged. At expiry the payoff of the call option will be

$$F(\hat{S},T) = F(Se^{-d_y},T) = (Se^{-d_y} - K)^+ = e^{-d_y}(S - Ke^{d_y})^+. \tag{11.23}$$

The payoff of the call option with underlying stock $\hat{S}$ is the same as the payoff of $e^{-d_y}$ call options on the underlying stock $S$ with strike price $Ke^{d_y}$. The boundary condition at $\hat{S} = 0$ remains the same as Eq. (10.7) and the boundary condition at infinity becomes

$$\lim_{\hat{S} \to \infty} F(\hat{S},t) = \hat{S} - Ke^{-r(T-t)} = e^{-d_y}\left[S - Ke^{d_y - r(T-t)}\right]. \tag{11.24}$$

Thus for times prior to the dividend date, the call option can be priced by equating it with the price of $e^{-d_y}$ call options having a strike price of $Ke^{d_y}$. For $t < t_d$,

$$C(S(t),K,T)) = e^{-d_y}\left(S(t)\Phi\left(d_1\right) - Ke^{d_y - r(T-t)}\Phi\left(d_2\right)\right). \tag{11.25}$$

For Eq. (11.25) the aggregate parameters $d_1$ and $d_2$ are calculated as

$$d_1 = \frac{\ln\frac{S(t)}{Ke^{d_y}} + \left(r + \frac{\sigma^2}{2}\right)(T-t)}{\sigma\sqrt{T-t}}$$

$$d_2 = d_1 - \sigma\sqrt{T-t}.$$

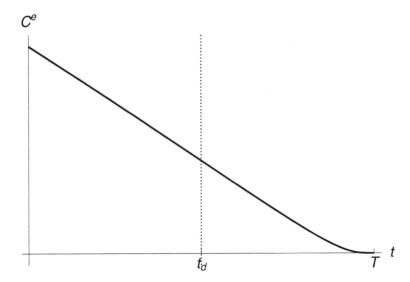

Fig. 11.4 The value of an at-the-money European call option on a stock paying a single discrete, proportional dividend at $t = t_d$.

The continuity of the value of the European call option across the dividend date can be verified in Fig. 11.4 which illustrates the value of an at-the-money call option over the interval $[0, T]$.

**Example 11.3.** Suppose a stock pays a discrete dividend of 2% of its value at $t_d = 1/2$. The risk-free interest rate is 8% per year and the volatility of the stock price is 20% per year. Find the price of a 1-year European at-the-money call option on the stock with a strike price of $100 at $t = 0$ and $t = 1/2$.

**Solution.** At $t = 0$ calculate

$$d_1 = \frac{\ln \frac{100}{100e^{0.02}} + \left(0.10 + \frac{(0.20)^2}{2}\right)(1 - 0)}{0.20\sqrt{1 - 0}} = 0.4$$

$$d_2 = 0.4 - 0.20\sqrt{1 - 0} = 0.2$$

$$C^e(S(0), 100, 1) = e^{-0.02}\left(100\Phi(0.4) - 100e^{0.02-0.08(1-0)}\Phi(0.2)\right)$$

$$= \$10.7719.$$

At $t = 1/2$ similar calculations yield

$$d_1 = \frac{\ln\frac{100}{100e^{0.02}} + \left(0.10 + \frac{(0.20)^2}{2}\right)(1 - 1/2)}{0.20\sqrt{1 - 1/2}} = 0.212132$$

$$d_2 = 0.212132 - 0.20\sqrt{1 - 1/2} = 0.070711$$

$$C^e(S(0), 100, 1/2) = e^{-0.02}\left(100\Phi(d_1) - 100e^{0.02 - 0.08(1 - 1/2)}\Phi(d_2)\right)$$

$$= \$6.4959.$$

Confirm there is no discontinuity in the premium for the European call by calculating the price at $t = 1/2$ using the post-dividend formula.

$$d_1 = \frac{\ln\frac{100e^{-0.02}}{100} + \left(0.10 + \frac{(0.20)^2}{2}\right)\left(1 - \frac{1}{2}\right)}{0.20\sqrt{1 - \frac{1}{2}}} = 0.212132$$

$$d_2 = 0.212132 - 0.20\sqrt{1 - 1/2} = 0.070711$$

$$C^e(S(\tfrac{1}{2})e^{-0.02}, 100, 1) = 100e^{-0.02}\Phi(d_1) - 100e^{-0.08(1 - \frac{1}{2})}\Phi(d_2)$$

$$= \$6.4959.$$

Figure 11.5 shows the value of the European call option at the dividend date as a function of $S(1/2)$. The post-dividend curve can be obtained by contracting the pre-dividend curve in the horizontal direction by the factor $e^{-d_y}$. The value of the European call on the vertical axis experiences no discontinuity due to the payment of the discrete dividend.

To summarize, the value of a European option written on a stock paying discrete, proportional dividends during the life of the option can be calculated using the Black–Scholes option pricing formulas (Eqs. (10.34) and (10.34)) by setting $\delta = 0$ and discounting the current value of the stock by the present value of the dividends to be paid.

**Example 11.4.** Find the value of the European put option corresponding to the call option in Example 11.3.

**Solution.** The premium for the European put can be determined from the value of the European call and the Put-Call Parity formula for options on stocks paying discrete dividends, Eq. (6.7). Adapting this formula to the situation of a single dividend payment gives

$$P^e(S(0), 100, 1) = C^e(S(0), 100, 1) + 100e^{-0.08(1 - 0)} - F^P_{0,1}(S(0)).$$

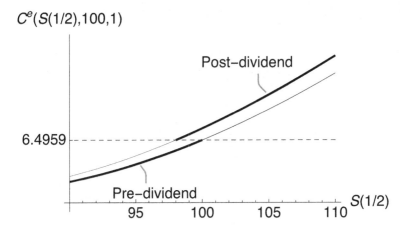

Fig. 11.5   The value of a European call option on the dividend date as a function of the underlying stock price. Note that while the value of the stock instantaneously decreases in value by 2% as the dividend is paid, the value of the call option is continuous.

Proceeding as in Example 11.3, $C^e(S(0), 100, 1) \approx \$10.7719$. The prepaid forward price of the stock is

$$F_{0,1}^P(S(0)) = 100e^{-0.02} = \$98.0199.$$

Thus premium for the European put is $P^e(S(0), 100, 1) = \$5.0637$. Recall that immediately prior to and after the dividend payment $C^e(S(1/2)) \approx \$6.4959$. Computing the price of the European put just before and after the dividend payment gives $P^e(S(1/2)) \approx \$4.5549$.

### Exercises

**11.3.1** Suppose a stock pays a dividend $D$ at the end of the year. At the end of the following year the stock will pay a dividend of $D(1 + g)$. The year after that, the stock will pay a dividend of $D(1 + g)^2$. If this continues until $k$ dividends have been paid, find a formula for the sum of the present values of the dividends assuming the interest rate is $r$ and is compounded annually.

**11.3.2** Find the sum of the present values of the dividends described in Exercise (11.3.1) as the number of dividends increases to infinity assuming $0 < g < r$.

**11.3.3** Starting with Eq. (11.16), derive Eq. (11.17).

**11.3.4** Show that $d_y = -\ln(1 - \delta)$.

**11.3.5** Show that if Eq. (11.20) does not hold, an arbitrage opportunity exists.

**11.3.6** Suppose the current price of a stock is $50 and the stock will pay a dividend of 3% of its value in 2 months. The risk-free interest rate is 2.5%, the strike price of a European call option is $50, the strike time is 5 months, and the volatility of the stock is 35%. Find the values of the call option at time $t = 0$ and at the dividend date.

**11.3.7** Find the values of a European put option at time $t = 0$ and at the dividend date for the stock described in Exercise 11.3.6.

**11.3.8** Suppose the dividend paid to the shareholder of a stock will be an absolute amount rather than an amount proportional to the value of the stock immediately before the dividend date. Suppose a stock is as described in Exercise 11.3.6 except that the stock pays a dividend of $2 after 2 months. Find the value at $t = 0$ of a European call option on the stock using the Black–Scholes call option price formula given in Eq. (10.34) where the current value of the stock has been discounted by the present value of the dividend amount to be paid and $\delta = 0$.

**11.3.9** For the stock described in Exercise 11.3.6, suppose two dividends are paid. At 2 months a dividend of $2 is paid and at 4 months another dividend of $2 is paid. Use the discounting procedure of Exercise 11.3.8 to approximate the value of the call option at time $t = 0$.

**11.3.10** Find the price a European put option on a stock described in Exercise 11.3.8.

**11.3.11** For the stock described in Exercise 11.3.7, suppose two dividends are paid. At 2 months a dividend of $2 is paid and at 4 months another dividend of $2 is paid. Use the discounting procedure of Exercise 11.3.10 to approximate the value of a European put option at time $t = 0$.

# Chapter 12

# Derivatives of Black–Scholes Option Prices

Now that the solution to the Black–Scholes equation is known, this chapter will investigate the sensitivity of the solution to changes in the independent variables. Knowledge of these sensitivities allows a portfolio manager to minimize changes in the value of a portfolio when underlying variables such as the risk-free interest rate change. The material of the present chapter will be used extensively in Chapter 13. An understanding of the sensitivity of an option's value to changes in its independent variables and parameters will also provide a new way of interpreting the Black–Scholes equation itself.

To mathematicians the word *derivative* means an instantaneous rate of change in a quantity. To a quantitative analyst a *derivative* is a financial entity whose value is derived from the value of some underlying asset. Hence a European call option is a derivative (in the quantitative analytical sense) whose value is a function of (among other things) the value of the security underlying the option. This chapter explores derivatives from the mathematician's perspective. In Chapter 13 these derivatives will be used in the manner of a quantitative analyst. Members of the quantitative financial profession refer to the subject matter of this chapter as the "Greeks" since a Greek letter is used to name each derivative (except for one which to be met in due time). Due to the typically large volume of options written in a contract, very accurate calculations of the Greeks are necessary to avoid round-off errors. Motivation for accuracy and accurate numerical techniques may be found in [Chawla (2006)] and [Chawla and Evans (2005)].

A particularly important relationship used throughout this chapter to simplify results is contained in the following lemma.

**Lemma 12.1.** *For European calls and puts which satisfy the Black–Scholes formulas in Eq. (10.34) and Eq. (10.35),*

$$S(t)e^{-\delta(T-t)}\varphi\left(d_1\right) = Ke^{-r(T-t)}\varphi\left(d_2\right), \qquad (12.1)$$

*where $\varphi\left(x\right)$ is the probability density function for the standard normal random variable.*

*Proof.* Recall from Eqs. (10.32) and (10.33) that

$$d_2 = d_1 - \sigma\sqrt{T - t}$$

$$d_2^2 = (d_1 - \sigma\sqrt{T - t})^2$$

$$d_2^2 - \sigma^2(T - t) = d_1^2 - 2\sigma\sqrt{T - t}d_1$$

$$\frac{-d_2^2 + \sigma^2(T - t)}{2} = \frac{-d_1^2}{2} + \ln\frac{S(t)}{K} + \left(r - \delta + \frac{\sigma^2}{2}\right)(T - t)$$

$$\frac{-d_2^2}{2} + \ln K - r(T - t) = \frac{-d_1^2}{2} + \ln S(t) - \delta(T - t).$$

Exponentiate both sides of the last equation.

$$Ke^{-r(T-t)}e^{-d_2^2/2} = S(t)e^{-\delta(T-t)}e^{-d_1^2/2},$$

$$Ke^{-r(T-t)}\frac{1}{\sqrt{2\pi}}e^{-d_2^2/2} = S(t)e^{-\delta(T-t)}\frac{1}{\sqrt{2\pi}}e^{-d_1^2/2}.$$

Thus the identity has been established.                                    □

## 12.1   Delta and Gamma

The value of an option is sensitive to changes in the price of the underlying stock. This rate of change is called **Delta**, $\Delta$. The astute reader may recall that $\Delta$ was introduced in Sec. 10.1 during the derivation of the Black–Scholes partial differential equation. Delta was chosen to be the partial derivative of the option with respect to the value of the underlying asset $S$ and was used to eliminate a stochastic term from Eq. (10.3). This clever choice of $\Delta$ produced a deterministic partial differential equation to solve rather than a stochastic differential equation. The Delta of a option is therefore defined to be the partial derivative of the option with respect to the underlying asset. For a European call whose value is given in Eq. (10.34),

$$\frac{\partial C^e}{\partial S} = e^{-\delta(T-t)}\Phi\left(d_1\right) + \left(Se^{-\delta(T-t)}\varphi\left(d_1\right) - Ke^{-r(T-t)}\varphi\left(d_2\right)\right)\frac{\partial d_{1,2}}{\partial S}.$$

Making use of Lemma 12.1 then the Delta of a European call is

$$\Delta_{C^e} = e^{-\delta(T-t)}\Phi\left(d_1\right). \qquad (12.2)$$

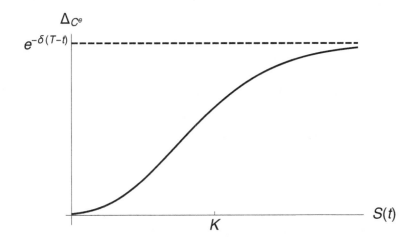

Fig. 12.1  The graph of the Delta for a European call option.

Figure 12.1 illustrates the graph of Delta for a typical European call. Note that $0 < \Delta_{C^e} < e^{-\delta(T-t)} < 1$. If a European call is far out-of-the-money the call's Delta is near zero while if a European call is far in-the-money its Delta asymptotically approaches $e^{-\delta(T-t)}$. Intuitively this can be justified by the realizing that if a call is far out-of-the-money it is unlikely to be exercised even if the price of the underlying asset changes by a small amount. Hence the price of the call is not very sensitive to changes in the price of the underlying asset and $\Delta_{C^e}$ is close to zero. If the European call is far in-the-money the payoff of the call is likely to be $S(T) - K$ and therefore the value of the call is approximately $Se^{-\delta(T-t)} - Ke^{-r(T-t)}$ implying that the Delta of the call is nearly $e^{-\delta(T-t)}$.

Delta for a European put option could be calculated directly from the definition for $P^e$ given in Eq. (10.35); however, the Put-Call Parity formula in Eq. (6.8) provides a convenient shortcut.

$$\frac{\partial}{\partial S}(P^e + S(t)e^{-\delta(T-t)}) = \frac{\partial}{\partial S}(C^e + Ke^{-r(T-t)})$$

$$\frac{\partial P^e}{\partial S} + e^{-\delta(T-t)} = \frac{\partial C^e}{\partial S}$$

$$\frac{\partial P^e}{\partial S} = e^{-\delta(T-t)}(\Phi(d_1) - 1)$$

$$\Delta_{P^e} = -e^{-\delta(T-t)}\Phi(-d_1). \tag{12.3}$$

Figure 12.2 illustrates the graph of Delta for a typical European put. Observe that $-1 < -e^{-\delta(T-t)} < \Delta_{P^e} < 0$. If a European put is far out-of-

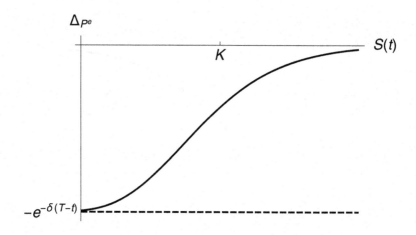

Fig. 12.2   The graph of the Delta for a European put option.

the-money (meaning $S(t)$ is much greater than $K$) the put's Delta is near zero while if a European put is far in-the-money ($S(t)$ is much lower than $K$) its Delta asymptotically approaches $-e^{-\delta(T-t)}$. Intuitively this can be justified by the realizing that if a put is far out-of-the-money it is unlikely to be exercised even if the price of the underlying asset changes by a small amount. Hence the price of the put is not very sensitive to changes in the price of the underlying asset and $\Delta_{P^e}$ is close to zero. If the European put is far in-the-money the payoff of the put is likely to be $K - S(T)$ and therefore the value of the put is approximately $Ke^{-r(T-t)} - Se^{-\delta(T-t)}$ implying that the Delta of the put is nearly $-e^{-\delta(T-t)}$.

**Example 12.1.** Suppose the price of a non-dividend-paying security is $100, the risk-free interest rate is 4% per annum, the annual volatility of the stock price is 23%, the strike price is $110, and the strike time is 6 months. Find the Deltas for European call and put options on the security. What do the Deltas imply about the sensitivity of the option premium to changes in the underlying security price?

**Solution.** Under these conditions $d_1 = -0.381747$ and

$$\Delta_{C^e} = 0.351325.$$

Using the same parameter values,

$$\Delta_{P^e} = 0.351325 - 1 = -0.648675.$$

So, as the European call option increases in value with an increase in the value of the underlying security, the put option decreases in value.

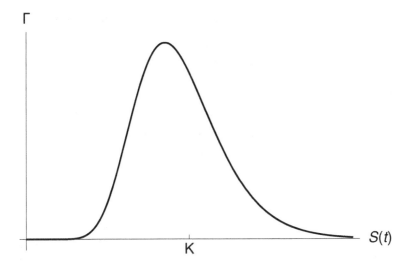

Fig. 12.3    Since $\Gamma^e > 0$ the option premium for a long European put (call) is decreasing (increasing) and concave upward.

The **Gamma**, $\Gamma$, of an option or collection of options is defined to be the second derivative of the option with respect to $S$, the price of the underlying security. Hence Gamma is the partial derivative of Delta with respect to $S$. Thus for a European call option $\Gamma$ is

$$\frac{\partial^2 C}{\partial S^2} = e^{-\delta(T-t)} \varphi(d_1) \frac{\partial d_1}{\partial S} = e^{-\delta(T-t)} \frac{e^{-d_1^2/2}}{\sigma S \sqrt{2\pi(T-t)}}.$$

According to Eq. (12.3) the Gamma for a European put is the same as that for a European call. Therefore without reference to whether the option is a European put or call, Gamma is defined as

$$\Gamma^e = \frac{e^{-\delta(T-t)}}{\sigma S \sqrt{T-t}} \varphi(d_1). \tag{12.4}$$

Pay close attention to the notation, $\varphi(x)$ is the probability density function for the standard normal random variable. Figure 12.3 illustrates the graph of $\Gamma^e$ for a European option. If all variables and parameters are held fixed except for $S$, then $\Gamma$ measures the convexity of the option. This will have important implications for the practice known as hedging which will be explored in Chapter 13. Recall that the Delta of a European put or call option asymptotically approaches 0 or $\pm e^{-\delta(T-t)}$ as the option is very much in-the-money or out-of-the-money. Consequently Gamma approaches 0 under the same circumstances.

**Example 12.2.** Find the Gamma for the European call and put options described in Example 12.1.

**Solution.** Since $S = 100$, $\delta = 0$, and $d_1 = -0.381747$, then $\Gamma^e = 0.022806$.

## Exercises

**12.1.1** The derivation of the simple formula for the Deltas of European calls and puts depended on the partial derivatives of $d_1$ and $d_2$ with respect to $S$ being equal. Show that $\partial d_1/\partial S = \partial d_2/\partial S$.

**12.1.2** Show that $\Delta_{C^e} - \Delta_{P^e} = e^{-\delta(T-t)}$.

**12.1.3** Find Delta for a European call option on a non-dividend-paying stock whose current value is \$150. The annual risk-free interest rate is 2.5%. The strike price is \$165 while the strike time is 5 months. The volatility of the stock is 22% annually.

**12.1.4** Find Delta for a European call option on a non-dividend-paying stock whose current value is \$50. The annual risk-free interest rate is 3.25%. The strike price is \$55 while the strike time is 4 months. The volatility of the stock is 27% annually.

**12.1.5** Find Delta for a European put option on a non-dividend-paying stock whose current value is \$125. The annual risk-free interest rate is 5.5%. The strike price is \$140 while the expiry date is 8 months. The volatility of the stock is 15% annually.

**12.1.6** Find Delta for a European put option on a non-dividend-paying stock whose current value is \$75. The annual risk-free interest rate is 6.5%. The strike price is \$81 while the expiry date is 7 months. The volatility of the stock is 33% annually.

**12.1.7** Calculate Gamma for a European put option for a non-dividend-paying security whose current price is \$180. The strike price is \$175, the annual risk-free interest rate is 3.75%, the volatility in the price of the security is 30% annually, and the strike time is 4 months.

**12.1.8** Calculate Gamma for a European call option for a non-dividend-paying security whose current price is \$205. The strike price is \$195, the annual risk-free interest rate is 4.65%, the volatility in the price of the security is 45% annually, and the strike time is 2 months.

**12.1.9** Suppose the current price of a stock is \$50. The stock pays a continuous dividend proportional to the price of the stock with a dividend yield of 1%. The expected value of the stock price after 6 months is 55 and the variance of the stock price is 140. The Delta of a 6-month, at-the-money European call option is 0.65. Find the price and the Gamma of the call option.

**12.1.10** Find formulas for the Delta and Gamma of asset-or-nothing put and call options.

**12.1.11** Find formulas for the Delta and Gamma of cash-or-nothing put and call options.

**12.1.12** The current price of a stock is $50, the stock pays dividends at a rate that is proportional to the stock price with a dividend yield of 2%, the risk-free, continuously compounded interest rate is 4%, and the volatility of the stock price is less than 20%. A 4-month, at-the-money European put option on the stock has a Delta of $-0.45$. Find the Gamma and price of the put.

**12.1.13** Refer to Sec. 11.2 and show that for options on futures,

$$F(t)\varphi(d_1) = K\varphi(d_2).$$

**12.1.14** Find the Delta and Gamma of a European call on a futures contract. You may use Eq. (11.14), differentiate with respect to $F$, and use the result of Exercise 12.1.13.

**12.1.15** Use the Put-Call parity formula Eq. (11.9) to find Delta and Gamma for put options on futures contracts.

## 12.2 Theta

The quantity **Theta**, denoted $\Theta$, is defined to be the rate of change of the value of an option with respect to $t$. This section will explore $\Theta$ for the European call and put options derived in Chapter 10. Since the valuation of European-style options depends heavily on the cumulative distribution function of the standard normal random variable and the aggregate variables $d_1$ and $d_2$, consider their partial derivatives. The derivative with respect to $t$ of quantity $d_{1,2}$ defined in Eq. (10.32) is given by

$$\frac{\partial d_{1,2}}{\partial t} = \frac{1}{2\sigma\sqrt{T-t}}\left(\frac{\ln(S/K)}{T-t} - r + \delta \mp \frac{\sigma^2}{2}\right). \qquad (12.5)$$

Hence using the Chain Rule [Stewart (1999), Chapter 3] for derivatives, the time rate of change in the value of a European call option (defined in Eq. (10.34)) is

$$\frac{\partial C^e}{\partial t} = \delta D e^{-\delta(T-t)}\Phi(d_1) - rKe^{-r(T-t)}\Phi(d_2)$$

$$+ Se^{-\delta(T-t)}\varphi(d_1)\frac{\partial d_1}{\partial t} - Ke^{-r(T-t)}\varphi(d_2)\frac{\partial d_2}{\partial t}$$

$$\Theta_{C^e} = Se^{-\delta(T-t)}\left(\delta\Phi(d_1) - \frac{\sigma\varphi(d_1)}{2\sqrt{T-t}}\right) - rKe^{-r(T-t)}\Phi(d_2). \qquad (12.6)$$

Alternatively, $\Theta$ can be determined from the Black–Scholes partial differential equation. Note that Eq. (10.4) can be re-written as

$$F_t = -(r - \delta)SF_S - \frac{1}{2}\sigma^2 S^2 F_{SS} + r\,F.$$

Replacing the dependent variable $F$ with the European call produces

$$\Theta_{C^e} = -(r - \delta)S\Delta_{C^e} - \frac{1}{2}\sigma^2 S^2 \Gamma^e + rC^e$$

$$= (\delta - r)Se^{-\delta(T-t)}\Phi(d_1) - \frac{\sigma Se^{-\delta(T-t)}\varphi(d_1)}{2\sqrt{T-t}}$$

$$+ rSe^{-\delta(T-t)}\Phi(d_1) - rKe^{-r(T-t)}\Phi(d_2)$$

$$= Se^{-\delta(T-t)}\left(\delta\Phi(d_1) - \frac{\sigma\varphi(d_1)}{2\sqrt{T-t}}\right) - rKe^{-r(T-t)}\Phi(d_2),$$

the same as before but without explicitly differentiating with respect to $t$.

Figure 12.4 shows a typical graph of $\Theta_{C^e}$. The case that $\Theta$ is negative for a call option is typical since these options lose value as the expiry date approaches. However, $\Theta_{C^e}$ for a European call can be positive if the option is very much in-the-money.

**Example 12.3.** Find Theta for a European call option on a non-dividend-paying stock whose current price is $250. The strike price is $245, the annual risk-free interest rate is 2.5%, the volatility of the stock price is 20%, and the strike time is 4 months.

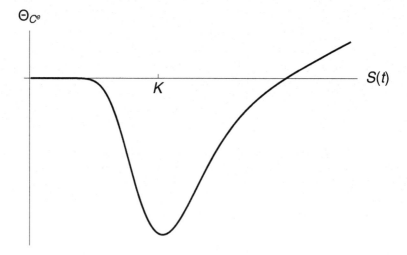

Fig. 12.4    Theta for a long European call is usually negative unless the option is significantly in-the-money.

**Solution.** Substituting $S = 250$, $K = 245$, $r = 0.025$, $\sigma = 0.20$, $T = 4/12$, and $t = 0$ into the formulas for $d_1$ and $d_2$ produce:

$$d_1 = \frac{\ln\frac{250}{245} + \left(0.025 + \frac{(0.20)^2}{2}\right)(4/12 - 0)}{0.20\sqrt{4/12 - 0}} = 0.304864$$

$$d_2 = 0.304864 - 0.20\sqrt{4/12 - 0} = 0.189394.$$

Using these values in Eq. (12.6) produces $\Theta_{C^e} = -19.9836$.

The calculation of $\Theta$ for a European put option can be carried out in a similar fashion to that for the call option or a somewhat simpler approach makes use of the Put-Call parity formula in Eq. (6.8).

$$\frac{\partial P^e}{\partial t} = \frac{\partial}{\partial t}\left[C^e + Ke^{-r(T-t)} - Se^{-\delta(T-t)}\right]$$

$$\Theta_{P^e} = rKe^{-r(T-t)}\Phi(-d_2) - Se^{-\delta(T-t)}\left(\delta\Phi(-d_1) + \frac{\sigma\varphi(d_1)}{2\sqrt{T-t}}\right). \quad (12.7)$$

The graph of the Theta for a European put is illustrated in Fig. 12.5. As was the case with the Theta for a call, this Theta is negative in most typical situations, but can be positive for puts which are very much out-of-the-money.

**Example 12.4.** Find Theta for a European call option on a non-dividend-paying stock whose current price is \$325. The strike price is \$330, the annual

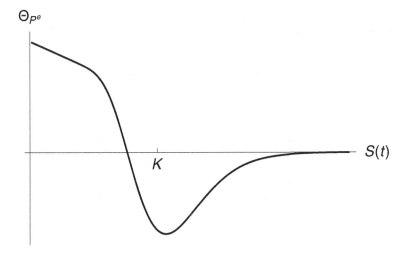

Fig. 12.5 Theta for a long European put is usually negative unless the option is significantly out-of-the-money.

risk-free interest rate is 3.5%, the volatility of the stock price is 27%, and the strike time is 3 months.

**Solution.** Substituting $S = 325$, $K = 330$, $r = 0.035$, $\sigma = 0.27$, $T = 3/12$, and $t = 0$ into the formulas for $d_1$ and $d_2$ produce:

$$d_1 = \frac{\ln \frac{325}{330} + \left(0.035 + \frac{(0.27)^2}{2}\right)(3/12 - 0)}{0.27\sqrt{3/12 - 0}} = 0.019222$$

$$d_2 = 0.019222 - 0.27\sqrt{3/12 - 0} = -0.115778.$$

Using these values in Eq. (12.7) produces

$$\Theta_{P^e} = -28.7484.$$

The derivative $\Theta$ is unique among the derivatives covered in this chapter since $t$ is the only non-stochastic variable among the list of independent variables upon which $C^e$ and $P^e$ depend. Since three of the most important Greeks have been introduced, the Black–Scholes partial differential equation can be recast to incorporate these quantities. If $F(S,t)$ satisfies Eq. (10.4) then

$$\Theta + (r - \delta)S\Delta + \frac{1}{2}\sigma^2 S^2 \Gamma = rF, \tag{12.8}$$

where $\Theta$, $\Delta$, and $\Gamma$ are partial derivatives of $F$ with respect to $t$, $S$, and $S$ twice respectively.

## Exercises

**12.2.1** Find Theta for a European call option on a non-dividend-paying stock whose current value is $300. The annual risk-free interest rate is 3%. The strike price is $310 while the strike time is 3 months. The volatility of the stock is 25% annually.

**12.2.2** Consider the Theta for a European call option on a non-dividend-paying stock whose current value is $100. The annual risk-free interest rate is 4.12%. The strike price of the option is $103 and expiry occurs in 12 months. The volatility of the stock price is 30% annually. Plot $\Theta_{C^e}$ as a function of the stock price at 12 months to expiry, six months to expiry, and 1 month to expiry.

**12.2.3** Find Theta for a European put option on a non-dividend-paying stock whose current value is $275. The annual risk-free interest rate is 2%. The strike price is $265 while the expiry date is 4 months. The volatility of the stock is 20% annually.

**12.2.4** Suppose the current price of a stock is $45, the stock pays a continuous dividend at a rate proportional to the stock price with a dividend yield of 2%, the volatility of the stock's return is less than 35%, a 3-month at-the-money European put on the stock has a Delta of $-0.4321$, and the continuously compounded risk-free interest rate is 6%.

(a) Find the premium charged for the European put.
(b) Find Gamma for the put.
(c) Find Theta for the put.

**12.2.5** A non-dividend-paying stock has a current price of $100 and a volatility of 30%. A European put on the stock has a price of $P^e = \$8.8846$ and a $\Theta_{P^e} = -2.0132$. A European call on the stock with the same strike price and time has a $\Delta_{C^e} = 0.6371$ and $\Gamma^e = 0.0125$. Find the continuously compounded, risk-free interest rate.

**12.2.6** Suppose the current price of a stock is $50 and the stock pays a continuous dividend at a rate proportional to the stock price with a dividend yield of 3%. The expected value of the stock's price in 6 months is $53.0918. The variance of the stock's price in 6 months is 157.736. The Delta of a 6-month, at-the-money European call option is 0.3456.

(a) Find the premium charged for the 6-month European call.
(b) Find Gamma for the call.
(c) Find Theta for the call.

**12.2.7** Consider a strangle on a non-dividend-paying stock whose price is currently $150 and whose volatility is 40%. After 1 year the payoff of the strangle will be

$$(S(1) - 160)^+ - (S(1) - 140)^+.$$

The continuously compounded, risk-free interest rate is 5%. Find the Theta for the strangle.

**12.2.8** Show that an equivalent formula for the Theta of a European put is

$$\Theta_{P^e} = rKe^{-r(T-t)}\Phi\left(-d_2\right) - \delta Se^{-\delta(T-t)}\Phi\left(-d_1\right) - \frac{\sigma Ke^{-r(T-t)}\varphi\left(d_2\right)}{2\sqrt{T-t}}.$$

## 12.3   Vega, Rho, and Psi

The **Vega**, $\mathcal{V}$, of an option is the change in the value of the option as a function of the volatility of the underlying security. Among the partial derivatives of the value of an option, Vega distinguishes itself as the only partial derivative given a non-Greek letter name. The name Vega comes through medieval Latin arising from Arabic.

For a European call option defined as in Eq. (10.34),

$$\frac{\partial C^e}{\partial \sigma} = Se^{-\delta(T-t)}\varphi\left(d_1\right)\frac{\partial d_1}{\partial \sigma} - Ke^{-r(T-t)}\varphi\left(d_2\right)\frac{\partial d_2}{\partial \sigma},$$

where, according to Exercise 12.3.1,

$$\frac{\partial d_{1,2}}{\partial \sigma} = \frac{-d_{2,1}}{\sigma}.$$

Substituting this into the partial derivative of the call pricing formula and using Lemma 12.1 yield

$$\frac{\partial C^e}{\partial \sigma} = Se^{-\delta(T-t)}\varphi\left(d_1\right)\left(-\frac{d_2}{\sigma} + \frac{d_1}{\sigma}\right)$$
$$\mathcal{V} = Se^{-\delta(T-t)}\varphi\left(d_1\right)\sqrt{T-t}.$$

Note that the value of the European call option increases with increasing volatility in the underlying security. It is easily seen from the Put-Call Parity formula in Eq. (6.8) that

$$\frac{\partial P^e}{\partial \sigma} = Se^{-\delta(T-t)}\varphi\left(d_1\right)\sqrt{T-t},$$

in other words the rate of change in the value of a European put option with respect to volatility is the same as the rate of change in the value of the corresponding European call option with respect to volatility. Thus Vega is unambiguously defined for a European-style option as

$$\mathcal{V} = Se^{-\delta(T-t)}\varphi\left(d_1\right)\sqrt{T-t}. \tag{12.9}$$

The general appearance of the graph of Vega is shown in Fig. 12.6.

**Example 12.5.** Find Vega for a European-style option (either call or put) on a non-dividend-paying security whose current value is \$160, whose strike price is \$150, and whose expiry date is 5 months. Assume the volatility of the security is 20% per year and the risk-free interest rate is 2.75% per year.

## Vega

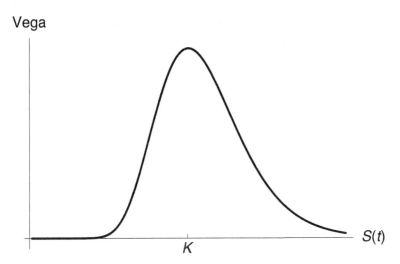

Fig. 12.6   Vega for long European calls and puts is positive.

**Solution.** Vega can be calculated using Eq. (12.9).

$$d_1 = \frac{\ln \frac{160}{150} + \left(0.0275 - 0 + \frac{(0.20)^2}{2}\right)(5/12 - 0)}{0.20\sqrt{5/12 - 0}} = 0.653219.$$

Then

$$\mathcal{V} = 160\varphi\left(0.653219\right)\sqrt{5/12} = 33.2866.$$

The **Rho**, $\rho$, of an option is the rate of change in the value of the option as a function of the risk-free interest rate $r$. Once again the European call option will be used to demonstrate the calculation of Rho. Using Eq. (10.34) it is seen that

$$\frac{\partial C^e}{\partial r} = Se^{-\delta(T-t)}\varphi\left(d_1\right)\frac{\partial d_1}{\partial r} - Ke^{-r(T-t)}\varphi\left(d_2\right)\frac{\partial d_2}{\partial r}$$
$$+ Ke^{-r(T-t)}\Phi\left(d_2\right)(T-t)$$
$$= Se^{-\delta(T-t)}\varphi\left(d_1\right)\left(\frac{\partial d_1}{\partial r} - \frac{\partial d_2}{\partial r}\right) + Ke^{-r(T-t)}\Phi\left(d_2\right)(T-t)$$
$$\rho_{C^e} = K(T-t)e^{-r(T-t)}\Phi\left(d_2\right). \tag{12.10}$$

The utility of Lemma 12.1 is demonstrated in the simplification of the formula for Rho. Note that for $\rho_{C^e} > 0$ implying that the value of a European call option increases with increasing interest rate.

**Example 12.6.** Consider the situation of a European call option on a non-dividend-paying security whose current value is \$225. The strike price is \$230, while the strike time is 4 months. The volatility of the price of the security is 27% annually and the risk-free interest rate is 3.25%. Find Rho for this option.

**Solution.** Using the given values of the parameters,

$$d_2 = \frac{\ln \frac{225}{230} + \left(0.0325 - 0 - \frac{(0.27)^2}{2}\right)(4/12 - 0)}{0.27\sqrt{4/12 - 0}} = -0.149441,$$

$$\rho_{C^e} = 230(4/12)e^{-0.0325(4/12)}\Phi(-0.149441) = 33.4156.$$

The Rho for a European put option can be determined from the Put-Call Parity formula of Eq. (6.8) and Eq. (12.10).

$$\frac{\partial P^e}{\partial r} = \frac{\partial C^e}{\partial r} - K(T - t)e^{-r(T-t)}$$

$$= K(T - t)e^{-r(T-t)}\Phi(d_2) - K(T - t)e^{-r(T-t)}$$

$$\rho_{P^e} = -K(T - t)e^{-r(T-t)}\Phi(-d_2). \tag{12.11}$$

The last equation above made use of the fact that $\Phi(x) = 1 - \Phi(-x)$ for the cumulative distribution function of a normally distributed random variable. Note that $\rho_{P^e} < 0$ implying that the value of a European put option decreases with increasing interest rate. Graphs of Rho for European calls and puts are shown in Fig. 12.7.

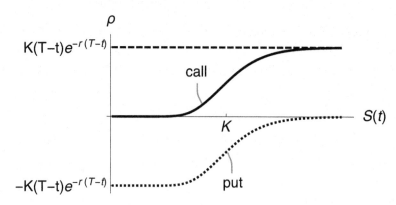

Fig. 12.7   Rho for long European calls is positive while rho for long European puts is negative. Rho approaches asymptotic values for options very much in-the-money and out-of-the-money.

**Example 12.7.** Consider the situation of a European put option on a security whose underlying value is \$150. The strike price is \$152, the volatility is 15% per annum, the expiry date is 3 months, and the risk-free interest rate is 2.5%. Find Rho for this option.

**Solution.** Using the given values of the parameters,

$$d_2 = \frac{\ln \frac{150}{152} + \left(0.025 - 0 - \frac{(0.15)^2}{2}\right)(3/12 - 0)}{0.15\sqrt{3/12 - 0}} = -0.13077,$$

$$\rho_{C^e} = -152(3/12)e^{-0.025(3/12)}\Phi(0.13077) = -20.8461.$$

The continuous dividend yield $\delta$ in the Black–Scholes model for option prices produces the final Greek, **Psi** denoted $\Psi$. For a European call option

$$
\begin{aligned}
\frac{\partial C^e}{\partial \delta} &= -(T-t)Se^{-\delta(T-t)}\Phi(d_1) + Se^{-\delta(T-t)}\varphi(d_1)\frac{\partial d_1}{\partial \delta} \\
&\quad - Ke^{-r(T-t)}\varphi(d_2)\frac{\partial d_1}{\partial \delta} \\
&= -(T-t)Se^{-\delta(T-t)}\Phi(d_1) + Se^{-\delta(T-t)}\varphi(d_1)\left(\frac{\partial d_1}{\partial \delta} - \frac{\partial d_1}{\partial \delta}\right) \\
\Psi_{C^e} &= -(T-t)Se^{-\delta(T-t)}\Phi(d_1).
\end{aligned}
$$

$$(12.12)$$

Since $\Psi_{C^e} < 0$ the value of the European call option on a stock paying constant, continuous dividends at rate $\delta$ is a decreasing function of $\delta$. Employing the Put-Call parity formula from Eq. (6.8) enables the formula for Psi for European puts to be determined.

$$
\begin{aligned}
\frac{\partial P^e}{\partial \delta} &= \frac{\partial}{\partial \delta}\left[C^e + Ke^{-r(T-t)} - Se^{-\delta(T-t)}\right] \\
&= \Psi_{C^e} + (T-t)Se^{-\delta(T-t)} \\
&= (T-t)Se^{-\delta(T-t)} - (T-t)Se^{-\delta(T-t)}\Phi(d_1) \\
\Psi_{P^e} &= (T-t)Se^{-\delta(T-t)}\Phi(-d_1).
\end{aligned}
$$

$$(12.13)$$

Figure 12.8 illustrates typical graphs of Psi for calls and puts.

*Exercises*

**12.3.1** Show that $\dfrac{\partial d_{1,2}}{\partial \sigma} = \dfrac{-d_{2,1}}{\sigma}$.

**12.3.2** Show that $\dfrac{\partial d_{1,2}}{\partial \delta} = \dfrac{-\sqrt{T-t}}{\sigma}$

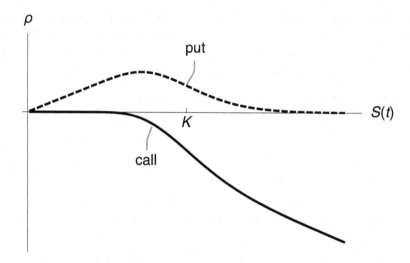

Fig. 12.8   Psi for long European calls is negative while Psi for long European puts is positive. Psi for a European call approaches 0 for options which are far out-of-the-money. Psi for a European put approaches 0 for options which are far in-the-money.

**12.3.3** Show that $\mathcal{V} = \Gamma^e \sigma S^2 (T - t)$.

**12.3.4** Calculate the $\mathcal{V}$ of a European-style option on a non-dividend-paying security whose current value is \$300, whose strike price is \$305, and whose expiry date is 6 months. Suppose the volatility of the security is 25% per year and the risk-free interest rate is 4.75% per year.

**12.3.5** Calculate the $\mathcal{V}$ of a European-style option on a non-dividend-paying security whose current value is \$123, whose strike price is \$125, and whose expiry date is 3 months. Suppose the volatility of the security is 35% per year and the risk-free interest rate is 5.15% per year.

**12.3.6** Calculate the Rho of a European style put option on a non-dividend-paying security whose current value is \$270, whose strike price is \$272, and whose strike time is 2 months. Suppose the volatility of the security is 15% per year and the risk-free interest rate is 3.75% per year.

**12.3.7** Calculate the Rho of a European style call option on a non-dividend-paying security whose current value is \$305, whose strike price is \$325, and whose expiry date is 4 months. Suppose the volatility of the security is 35% per year and the risk-free interest rate is 2.55% per year.

**12.3.8** Suppose a stock is currently priced at \$75 and pays a continuous dividend with a yield of 2.5% annually. The volatility of the stock is 50% per annum and the expected return of the stock is 18%. The continuously compounded, risk-free interest rate is 5%. Find the Vega, Rho, and Psi for a 6-month European put option on the stock with a strike price of $70.

**12.3.9** Suppose an investor purchases a 45-strike put and sells a 50-strike put (a bull spread). The stock underlying the spread has a volatility of 30%, a dividend yield of 1%, and the spread expires in 3 months. If the continuously compounded, risk-free interest rate is 6% and the current price of the stock is $47.50, find the Vega, Rho, and Psi for the spread.

## 12.4  Applications of the Greeks

The Greeks together with linear and quadratic approximation can be used to produce approximations to European option prices when one or more of the parameters of the Black–Scholes model change by a small amount. To keep the discussion generally applicable to both calls and puts, the value of the option will be denoted $V(S(t), \delta, \sigma, r, t)$ where the variables $S(t)$, $\delta$, $\sigma$, $r$, and $t$ have the traditional interpretations.

If the price of the asset underlying the European option changes by a small amount, an investor may find the new price of the option by repeating all the work done to find the original price of the option – or may save some effort by using the original option price and one or more Greeks associated with the change in $S$, namely Delta and Gamma. The simplest approximation formula is the Delta approximation which is a linear approximation to the option price.

$$V(S(t) + dS, \delta, \sigma, r, t) \approx V(S(t), \delta, \sigma, r, t) + \Delta_V dS \qquad (12.14)$$

The quantity $\Delta_V$ is the Delta of the option. A generally more accurate approximation formula includes an additional term involving the Gamma of the option. This formula is commonly known as the Delta/Gamma approximation.

$$V(S(t) + dS, \delta, \sigma, r, t)$$
$$\approx V(S(t), \delta, \sigma, r, t) + \Delta_V dS + \frac{1}{2}\Gamma_V (dS)^2. \qquad (12.15)$$

**Example 12.8.** For the call and put options described in Example 12.1 approximate the prices of the options using the Delta approximations and the Delta/Gamma approximation if the price of the underlying stock increases by $5.

**Solution.** As originally described in Example 12.1 the premiums for the call and put are respectively $C^e(100) = 3.5310$ and $P^e(100) = 11.3529$. The

Deltas for the calla and put are $\Delta_{C^e} = 0.351325$ and $\Delta_{P^e} = -0.648675$. Using the Delta approximation of Eq. (12.14) gives the new prices of the options as

$$C^e(105) \approx C^e(100) + \Delta_{C^e}(5)$$
$$= 3.5310 + (0.351325)(5) = 5.2876$$
$$P^e(105) \approx P^e(100) + \Delta_{P^e}(5)$$
$$= 11.3529 + (-0.648675)(5) = 8.1095.$$

The Gamma for these options was calculated in Example 12.2 and found to be $\Gamma^e = 0.022806$. The Delta/Gamma approximation in Eq. (12.15) gives new prices for the call and put as

$$C^e(105) \approx C^e(100) + \Delta_{C^e}(5) + \frac{1}{2}\Gamma^e(5)^2 = 5.2876 + \frac{25}{2}(0.022806) = 5.5727$$
$$P^e(105) \approx P^e(100) + \Delta_{P^e}(5) + \frac{1}{2}\Gamma^e(5)^2 = 8.1095 + \frac{25}{2}(0.022806) = 8.3946.$$

For the sake of comparison the exact prices of the call and put when the value of the underlying stock is \$105 are $C^e(105) = 5.5769$ and $P^e(105) = 8.3987$, hence both approximation formulas yield good results.

One interpretation of the Delta approximation is that the change in the price of the option can be expressed as the change in the price of the underlying asset multiplied by the Delta for the option, or

$$V(S + dS) - V(S) = \Delta_V dS.$$

Thus for every unit change in the price of the underlying asset, the option price changes by $\Delta_V$. Rather than thinking of this change in absolute terms, think of it as the relative change.

$$\frac{\Delta_V dS}{V(S)} = \frac{V(S + dS) - V(S)}{V(S)}$$
$$\frac{\Delta_V S}{V(S)} = \frac{(V(S + dS) - V(S)) S}{V(S) dS}$$
$$\Omega = \frac{(V(S + dS) - V(S))/V(S)}{dS/S}.$$

The quantity $\Omega = \Delta_V S/V$ is the relative change in the price of the option divided by the relative change in the price of the underlying asset. This may be interpreted as stating that for every 1% change in the price of the underlying asset, the option will change in value by $\Omega\%$. The quantity $\Omega$ is called the **option elasticity**. For European calls $\Omega \geq 1$ and for puts $\Omega \leq 0$

(see Exercise 12.4.9). Since the price of an option depends on the price of the underlying asset, then define the volatility of the option (denoted $\sigma_V$) using the option elasticity and the volatility of the underlying asset, $\sigma$.

$$\sigma_V = \sigma|\Omega|. \tag{12.16}$$

Since volatility is positive the absolute value of option elasticity must be used.

Recall that volatility is a measure of the risk in a security. Thus when $|\Omega| > 1$ options are riskier than their underlying assets. According to the discussion in Sec. 10.1 an option is equivalent to a portfolio consisting of a position in Delta shares of the underlying stock and risk-free borrowing or lending of an amount $B$. The stock has an expected rate of return $\alpha$ and borrowing/lending takes place at the continuously compounded risk-free rate $r$. Thus an option possesses an expected rate of return which is a weighted average of $\alpha$ and $r$. Let the option $V = (\Delta)S + B$ and denote its expected rate of return compounded continuously as $\gamma$.

$$V e^{\gamma\,dt} = (\Delta)S e^{\alpha\,dt} + B e^{r\,dt}$$

$$e^{\gamma\,dt} = \frac{(\Delta)S e^{\alpha\,dt} + (V - (\Delta)S)e^{r\,dt}}{V}$$

$$= \Omega e^{\alpha\,dt} + (1 - \Omega)e^{r\,dt}.$$

If $dt$ is an infinitesimal interval of time then define

$$\gamma = \Omega\alpha + (1 - \Omega)r. \tag{12.17}$$

The Delta/Gamma approximation can be extended further by incorporating a term accounting for the passage of time. This is known as the Delta/Gamma/Theta approximation.

$$V(S(t) + dS, \delta, \sigma, r, t + dt)$$

$$\approx V(S(t), \delta, \sigma, r, t) + \Delta_V dS + \frac{1}{2}\Gamma_V(dS)^2 + \Theta_V dt. \tag{12.18}$$

**Example 12.9.** For the European call and put options described in Example 12.8 assume their Thetas are respectively $\Theta_{C^e} = -7.29629$ and $\Theta_{P^e} = -2.98342$. Approximate the premiums for the call and put option if $S(1/12) = 105$.

**Solution.** Using Eq. (12.18) the call and put premiums are approximately,

$$C^e(105, 1/12) \approx 5.5727 + (-7.29629)(1/12) = 4.9647,$$

$$P^e(105, 1/12) \approx 8.3946 + (-2.98342)(1/12) = 8.1459.$$

Again these values can be compared against the exact values which are respectively $C^e(105, 1/12) = 4.8434$ and $P^e(105, 1/12) = 8.0252$.

The price of an option also depends on the volatility of the underlying asset, the risk-free interest rate, and the dividend yield of the asset. An approximation formula based on changes in these parameters is called the Vega/Rho/Psi approximation and can be expressed as

$$V(S, \delta + d\delta, \sigma + d\sigma, r + dr, t)$$
$$\approx V(S(t), \delta, \sigma, r, t) + \mathcal{V} d\sigma + \rho_V dr + \Psi_V d\delta. \tag{12.19}$$

**Example 12.10.** For the call option described in Example 12.1 approximate the following prices of the option using the Vega/Rho/Psi approximation.

(i) Suppose the risk-free interest rate changes to 4.5%.
(ii) Suppose the stock pays a dividend at a yield rate of 1%.
(iii) Suppose the volatility of the stock decreases to 20%.
(iv) Suppose the risk-free interest rate changes to 2.5%, the stock pays a dividend at a yield rate of 2%, and volatility of the stock increases to 25%.

**Solution.** First calculate the Vega, Rho, and Psi for the European call option as $\mathcal{V} = 26.2271$, $\rho_{C^e} = 15.8007$, and $\Psi_{C^e} = -17.5662$.

$$C^e(r = 4.5\%) \approx 3.5310 + (15.8007)(0.005) = 3.6100$$
$$C^e(\delta = 1\%) \approx 3.5310 + (-17.5662)(0.01) = 3.3553$$
$$C^e(\sigma = 20\%) \approx 3.5310 + (26.2271)(-0.03) = 2.7442$$
$$C^e(r = 2.5\%, \delta = 2\%, \sigma = 25\%) \approx 3.5310 + (15.8007)(-0.015)$$
$$+ (-17.5662)(0.02) + (26.2271)(0.02)$$
$$= 3.4672.$$

For comparison the exact prices (rounded to four decimal places) would be

$$C^e(r = 4.5\%) = 3.6100$$
$$C^e(\delta = 1\%) = 3.3586$$
$$C^e(\sigma = 20\%) = 2.7564$$
$$C^e(r = 2.5\%, \delta = 2\%, \sigma = 25\%) = 3.4797.$$

*Exercises*

**12.4.1** The current price of a stock is $25, an option on the stock is sold for a premium of $10.8272, and the Gamma of the option is 0.0475. When the price of the stock increases by $1 the price of the option decreases by 1.0061. Approximate the price of the option when the price of the stock is $24.

**12.4.2** An option written on a particular stock has a Delta of 1.9565 and Gamma of 0.008468. The price of the option decreases by $0.20 when the price of the stock changes by $x$. Find $x$.

**12.4.3** The current price of a stock is $27.50/share and the price of a 4-month, asset-or-nothing call option is $14.8753 with a Delta of 2.0567. Approximate the price of the option if the stock price is $28/share.

**12.4.4** Suppose the price of a non-dividend paying stock is $54 per share with volatility 23% per annum and the risk-free interest rate is 4.5% per year. Calculate the price of a European call option with a strike price of $57 per share and a strike time of 5 months. Use Delta and a linear approximation to determine the price of a 4-month call option on a stock whose current price is $54.75 per share (all other features of the stock and option are the same as before). Compare the linear approximation with the Black–Scholes value of the second option.

**12.4.5** The linear approximation in Exercise (12.4.4) over estimates the value of the second call option since it does not take into account the change in the value of the option due to the passage of time. Repeat the previous linear approximation including both Delta and Theta for the option.

**12.4.6** The linear approximation calculated in Exercise (12.4.5) is closer to the exact value of the second call option than the approximation found in Exercise (12.4.4). Including the price correction associated with Gamma will produce an even more accurate approximation. Repeat the previous approximation including Delta, Gamma, and Theta for the option.

**12.4.7** Calculate the option elasticity of a European style call option on a non-dividend-paying security whose current value is $425, whose strike price is $435, and whose expiry date is 6 months. Suppose the volatility of the security is 17% per year and the risk-free interest rate is 5.25% per year.

**12.4.8** Calculate the option elasticity of a European style put option on a non-dividend-paying security whose current value is $315, whose strike price is $310, and whose expiry date is 4 months. Suppose the volatility of the security is 24% per year and the risk-free interest rate is 4.75% per year.

**12.4.9** Show that for European calls $\Omega \geq 1$ and for puts $\Omega \leq 0$.

**12.4.10** Find the volatility of the European call option described in Exercise 12.4.7.

**12.4.11** Estimate the change in the value of the option in Exercise 12.3.5 if the volatility of the security is 30% per year.

**12.4.12** Suppose the current price of a stock is $75/share and the stock pays a continuous dividend with a yield of 2% per annum. The volatility of the stock is 35% and the continuously compounded, risk-free interest rate is 5.5%. Find the volatility of a 6-month, at-the-money European put option on the stock.

**12.4.13** Suppose a stock has an expected appreciation rate of 12%, and pays a continuous dividend with a yield of 3% per annum. The volatility of the stock is 45% and the continuously compounded, risk-free interest rate is 6.5%. Find the expected return of a 3-month, at-the-money European call option on the stock.

**12.4.14** Suppose the current price of a stock is $55/share and the stock pays a continuous dividend with a yield of 1.5% per annum. The volatility of the stock is 25% and the Delta for a 6-month European call option on the stock is 0.75. Find the volatility of the option.

**12.4.15** The option elasticity of a portfolio of options all based on the same underlying asset is the value-weighted average of the option elasticities of the component options of the portfolio. Suppose an investor takes a position in $w_i$ units of the $i$th option whose price is $V_i$ and whose option elasticity is $\Omega_i$ for $i = 1, 2, \ldots, n$. Show that the option elasticity of the portfolio,

$$\Omega_P = \sum_{i=1}^{n} \frac{w_i V_i \Omega_i}{P}, \qquad (12.20)$$

where $P$ is the total value of the portfolio.

**12.4.16** Calculate the volatility of the strangle described in Exercise 12.2.7 using the result of Exercise 12.4.15.

Table 12.1 summarizes the option value derivatives discussed in this chapter and lists some of their relevant properties. This table is provided as a convenient place to find the definitions of these quantities in the future.

In the next chapter the practice of hedging will be described. Hedging is performed on a **portfolio** of securities, a collection which may include stocks, put and call options, cash in savings accounts, *etc.* The mathematical aspect of hedging is made possible by the linearity property of the Black–Scholes partial differential equation. An operator $L[\cdot]$ is linear if $L[aX+Y] = aL[X] + L[Y]$ for all scalars $a$ and vectors $X, Y$ in the domain of $L$. In the case of the Black–Scholes PDE the vectors are solution functions. Define the operator $L[\cdot]$ to be

$$L[X] = \frac{\partial X}{\partial t} + (r - \delta)S\frac{\partial X}{\partial S} + \frac{1}{2}\sigma^2 S^2 \frac{\partial^2 X}{\partial S^2} - rX,$$

Table 12.1 Listing of the derivatives of European option values and some of their properties.

| Name | Symbol | Definition | Property |
|---|---|---|---|
| Delta | $\Delta_{C^e}$ | $e^{-\delta(T-t)}\Phi(d_1)$ | $\Delta_{C^e} > 0$ |
| | $\Delta_{P^e}$ | $-e^{-\delta(T-t)}\Phi(-d_1)$ | $\Delta_{P^e} < 0$ |
| Gamma | $\Gamma^e$ | $\dfrac{e^{-\delta(T-t)}\varphi(d_1)}{\sigma S\sqrt{T-t}}$ | $\Gamma^e > 0$ |
| Theta | $\Theta_{C^e}$ | Eq. (12.6) | |
| | $\Theta_{P^e}$ | Eq. (12.7) | |
| Vega | $\mathcal{V}$ | $Se^{-\delta(T-t)}\varphi(d_1)\sqrt{T-t}$ | $\mathcal{V} > 0$ |
| Rho | $\rho_{C^e}$ | $K(T-t)e^{-r(T-t)}\Phi(d_2)$ | $\rho_{C^e} > 0$ |
| | $\rho_{P^e}$ | $-K(T-t)e^{-r(T-t)}\Phi(-d_2)$ | $\rho_{P^e} < 0$ |
| Psi | $\Psi_{C^e}$ | $-(T-t)Se^{-\delta(T-t)}\Phi(d_1)$ | $\Psi_{C^e} < 0$ |
| | $\Psi_{P^e}$ | $(T-t)Se^{-\delta(T-t)}\Phi(-d_1)$ | $\Psi_{P^e} > 0$ |

then any solution $X$, to the Black–Scholes partial differential equation, Eq. (10.4), will satisfy $L[X] = 0$. The Greeks will be used in the process of hedging to reduce risk in portfolios.

# Chapter 13

# Hedging

**Hedging** is the practice of making a portfolio of investments less sensitive to changes in market variables such as the prices of securities and interest rates. Market makers, those who create places where options can be bought and sold, are exposed to the changes in the values of options brought on by changes in the prices of the underlying securities. This chapter will describe techniques market makers use to reduce their risk of financial loss due to adverse price movements. The techniques will depend heavily on the use of some of the Greeks presented in Chapter 12, specifically Delta, Gamma, and Theta. This chapter will start with a discussion of hedging practices making use of only Delta, then later extend to the use of Gamma and Theta. Recall that if $F(S,t)$ is a solution to the Black–Scholes partial differential equation Eq. (10.4), the quantity $\Delta = F_S$ is central to hedging a portfolio. In fact this same quantity $\Delta$ was used to derive the Black–Scholes partial differential equation. The term "Delta" may be used in two senses in this chapter. In some cases $\Delta$ will refer to the instantaneous rate of change of a financial derivative with respect to the value of the underlying security. This is the mathematical Delta. The term may also be used as part of the phrase "Delta-neutral" used to describe a portfolio consisting of the underlying security and financial derivatives. The portfolio $\mathcal{P}$ made up of one option $F$ and $\Delta$ shares of the underlying security $S$ is Delta-neutral if $\mathcal{P}_S = 0$. This is the quantitative analytical sense of "Delta". The Black–Scholes equation can be thought of as the statement that when the $\Delta$ of a portfolio is zero, the rate of return from a portfolio consisting of ownership of the underlying security and sale of the option (or the opposite positions) should be the same as the rate of return at the risk-free interest rate on the same net amount of cash.

## 13.1   General Principles

Before launching into a discussion of hedging, recall what happens when a market maker sells a call option on a stock to an investor. The market maker has promised that the investor will be able to purchase the stock for no more than the call option's strike price on or prior to the call option's expiry. The stock must be available to the investor at the strike price even if the market value of the stock exceeds the strike price. The market maker is in a favorable position if, at the strike time, the market price of the stock is below the strike price set when the call option was sold to the investor. In this case the call option will not be exercised and the market maker keeps the premium it received when the option was sold. Conversely, if the market price of the stock exceeds the strike price of the option, the market maker must ensure the investor can find a seller of the stock who will agree to accept the strike price. One way to accomplish this would be for the market maker to be the seller of the stock to the investor.

Consider this scenario: a market maker sells 100 European call options on a non-dividend-paying stock whose current price is $50 while the strike price is $52, the risk-free interest rate is 2.5% per annum, the expiry date is 4 months, and the volatility of the stock price is 22.5% per annum. According to the Black–Scholes option pricing formula, the value of this European call option is $1.91965. The market maker has a positive cash flow of $191.965. At two extremes, the market maker could wait until the expiry date to purchase the 100 shares of the stock, or they could purchase the shares at the time the investor bought the call options. The former strategy is called a **naked** position since the market maker's potential responsibility for providing the stock at the strike price is exposed to the risk that the stock price will increase. Potential losses in this position are unlimited. The latter strategy is called a **covered** position since the market maker is now protected against increases in the price of the security prior to the expiration of the option. However, the portfolio of stocks held by the market maker is exposed to the risk that the stock price falls. In contrast to the naked position, the potential losses in the covered position are bounded below. Suppose the market maker takes the covered position, and purchases 100 shares of stock using the proceeds of the option sale and funds borrowed at the risk-free rate. At expiry the market maker will experience a net cash flow of

$$100 \min\{S(4/12), 52\} - 100(50 - 1.91965)e^{0.025(4/12)}.$$

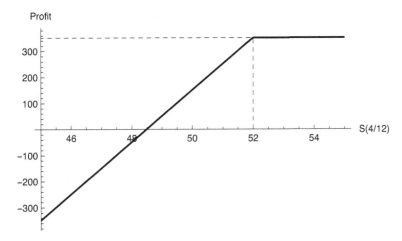

Fig. 13.1 The line shows the profit or loss to the seller of a European call option who adopts a covered position by purchasing the underlying security at the time they sell the option.

The graph of the profit is illustrated in Fig. 13.1. The minimum profit is $-\$4848.27$, the maximum profit is $\$351.731$ (and occurs when $S(4/12) = \$52$), and the market maker breaks even when $S(4/12) = \$48.4827$.

If the market maker adopts a naked position the profit/loss at expiry will be

$$-100(S(4/12) - 52)^+ + 100(1.91965)e^{0.025(4/12)}.$$

Note that the market maker is investing the initial cash flow from the sold options at the risk-free rate. The profit is zero when $S(4/12)$ is approximately $\$53.9357$. As long as the stock price remains below the strike price the market maker keeps the initial cash flow plus interest from the sale of the options. The maximum profit is $\$193.571$. However the market maker's losses are unlimited as $S(4/12)$ increases, which is always a possibility. Figure 13.2 illustrates the profit to the seller of the European call option.

Neither of these schemes are practical for hedging portfolios in the financial world due to their potentially large costs. The naked and the covered positions represent two extremes, one in which no stock is held and the other in which all the potentially needed shares are held from the moment the option is sold until expiry. A better strategy may be a compromise between these extremes. Perhaps some fraction of the total, potential number of necessary shares should be held by the market maker. The optimal number would be the number which reduces the potential loss to the market maker. In the next section a simple method for hedging is explored.

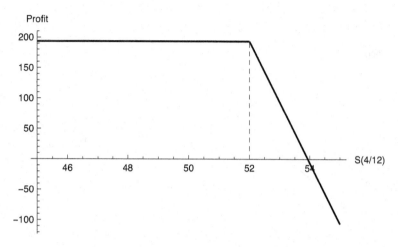

Fig. 13.2    The line shows the profit or loss to the seller of a European call option who adopts a naked position by purchasing the underlying security at the expiry date of the option.

## *Exercises*

**13.1.1** The current price of a non-dividend-paying stock is \$65/share. The volatility of the return on the stock is 35% and the continuously compounded risk-free interest rate is 5%. A market maker has sold 100 6-month \$70-strike call options on the stock and wishes to hedge their position. How much should be borrowed at the risk-free rate to create a portfolio consisting of the proceeds from the option sale and 50 shares of the underlying stock?

**13.1.2** The current price of a non-dividend-paying stock is \$75/share. The volatility of the return on the stock is 45% and the continuously compounded risk-free interest rate is 3.5%. A market maker has sold 100 9-month \$70-strike put options on the stock and wishes to hedge their position by shorting 20 shares of the underlying stock. The proceeds of the option and short sales are invested at the risk-free rate. What is the market maker's profit or loss if $S(9/12) = \$67$?

**13.1.3** The current price of a stock paying a continuous dividend at yield 3% is \$50/share. The volatility of the return on the stock is 25% and the continuously compounded risk-free interest rate is 4%. A market maker has sold 100 6-month \$55-strike call options on the stock and wishes to hedge their position. How much should be borrowed at the risk-free rate to create a portfolio consisting of the proceeds from the option sale and 35 shares of the underlying stock?

**13.1.4** The current price of a stock is \$100/share. The stock pays a continuous dividend at a rate of 2% per annum. The volatility of the return on the stock is 30% and the continuously compounded risk-free interest rate is 4.5%. A market maker has sold 100 4-month \$95-strike put options on the stock and wishes to hedge their position by shorting 40 shares of the underlying stock. The proceeds of the option and short sales are invested at the risk-free rate. What is the market maker's profit or loss if $S(4/12) = \$99$?

**13.1.5** The current price of a stock is \$315/share. The stock pays a continuous dividend at a rate of 3.5% per annum. The volatility of the return on the stock is 45% and the continuously compounded risk-free interest rate is 6.25%. A market maker has sold 100 4-month \$325-strike call options on the stock and wishes to hedge their position by shorting 50 shares of the underlying stock. What is the Delta of the market maker's hedge portfolio?

## 13.2 Delta Hedging

The purpose of hedging is to eliminate or at least minimize the change in the value of a portion of an investor's or an institution's portfolio of investments as conditions change in the financial markets. The hedged portion of a portfolio may represent the value of some future financial obligation, for example a pension intended to fund retirement. In general, people and institutions invest in stocks to earn a greater rate of return per dollar invested than can be earned in low risk government-issued bonds. The higher rate of return can be thought of a payment for accepting the greater risk. Hedging is a method of reducing the risk at the cost of some of the reward. The value of a portfolio containing stocks and options may change due to changes in the value of the stock underlying the option or simply as a result of the approaching expiry date. The reader will recall from the previous chapter that Delta is the partial derivative of the value of an option with respect to the value of the underlying security. In Sec. 12.1, expressions for Delta for European call and put options were derived. One can think of Delta in the following way: for every unit change in the value of the underlying security, the value of the option changes by $\Delta$. Delta is the marginal option value of the underlying security. A portfolio consisting of securities and options is called **Delta-neutral** if for every option sold, $\Delta$ units of the security are bought.

Suppose at time $t = 0$ a market maker sells a European call option and wishes to reduce their risk by establishing a hedging, synthetic long call to offset the short position. Suppose the short call has a Delta of $-\Delta_{C^e}$ where the minus sign has been attached since the European call was sold. If the market maker purchases $\Delta_{C^e}$ shares of the underlying security by borrowing the amount

$$\mathcal{P} = S(0)\Delta_{C^e} - C^e,$$

at the risk-free rate. Note that $\partial \mathcal{P}/\partial S = 0$, so the portfolio can be described as Delta-neutral.

**Example 13.1.** Consider the case of a non-dividend-paying security whose current price is \$100, while the risk-free interest rate is 4% per annum, the annual volatility of the stock price is 23%, the strike price is set at \$105, and the expiry date is 3 months hence. Under these conditions $d_1 = -0.279806$, the value of a European call option is $C^e(100, 3/12, 105) = 2.96155$ and Delta for the option is

$$\Delta_{C^e} = \frac{\partial C}{\partial S} = 0.389813.$$

Thus if a market maker sold an investor European call options on ten thousand shares of the security[1], the firm would receive \$29,615.50 and purchase shares worth

$$(10{,}000)S(0)\Delta_{C^e} = (10{,}000)(100)0.389813 = \$389{,}813.$$

To enable the purchase, the market maker would borrow

$$389{,}813 - (10{,}000)(2.96155) = \$360{,}198,$$

at the risk-free rate.

The firm may choose to do nothing further until the strike time of the option. In that case, this is referred to as a "hedge and forget" scheme. On the other hand, since the price of the security is dynamic, the firm may choose to make periodic adjustments to the number of shares of the security it holds. This strategy is known as **rebalancing** the portfolio. Suppose 1 day passes and the price of the stock increases to \$101/share. The value of the call options sold will have changed due to the change in stock price and the passage of time. The reader can check that

$$C^e(101, 3/12 - 1/365, 105) = \$3.33882.$$

---

[1] Typically options are sold in blocks of 100.

If the market maker wants to unwind their position by selling the hedging portfolio and repurchasing the options, the 1-day profit or loss would be affected by the change in the stock price, change in the option price, and interest due on the funds borrowed. The net cash flow after 1 day would be

$$-10,000(3.33882) + (10,000)(101)0.389813 - 360,198e^{0.04(1/365)} = \$85.99.$$

Thus the market maker made a profit.

If the market maker does nothing the hedging portfolio is no longer Delta-neutral since the Delta of the call options is now 0.422328, indicating that a Delta-neutral hedging portfolio should contain 4223.28 shares of the underlying stock. To rebalance the portfolio the market maker must purchase

$$10000(0.422382 - 0.389813) = 325.15 \text{ additional shares.}$$

This will require the market maker to borrow an additional $(101)(325.15) = \$32,840.10$. In this fashion the market maker may continue marking-to-market daily (or on some other schedule) until liquidation of the portfolio or expiry of the sold options.

Consider the scenario of a market maker selling 100 7-day, \$105-strike European call options on a non-dividend-paying stock whose current price is \$100. The volatility of the stock's return is 33% per annum and the risk-free interest rate is 5%. Assume that the value of the security follows the random walk illustrated in Fig. 13.3. Consider the actions the market maker must take to maintain a Delta-neutral hedging portfolio. On day 0 the market maker sells 100 call options at a price of \$0.356489 each and creates the hedging portfolio by purchasing $100(0.152968) = 15.2968$ shares. In order to accomplish this, the market maker borrows

$$100[(0.152968)(100) - 0.356489] = \$1,494.04,$$

at the risk-free rate.

On day 1 the stock price has increased to \$101. The value of the European call options is now \$0.438508 each and their Delta is $\Delta_{C^e} = 0.190134$. If the market maker liquidates the hedging portfolio and repurchases the options on day 1, the profit is

$$100[(0.152968)(101) - 0.438508] - 1,494.04e^{0.05(1/365)} = \$6.89021.$$

If the market maker rebalances the portfolio they must increase the number of shares in the hedging portfolio to $100(0.190134) = 19.0134$. To do this they must borrow an additional

$$(19.0134 - 15.2968)(101) = \$375.377,$$

in order to purchase 3.7166 shares.

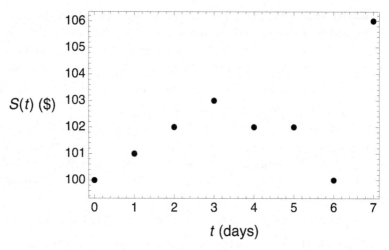

Fig. 13.3    A realization of a random walk taken by the value of a security.

On day 2 the stock price has increased to $102. The value of the European call options is now $0.539825 each and their Delta is $\Delta_{C^e} = 0.237778$. If the market maker liquidates the hedging portfolio and repurchases the options on day 2, the profit is

$$100[(0.190134)(102) - 0.539825] - 1494.04e^{0.05\left(\frac{2}{365}\right)} - 375.377e^{0.05\left(\frac{1}{365}\right)}$$
$$= \$15.5159.$$

If the market maker rebalances the portfolio on day 2 they must again purchase additional shares of the stock. The new number of shares in the hedging portfolio is $100(0.237778) = 23.7778$. To do this they must borrow an additional

$$(23.7778 - 19.0134)(102) = \$485.963,$$

in order to purchase 4.76435 shares. Continuing on until expiry the results of rebalancing can be summarized in Table 13.1.

The price of the underlying security drops for the first time in this example on day 4. On that day the price of the options becomes $0.280664 with a Delta of 0.173554. If the market maker liquidates the hedging portfolio on day 4 their profit will be

$$100[(0.300294)(102) - 0.280664] - 3000.08e^{0.05(1/365)} = \$34.4466.$$

The market maker's Delta-neutral hedging portfolio should hold 17.3554 shares of the stock, so $30.0294 - 17.3554 = 12.674$ shares must be sold.

Table 13.1 Daily marking-to-market of a Delta-neutral portfolio. The quantities listed in the "Profit" column would be the market maker's profit on that day if the market maker closed out their position by selling any held shares, repurchasing any sold options, and repaying the principal and interest on any borrowing.

| Time (days) | S(t) ($) | $C^e$ ($) | Shares ($100\Delta_{C^e}$) | Borrowing ($) | Profit ($) |
|---|---|---|---|---|---|
| 0 | 100 | 0.356489 | 15.2968 | 1494.04 | — |
| 1 | 101 | 0.438508 | 19.0134 | 1869.62 | 6.8902 |
| 2 | 102 | 0.539825 | 23.7778 | 2355.84 | 15.5159 |
| 3 | 103 | 0.666164 | 30.0294 | 3000.08 | 26.3370 |
| 4 | 102 | 0.280664 | 17.3554 | 1707.74 | 34.4466 |
| 5 | 102 | 0.149090 | 12.2367 | 1185.86 | 47.3701 |
| 6 | 100 | 0.001273 | 0.249186 | −12.72 | 37.5159 |

The proceeds from the sale can be used to repay a portion of the borrowing undertaken to finance the hedging portfolio.

Thus to mark-to-market daily the market maker must keep track of interest paid on the previous day's borrowing (or lending), the cost of the purchase (or sale) of shares of the underlying stock necessary to maintain a Delta-neutral portfolio, and any borrowing (or lending) necessary to finance the rebalancing. Rebalancing can be done as often as the market maker wishes, so to develop a flexible, general formula assume that rebalancing takes place at regular intervals spaced $h$ apart in time. Let $S(ih)$ be the stock price at time $ih$ and similarly let $C^e(ih)$ and $\Delta_{C^e}(ih)$ be the cost and Delta respectively of the European call option at time $ih$. The Delta-neutral portfolio at time $ih$ will be denoted $\mathcal{P}(ih)$. At inception $\mathcal{P}(0) = \Delta_{C^e}(0)S(0) - C^e(0)$. If $\mathcal{P}(0) > 0$ the amount is borrowed at the risk-free rate while if $\mathcal{P}(0) < 0$ the amount is lent at the risk-free rate. The profit or loss to the market maker liquidating the Delta-neutral portfolio at time $ih$ is

$$\text{P/L} = \Delta_{C^e}((i-1)h)S(ih) - C^e(ih) - \mathcal{P}((i-1)h)e^{rh}.$$

If the portfolio is rebalanced

$$\mathcal{P}(ih) = (\Delta_{C^e}(ih) - \Delta_{C^e}((i-1)h))\, S(ih) + \mathcal{P}((i-1)h)e^{rh}.$$

Throughout it is assumed the market maker can borrow sufficiently at the risk-free rate to finance the hedging portfolio. At any given time the market value of the portfolio is $\Delta_{C^e}(ih)S(ih) - C^e(ih)$. These formulas can be scaled by the number of options sold and can be generalized to other types of options such as European puts. The reader will explore several of these issues in the exercises at the end of the section.

Rebalancing of the portfolio can take place either more or less frequently than daily as was done in the previous extended example. The quantity discussed in Sec. 12.1 known as Gamma ($\Gamma$) is the second partial derivative of the portfolio with respect to the value of the underlying security. In other words, Gamma is the rate of change of Delta with respect to $S$. If $|\Gamma|$ is large then $\Delta$ changes rapidly with relatively small changes in $S$. In this case frequent rebalancing of the investment firm's position may be necessary. If $|\Gamma|$ is small then $\Delta$ is relatively insensitive to changes in $S$ and hence rebalancing may be needed only infrequently. Thus a market maker can monitor Gamma in order to determine the frequency at which the hedging portfolio should be rebalanced.

## *Exercises*

**13.2.1** Use Eq. (10.32) to show that $\Delta_{C^e} \to 1$ as $t \to T^-$ for an in-the-money European call option.

**13.2.2** Use Eq. (10.32) to show that $\Delta_{C^e} \to 0$ as $t \to T^-$ for an out-of-the-money European call option.

**13.2.3** Find $\lim_{t \to T^-} \Delta_{C^e}$ for an at-the-money European call option.

**13.2.4** Consider 100 sold European calls on a non-dividend-paying security whose current value is $60. The risk-free interest rate is 5.65% and the volatility of the security is 45% per year. The strike price of the call is $60 and the strike time is 90 days hence. If a Delta-neutral portfolio is created determine the profit or loss after one day if the price of the security the next day is $60.50.

**13.2.5** For the Delta-neutral portfolio described in Exercise 13.2.4 plot the profit or loss after one day if the price of the security is $S(1/365)$.

**13.2.6** Suppose a non-dividend-paying security follows the Black–Scholes model. The time $t = 0$ price of the security is $65. The volatility of the security is 40%. The continuously compounded, risk-free interest rate is 5%. An investor has just purchased 100 units of a 9-month, 65-strike European call option. To hedge the risk, the investor hedges their position by creating a Delta-neutral hedge portfolio. Find the stock and cash components of the hedge portfolio.

**13.2.7** For the situation described in Exercise 13.2.6, what is the change in the value of the investor's hedging portfolio 1 month later? Assume that $S(1/12) = \$65$.

**13.2.8** For the situation described in Exercise 13.2.6, what is the investor's profit or loss if the investor liquidates the hedging portfolio 1 month later? Assume that $S(1/12) = \$65$.

**13.2.9** Suppose that a security follows the Black–Scholes model. The time $t = 0$ price of the security is \$75. The volatility of the security is 35% per annum. The continuously compounded, risk-free interest rate is 6% per annum. The security pays a dividend continuously at rate 4% per annum. An investor has bought 100 units of a 6-month, 70-strike put option. To hedge the risk, the investor hedges their position by creating the Delta-neutral hedge portfolio. Find the stock and cash components of the hedge portfolio at $t = 0$.

**13.2.10** For the situation described in Exercise 13.2.9, what is the change in the value of the investor's hedging portfolio 1 month later? Assume that $S(1/12) = \$72$.

**13.2.11** For the situation described in Exercise 13.2.9, what is the investor's profit or loss if the investor liquidates the hedging portfolio 1 month later? Assume that $S(1/12) = \$72$.

**13.2.12** Suppose the continuously compounded, risk-free interest rate is 3.75% and an investor creates a bull spread on a stock by purchasing a 2-month, 100-strike European call option and sells a 2-month, 110-strike European call option. The stock does not pay dividends and has a volatility of 45% per annum. The current price of the stock is \$99. What actions should the investor take to create a Delta-neutral hedging portfolio?

**13.2.13** An investor creates a butterfly spread using puts. European puts with strike prices of \$45 and \$55 are purchased while two puts with strike prices of \$50 are sold. All options expire in 3 months. The risk-free interest rate is 5%, the current price of the underlying stock is \$50/share, the volatility of the stock is 33%, and the stock pays a continuous dividend with a yield of 2% per annum. What actions should the investor take to create a Delta-neutral hedging portfolio?

**13.2.14** Suppose the current price of a non-dividend-paying stock is \$45 per share with volatility $\sigma = 0.20$ per annum and the risk-free interest rate is 4.5% per year. Create a table similar to Table 13.1 for European call options sold on 1000 shares of the stock with a strike price of \$47 per share and a strike time of 15 weeks. Rebalance the portfolio weekly assuming the weekly prices of the stock are as follows.

| Week | 0 | 1 | 2 | 3 | 4 | 5 | 6 | 7 |
|------|------|------|------|------|------|------|------|------|
| $S$  | 45.00 | 44.58 | 46.55 | 47.23 | 47.62 | 47.28 | 49.60 | 50.07 |

| Week | 8 | 9 | 10 | 11 | 12 | 13 | 14 | 15 |
|------|------|------|------|------|------|------|------|------|
| $S$  | 47.79 | 48.33 | 48.81 | 51.36 | 52.06 | 51.98 | 54.22 | 54.31 |

**13.2.15** Repeat Exercise 13.2.14 for the following scenario in which the call option finishes out-of-the-money.

| Week | 0 | 1 | 2 | 3 | 4 | 5 | 6 | 7 |
|------|------|------|------|------|------|------|------|------|
| $S$  | 45.00 | 44.58 | 45.64 | 44.90 | 43.42 | 42.23 | 41.18 | 41.52 |
| Week | 8 | 9 | 10 | 11 | 12 | 13 | 14 | 15 |
| $S$  | 41.94 | 42.72 | 44.83 | 44.93 | 44.19 | 41.77 | 39.56 | 40.62 |

**13.2.16** Repeat Exercise 13.2.14 assuming this time that the market maker has sold 1000 European put options for the stock.

**13.2.17** Repeat Exercise 13.2.15 assuming this time that the market maker has issued sold 1000 European put options for the stock.

## 13.3   Self-Financing Portfolios

Returning to the beginning of the multi-day rebalancing example, suppose a call option is sold and a Delta-neutral portfolio is immediately created. Figure 13.4 shows there is an interval of stock prices on the following day for which the market maker has a positive profit. At the endpoints of this interval the market maker breaks even and the portfolio is described as **self-financing**. This section will explore the stock price movements which lead to self-financing portfolios. The reader will discover that these movements are not random or stock specific, but instead are known to the market maker through the statistical parameters describing the stock.

Recall the Delta/Gamma/Theta approximation to option prices given in Eq. (12.18). This approximation takes into account changes in option

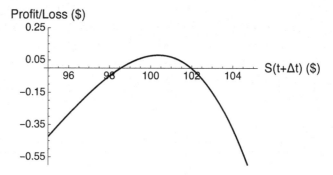

Fig. 13.4   There is an interval of stock prices for which the market maker makes a positive profit on a Delta-neutral portfolio.

prices due to small changes in the price of the underlying asset and changes in time, precisely the variables that are used in the rebalancing examples and exercises considered so far. Suppose a market maker sells an option worth $V(S(t), t)$ on a non-dividend-paying stock whose price is $S(t)$ and volatility is $\sigma$. The risk-free interest rate is denoted $r$. Suppose the market maker Delta hedges with $\Delta = \partial V / \partial S$ (evaluated at time $t$) shares of the underlying stock and cash in the amount of $B = \Delta S(t) - V(S(t), t)$. If time $h$ passes the profit/loss on this portfolio is the gain/loss on the sold option, the purchased stock, and the gain/loss due to interest on the cash.

$$P/L = \Delta S(t+h) - V(S(t+h), t+h) - Be^{rh}. \tag{13.1}$$

To simplify this expression for profit/loss, replace the value of the option at time $t + h$ with the Delta/Gamma/Theta approximation and assume $h$ is small enough that the approximation $e^{rh} \approx 1 + rh$ is valid. The reader will be asked to show in Exercise 13.3.1 that

$$P/L = -\frac{1}{2}\Gamma(S(t+h) - S(t))^2 - (\Theta + Br)h. \tag{13.2}$$

The symbol $\Theta$ can be replaced using the recast form of the Black–Scholes partial differential equation shown in Eq. (12.8). The single time period profit/loss expression can then be approximated as

P/L

$$= -\frac{\Gamma}{2}(S(t+h) - S(t))^2 - \left(r(V(S(t), t) - \Delta S(t)) - \frac{\Gamma}{2}\sigma^2(S(t))^2 + Br\right)h$$

$$= -\frac{1}{2}\Gamma(S(t+h) - S(t))^2 + \frac{1}{2}\Gamma\sigma^2(S(t))^2 h.$$

This is a quadratic expression in $S(t + h)$ and the profit/loss will be zero when

$$\frac{1}{2}\Gamma(S(t+h) - S(t))^2 = \frac{1}{2}\Gamma\sigma^2(S(t))^2 h$$

$$(S(t+h) - S(t))^2 = h\sigma^2(S(t))^2.$$

Thus when $S(t+h) = S(t)(1 \pm \sigma\sqrt{h})$ the hedging portfolio is self-financing. By assumption $S(t + h) \sim \mathcal{LN}(\ln S(t) + \mu h, \sigma^2 h)$ so the hedging portfolio is self-financing whenever the underlying stock price moves by a factor of one standard deviation, $\sigma\sqrt{h}$.

## Exercises

**13.3.1** Use the Delta/Gamma/Theta approximation found in Eq. (12.18) to derive Eq. (13.2) from Eq. (13.1).

**13.3.2** Assume that a non-dividend-paying security follows the Black–Scholes model. The time $t = 0$ price of the security is \$65. The volatility of the security is 45%. The continuously compounded, risk-free interest rate is 5%. An investor has just sold 100 units of a 6-month, 60-strike European call option. To hedge the risk, the investor hedges their position by creating the hedge portfolio. Suppose the security price is \$60 after 1 month. Determine if this hedge portfolio is self-financing.

**13.3.3** Assume that a non-dividend-paying security follows the Black–Scholes model. The time $t = 0$ price of the security is \$75. The volatility of the security is 50%. The continuously compounded, risk-free interest rate is 8%. An investor has just purchased 100 units of a 3-month, 80-strike European put option. To hedge the risk, the investor hedges their position by creating a hedged portfolio. Sketch the profit/loss curve as a function of the security price at a time of 1 month.

**13.3.4** Determine if the hedging portfolio of Exercise 13.3.3 is self-financing if the security price is \$70 after 1 month.

**13.3.5** Suppose a security follows the Black–Scholes model. A market maker who Delta hedges, sells a 4-month at-the-money European call option on a non-dividend-paying security. The current security price is \$60. The continuously compounded, risk-free interest rate is 6%. The Delta for the call option is 0.655422. If after 1 week the market maker has zero profit or loss, determine the security price movement over the week.

**13.3.6** Consider the Delta-neutral portfolio described in Exercise 13.2.4. Find the day 1 prices of the security so that the profit/loss is zero.

## 13.4     Gamma-Neutral Portfolios

Gamma for a portfolio is the second partial derivative of the portfolio with respect to $S$. The portfolio cannot be made Gamma-neutral using a linear combination of just an option and its underlying security, since the second derivative of $S$ with respect to itself is zero. As explained in Sec. 13.2, a portfolio can be made Delta neutral with the appropriate combination of the option and the underlying security. The portfolio can be made Gamma-neutral by manipulating a component of the portfolio which depends non-linearly on $S$. Options are financial instruments whose values depend non-linearly on $S$; however, as mentioned earlier, the portfolio cannot contain only an option and its underlying security. One way to achieve the goal of a Gamma-neutral portfolio is to include in the portfolio two or more different types of options dependent on the same underlying security.

For example, suppose the portfolio contains options with two different strike times or two different strike prices (or both) written on the same stock. In concrete terms, a market maker may sell a number of European call options with expiry 3 months hence and buy some other number of the European call options on the same stock with expiry arriving in 6 months. Let $\Gamma_3$ be the Gamma of a sold 3-month call option, $\Gamma_6$ be the Gamma of a purchased 6-month call option, $x$ be the number of the 3-month strike options sold, and $y$ be the number of 6-month strike options purchased. The Gamma of the portfolio would be

$$\Gamma_{\mathcal{P}} = x\Gamma_3 - y\Gamma_6.$$

The numbers $x$ and $y$ can be chosen so that $\Gamma_{\mathcal{P}} = 0$. Once the portfolio is made Gamma-neutral, the underlying security can be added to the portfolio in such a way so as to make the portfolio Delta-neutral. The reader should keep in mind that the underlying security will have a $\Gamma = 0$ and thus Gamma-neutrality for the portfolio will be maintained while Delta-neutrality is achieved. Thus for a Delta- and Gamma-neutral portfolio $\mathcal{P}$ the Delta/Gamma/Theta approximation in Eq. (12.18) portfolio reduces to

$$\mathcal{P}(S(t) + dS, t + \Delta t) \approx \mathcal{P}(S(t), t) + \Theta_{\mathcal{P}}(\Delta t). \qquad (13.3)$$

Equation (13.3) implies that the change in the value of a simultaneously Gamma- and Delta-neutral portfolio is proportional to the size of the time step.

**Example 13.2.** Suppose a non-dividend-paying stock's current value is $100 while its volatility is $\sigma = 0.22$ and the risk-free interest rate is 2.5% per year. An market maker sells 100 European call options on this stock with a strike time of 3 months and a strike price of $102. The firm buys European call options on the same stock with the same strike price but with a strike time of 6 months. Find the proper number of 6-month strike calls to purchase and the position in the underlying stock the market maker should take to create a Delta- and Gamma-neutral hedging portfolio.

**Solution.** The following table summarizes the features of the stock and options.

| Instrument | Quantity | Price | Delta | Gamma |
|---|---|---|---|---|
| 3-month call | −100 | $3.76727 | 0.472811 | 0.0361832 |
| 6-month call | $y$ | $5.86293 | 0.512301 | 0.0256328 |
| Stock | $z$ | $100 | 1 | 0 |

The values in the Gamma column are determined using Eq. (12.4). To make the hedging portfolio Gamma-neutral the market maker must take the position $y$ in the 6-month call option, where

$$(-100)(0.0361832) + y(0.0256328) = 0 \implies y = 141.16 \text{ units.}$$

The market maker should take a long position in 141.16 6-month call options. To make the hedging portfolio Delta-neutral the market maker must take a position in $z$ units of the underlying stock where,

$$(-100)(0.472811) + (141.16)(0.512301) + z = 0 \implies z = -25.0352 \text{ units.}$$

Since $z < 0$ the market maker should short the underlying stock. The net cash flow required to create the Delta/Gamma-neutral hedging portfolio is

$$(100)(3.76727) - (141.16)(5.86293) + (25.0352)(100) = \$2052.64.$$

This amount can be invested at the risk-free rate.

In general when two different options (say $V_1$ and $V_2$) with the same underlying stock are used to create a Gamma-neutral hedge

$$x\Gamma_{V_1} + y\Gamma_{V_2} = 0 \iff -\frac{x}{y} = \frac{\Gamma_{V_2}}{\Gamma_{V_1}}.$$

The ratio of the Gamma's of the two options equals the negative of the ratios of the number of units of the two options in the portfolio. For every $y$ units of option $V_2$ bought (sold), $x$ units of option $V_1$ must be sold (bought) to achieve Gamma-neutrality. Note that in Example 13.2 the Gamma ratio was

$$\frac{\Gamma_3}{\Gamma_6} = \frac{0.0361832}{0.0256328} = 1.4116.$$

Thus for every 3-month call option sold, 1.4116 6-month call options should be bought.

Delta-neutrality preserves the value of a portfolio when the price of the stock in the portfolio changes by a small amount. Gamma- and Delta-neutrality preserves the value of a portfolio for relatively larger changes in the price of the stock in the portfolio. Figure 13.5 shows that over a fairly wide range of values of the underlying stock the value of the Delta/Gamma-neutral portfolio $\mathcal{P}$ of Example 13.2 and Delta-neutral portfolio remains nearly the same. In the figure the 1-day profit/loss on a Delta-neutral portfolio in which 100 3-month, 102-strike call options are sold is shown alongside the 1-day profit/loss for the Delta/Gamma-neutral portfolio of

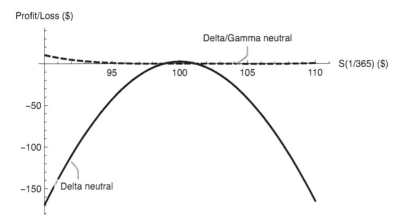

Fig. 13.5   The 1-day profit/loss on related Delta-neutral and Delta/Gamma-neutral portfolios. In both portfolios 100 units of 3-month, 102-strike European call options are sold. The profit/loss Delta/Gamma-neutral hedging portfolio is less sensitive to movements in the underlying stock price than the merely Delta-neutral portfolio.

Example 13.2. The Delta-neutral portfolio consists of a short position in 100 units of the 3-month European call option and a long position in 47.2811 shares of the stock. This initial cash flow required to create the Delta-neutral hedging portfolio is $4,351.38 which is borrowed at the risk-free rate.

This discussion of hedging is far from complete. The market maker who sells options to an investor and who wishes to create a Gamma-neutral hedging portfolio must buy related options with the same underlying stock. To do so, the market maker must buy the options from *another* market maker. Paying the bid/ask spread to the other market maker reduces the potential profit of the first market maker.

This chapter has focused mainly on making a portfolio's value resistant to changes in the value of a security. In reality, the risk-free interest rate and the volatility of the stock also affect the value of a portfolio. The quantities Rho and Vega discussed in Chapter 12 can be used to set up portfolios hedged against changes in the interest rate and volatility. In this chapter it was also assumed that the necessary options and securities could be bought so as to form the desired hedge. In practice this may not always be possible. For example, a market maker may not be able to purchase sufficient quantities of a stock to make their portfolio Delta- or Gamma-neutral. In this case they may have to substitute a different, but related security or other financial instrument in order to set up the hedge.

### Exercises

**13.4.1** Set up a Gamma-neutral portfolio consisting of European call options on a non-dividend-paying security whose current value is $85. Suppose the risk-free interest rate is 5.5% and the volatility of the security is 17% per year. The portfolio should contain a short position in 500 4-month options with a strike price of $88 and a long position in 6-month options with the same strike price.

**13.4.2** For the portfolio described in Exercise 13.4.1 add a position in the underlying security so that the portfolio is also Delta-neutral.

**13.4.3** Set up a Gamma-neutral portfolio consisting of European call options on a non-dividend-paying security whose current value is $95. Suppose the risk-free interest rate is 4.5% and the volatility of the security is 23% per year. The portfolio should contain a short position in 100 3-month options with a strike price of $97 and a long position in 5-month options with a strike price of $98.

**13.4.4** For the portfolio described in Exercise 13.4.3 add a position in the underlying security so that the portfolio is also Delta-neutral.

**13.4.5** Set up a Gamma-neutral portfolio consisting of positions in 3-month European call options on a security whose current value is $60. The security pays a continuous dividend with yield 2% per annum. Suppose the risk-free interest rate is 5.65% and the volatility of the security is 45% per year. The portfolio should contain 100 sold options with a strike price of $62 and a long position in options with a strike price of $65.

**13.4.6** For the portfolio described in Exercise 13.4.5 add a position in the underlying security so that the portfolio is also Delta-neutral.

**13.4.7** Using the results of Exercises 13.4.5 and 13.4.6 determine the profit or loss after one day if the price of the security the next day is $60.50.

**13.4.8** For the Delta/Gamma-neutral portfolio described in Exercise 13.4.7 plot the profit or loss after one day if the price of the security is $S$.

# Chapter 14

# Exotic Options

Previous chapters have developed and explored the pricing formulas and features of plain vanilla or standard options. These are options which can be priced and hedged using the Black–Scholes equation and its associated Greeks. Financial firms and investors may have need of options with features that do not satisfy the Black–Scholes model assumptions. This chapter will present several such examples of **exotic options**, though a more accurate label for them may be **non-standard options**. The reader may well wonder how these exotic options can be priced and hedged when the options do not fit the Black–Scholes framework. In many cases the non-standard options can be decomposed into combinations of standard options and the usual pricing formulas applied. In other cases (for example, the Asian options) the numerical technique provided by the binomial tree pricing method of Chapter 7 can be used. In other cases the no arbitrage principle provides a price. In other words the time $t = 0$ price of an option should be the present value of the risk-neutral expected value of its payoff.

The exotic options explored in this chapter do not exhaust all the existing non-standard options (since new types are created continually) but the approach outlined in this chapter should give readers a way to start analyzing unfamiliar options when encountered in the future.

## 14.1  Gap Options

Gap options are quite similar to standard options differing in an extra feature present in their payoff formula not present in the plain vanilla call and put options. In addition to the strike price, gap options also have a payment trigger in their payoff expression. Hence in this section the strike price of a gap option will be denoted $K_1$ and the payment trigger price will be denoted $K_2$. There are several variations to consider depending on whether the gap option is a call or a put and whether the strike price is less than or greater than the trigger price. In either case for a gap call option the payoff can be represented by the formula,

$$\text{payoff} = \begin{cases} 0 & \text{if } S(T) \leq K_2 \\ S(T) - K_1 & \text{if } S(T) > K_2. \end{cases}$$

For a gap put, the payoff is given by the following piecewise defined function.

$$\text{payoff} = \begin{cases} K_1 - S(T) & \text{if } S(T) < K_2 \\ 0 & \text{if } S(T) \geq K_2 \end{cases}$$

A gap call must be exercised when $S(T) > K_1$ while a gap put must be exercised when $S(T) < K_1$. The graphs of the payoff functions for gap calls and puts are shown in Fig. 14.1. Thus the payoff of a gap option can be negative. All of the plain vanilla call and put options studied previously had non-negative payoffs and thus *a priori* the premiums for the options were non-negative. Since a gap option can have a negative payoff, there are situations in which they will also carry negative premiums, in other words the option holder may be paid to hold the gap option. The reader should

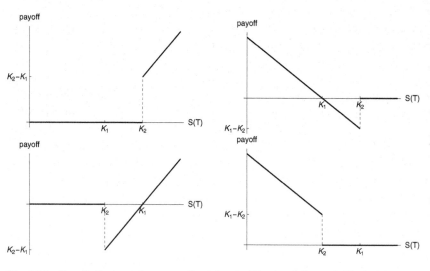

Fig. 14.1  Payoff diagrams for gap calls and puts. The top left graph illustrates a gap call option with strike price $K_1 < K_2$ the trigger price. The top right graph illustrates a gap put option with strike price $K_1 < K_2$ the trigger price. The bottom left graph illustrates a gap call option with strike price $K_1 > K_2$ the trigger price. The bottom right graph illustrates a gap put option with strike price $K_1 > K_2$ the trigger price.

also note that if $K_1 = K_2$ then gap calls and puts have the same payoff as standard calls and puts and thus would carry the same premiums.

Pricing gap calls and puts when $K_1 \neq K_2$ is relatively straightforward. Consider the gap call option for a stock whose current price is $S(t)$, volatility is $\sigma$, and dividend yield is $\delta$. The strike time of the option is $T > t$ and the continuously compounded risk-free interest rate is $r$. At the strike time, the gap call will pay

$$C^g(S(T); T) = (S(T) - K_1)\, \mathbb{1}_{(S(T)>K_2)} = S(T)\, \mathbb{1}_{(S(T)>K_2)} - K_1\, \mathbb{1}_{(S(T)>K_2)}.$$

Recall that $\mathbb{1}_{(X)}$ is the indicator function which is 1 when event $X$ occurs and zero otherwise. This payoff can be interpreted as the difference in the payoffs of an asset-or-nothing call with strike price $K_2$ and $K_1$ cash-or-nothing calls with strike price $K_2$. The pricing formulas for these types of binary options are found in Eqs. (10.36) and (10.37). Combining those formulas produces the following formula for the time $t$ price of a gap call.

$$d_1 = \frac{\ln \frac{S(t)}{K_2} + \left(r - \delta + \frac{\sigma^2}{2}\right)(T-t)}{\sigma\sqrt{T-t}} \tag{14.1}$$

$$d_2 = \frac{\ln \frac{S(t)}{K_2} + \left(r - \delta - \frac{\sigma^2}{2}\right)(T-t)}{\sigma\sqrt{T-t}} \tag{14.2}$$

$$C^g(S(t); T, K_1, K_2) = S(t)e^{-\delta(T-t)}\Phi(d_1) - K_1 e^{-r(T-t)}\Phi(d_2). \tag{14.3}$$

The reader should note the similarity between this formula and the Black–Scholes formula for an ordinary call option. Specifically compare Eq. (14.1) with Eq. (10.32), compare Eq. (14.2) with Eq. (10.33), and compare Eq. (14.3) with Eq. (10.34). The formulas are nearly identical except for the trigger price $K_2$ being used in place of the strike price in the expressions for $d_1$ and $d_2$.

Now consider a gap put option for the stock described above. At the strike time, the gap put will pay

$$P^g(S(T); T) = (K_1 - S(T))\, \mathbb{1}_{(S(T)<K_2)} = K_1\, \mathbb{1}_{(S(T)<K_2)} - S(T)\, \mathbb{1}_{(S(T)<K_2)}.$$

This payoff can be interpreted as the difference in the payoffs of $K_1$ cash-or-nothing puts with strike price $K_2$ and an asset-or-nothing put with strike price $K_2$. The pricing formulas for these types of binary options are found in Eqs. (10.38) and (10.39). Combining those formulas produces the following formula for the time $t$ price of a gap put,

$$P^g(S(t); T, K_1, K_2) = K_1 e^{-r(T-t)}\Phi(-d_2) - S(t)e^{-\delta(T-t)}\Phi(-d_1), \tag{14.4}$$

where $d_1$ and $d_2$ are calculated as in Eqs. (14.1) and (14.2).

**Example 14.1.** Suppose the current price of a security is \$150/share, the security pays a continuous dividend with a yield 2%, and the volatility of the security is 35%. The continuously compounded risk-free interest rate is 5%. Find the premium for a 3-month strike gap call option that pays $S(3/12) - 150$ if $S(3/12) > 160$.

**Solution.** Using Eqs. (14.1), (14.2), and (14.3),

$$d_1 = \frac{\ln\frac{150}{160} + \left(0.05 - 0.02 + \frac{(0.35)^2}{2}\right)(3/12 - 0)}{0.35\sqrt{3/12 - 0}} = -0.238434$$

$$d_2 = d_1 - 0.35\sqrt{3/12 - 0} = -0.413434$$

$$C^g = 150e^{-0.02(3/12-0)}\Phi(-0.238434) - 150e^{-0.05(3/12-0)}\Phi(-0.413434)$$

$$= \$10.2485.$$

Readers will demonstrate in Exercise 14.1.3 an alternative way to express the price of a gap call as

$$C^g(S(t); T, K_1, K_2) = C^e(S(t); T, K_2) + (K_2 - K_1)e^{-r(T-t)}\Phi(d_2), \quad (14.5)$$

where $C^e(S(t); T, K_2)$ is an ordinary European $K_2$-strike call and $d_2$ is calculated as in Eq. (14.2). Likewise the gap put option can be priced as a combination of the ordinary European put option and a binary option.

$$P^g(S(t); T, K_1, K_2) = P^e(S(t); T, K_2) - (K_2 - K_1)e^{-r(T-t)}\Phi(-d_2). \quad (14.6)$$

This makes calculation of the Greeks for gap options more straightforward. In particular,

$$\Delta_{C^g} = e^{-\delta(T-t)}\Phi(d_1) + \frac{(K_2 - K_1)e^{-r(T-t)}}{\sigma S\sqrt{T-t}}\varphi(d_2), \quad (14.7)$$

$$\Delta_{P^g} = -e^{-\delta(T-t)}\Phi(-d_1) + \frac{(K_2 - K_1)e^{-r(T-t)}}{\sigma S\sqrt{T-t}}\varphi(d_2), \quad (14.8)$$

$$\Gamma^g = \frac{e^{-\delta(T-t)}\varphi(d_1)}{\sigma S\sqrt{T-t}} - \frac{(K_2 - K_1)e^{-r(T-t)}}{\sigma S^2\sqrt{T-t}}\varphi(d_2)\left(1 + \frac{d_2}{\sigma\sqrt{T-t}}\right). \quad (14.9)$$

The reader must keep in mind that $d_1$ and $d_2$ are calculated using the trigger price $K_2$ rather than the strike price $K_1$.

**Example 14.2.** Suppose a market maker has sold 100 gap calls on a stock whose current price is \$200/share. The stock pays a continuous dividend with yield 1% per annum and has a volatility of 30%. The options expire in

3 months and have strike and trigger prices of \$195 and \$200 respectively. The continuously compounded risk-free interest rate is 4% per annum. Find the price of the gap call option and determine the market maker's position in the underlying stock to create a Delta-neutral hedging portfolio.

**Solution.** Using Eqs. (14.1), (14.2), and (14.3),

$$d_1 = \frac{\ln \frac{200}{200} + \left(0.04 - 0.01 + \frac{(0.30)^2}{2}\right)(3/12 - 0)}{0.30\sqrt{3/12 - 0}} = 0.125$$

$$d_2 = d_1 - 0.30\sqrt{3/12 - 0} = -0.025$$

$$C^g = 200e^{-0.01(3/12-0)}\Phi(0.125) - 195e^{-0.04(3/12-0)}\Phi(-0.025) = \$15.0686.$$

The Delta for the gap call can be found using Eq. (14.7).

$$\Delta_{C^g} = e^{-0.01(3/12-0)}\Phi(0.125) + \frac{(200-195)e^{-0.04(3/12-0)}}{0.30(200)\sqrt{3/12 - 0}}$$

$$= 0.614174.$$

Thus the market maker should have a long position in 61.4174 shares of the underlying stock. The net cost to the market maker of the hedging portfolio is

$$B = 61.4174(200) - 100(15.0686) = \$10,776.60.$$

*Exercises*

**14.1.1** Find the price of a gap put option that pays $150 - S(3/12)$ if $S(3/12) < 160$ for the stock described in Example 14.1.

**14.1.2** A gap put option on the stock described in Example 14.1 pays $K_1 - S(3/12)$ if $S(3/12) < 160$ and zero otherwise. The current price of the gap put is \$0. Find the strike price $K_1$ of the gap put.

**14.1.3** Show that a gap call option can be priced using the formula shown in Eq. (14.5).

**14.1.4** Suppose the market maker of Example 14.2 wishes for the hedging portfolio to be both Delta- and Gamma-neutral. To establish Gamma-neutrality the market maker will take a position in ordinary European call options with the same features as the gap calls (except for the trigger price). Find the positions in the European call options and the underlying stock the market maker must take so as to create a Delta- and Gamma-neutral hedging portfolio.

**14.1.5** Starting with the Put-Call Parity formula for ordinary European puts and calls, find a Put-Call Parity formula for gap puts and calls.

**14.1.6** Suppose the current price of a security is \$90 and a time $T$-strike gap option on this security will have a payoff of $S(T) - 100$ is $S(T) > 110$ and a zero payoff otherwise. The price of an 110-strike European call option on the security is \$5. The Delta of the European call option is 0.25. Find the price of the gap option.

## 14.2 Power Options

A **power option** is a type of financial derivative whose payoff is related to a power of the underlying asset. $K$-strike call and put power options have time $T$ payoffs which can be expressed as $((S(T))^a - K)^+$ and $(K - (S(T))^a)^+$ respectively where $a > 0$ is a real number. Figure 14.2 illustrates the payoff for values of $a$ in the intervals $0 < a < 1$ and $a > 1$. This section will explore the pricing of power options and illustrate how the Black–Scholes framework for option pricing can be adapted to this scenario. The main tools used will be Itô's result of Lemma 9.1 and properties of lognormally distributed random variables.

Suppose the time $t$ value of the asset underlying a power option is $S(t)$ and it follows the stochastic process given in Eq. (10.1) and repeated here for convenience.

$$dS(t) = (\alpha - \delta)S(t)\,dt + \sigma S(t)\,dW(t).$$

If $Y(t) = a \ln S(t)$ then by Lemma 9.1 $Y(t)$ follows the stochastic process,

$$dY(t) = a\left(\alpha - \delta - \frac{\sigma^2}{2}\right)dt + a\sigma\,dW(t). \qquad (14.10)$$

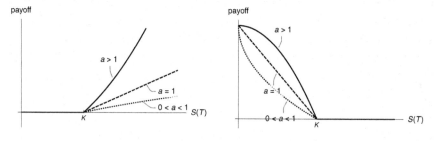

Fig. 14.2 Payoff diagrams for power calls and puts. The graph on the left illustrates the payoffs of power call options with strike price $K$ and powers $0 < a < 1$, $a = 1$ (an ordinary call), and $a > 1$. The graph on the right illustrates the payoffs of power put options for the same intervals of the power $a$.

The solution to this stochastic differential equation is

$$Y(t) = Y(0) + a\left(\alpha - \delta - \frac{\sigma^2}{2}\right)t + a\sigma W(t).$$

The random variable $Y(t) = a\ln S(t)$ is normally distributed and thus $(S(t))^a$ is lognormally distributed.

$$(S(t))^a = (S(0))^a e^{a(\alpha-\delta-\sigma^2/2)t + a\sigma W(t)}.$$

The expected value of this random variable is

$$\mathbb{E}\left((S(t))^a\right) = (S(0))^a e^{a(\alpha-\delta-\sigma^2/2)t + a^2\sigma^2 t/2} = (S(0))^a e^{[a(\alpha-\delta)+a(a-1)\sigma^2/2)]t}.$$

Based on the work done to this point, the time $t$ risk-neutral price of a prepaid forward which delivers $(S(T))^a$ at time $T$ is

$$F_{t,T}^P((S(t))^a) = e^{-r(T-t)}(S(t))^a e^{[a(r-\delta)+a(a-1)\sigma^2/2)](T-t)}. \qquad (14.11)$$

Note that the risk-neutral interest rate $r$ is used in place of the expected growth rate $\alpha$, which the asset has in the real world.

The risk-neutral expected value of $(S(t))^a$ is expressed as

$$\mathbb{E}^*\left((S(t))^a\right) = (S(0))^a e^{[a(r-\delta)+a(a-1)\sigma^2/2)]t},$$

thus if $\delta^*$ is the annual dividend yield of $S^a$ then

$$r - \delta^* = a(r - \delta) + a(a-1)\frac{\sigma^2}{2},$$

or equivalently

$$-\delta^* = (a-1)\left(r + \frac{a\sigma^2}{2}\right) - a\delta. \qquad (14.12)$$

The ordinary European call and put option formulas (see Eqs. (10.34) and (10.35)) can be applied to an underlying security $(S(t))^a$ with dividend yield $\delta^*$ and volatility $a\sigma$. Consequently a $K$-strike, $T$-expiry European call power option with power $a$ on a security whose current price is $S(t)$, dividend yield is $\delta$ per annum, and volatility is $\sigma$ when the risk-free continuously compounded interest rate is $r$ can be expressed as

$$C^p(S(t); a, T, K) = (S(t))^a e^{-\delta^*(T-t)}\Phi(d_1) - Ke^{-r(T-t)}\Phi(d_2) \qquad (14.13)$$

where Eqs. (10.32) and (10.33) can be simplified to

$$d_1 = \frac{\ln\frac{S(t)}{K^{1/a}} + \left(r - \delta + (a - \frac{1}{2})\sigma^2\right)(T-t)}{\sigma\sqrt{T-t}} \qquad (14.14)$$

$$d_2 = \frac{\ln\frac{S(t)}{K^{1/a}} + \left(r - \delta - \frac{\sigma^2}{2}\right)(T-t)}{\sigma\sqrt{T-t}}. \qquad (14.15)$$

The pricing formula for a European power put is

$$P^p(S(t); a, T, K) = Ke^{-r(T-t)}\Phi(-d_2) - (S(t))^a e^{-\delta^*(T-t)}\Phi(-d_1). \qquad (14.16)$$

**Example 14.3.** Suppose the current price of a security is \$90, the volatility of the security is 45%, and the security pays a continuous dividend with yield 1.5% per annum. The risk-free interest rate is 5% per annum. Find the prices of 6-month, \$95-strike European call and put power options on the security with payoffs of $((S(6/12))^{3/2} - 95)^+$ and $(95 - (S(6/12))^{3/2})^+$ respectively.

**Solution.** Using Eqs. (14.14) and (14.15)

$$d_1 = \frac{\ln\frac{90}{95^{2/3}} + \left(0.05 - 0.015 + (3/2 - \frac{1}{2})(0.45)^2\right)(6/12 - 0)}{0.45\sqrt{6/12 - 0}} = 4.97376$$

$$d_2 = \frac{\ln\frac{90}{95^{2/3}} + \left(0.05 - 0.015 - \frac{(0.45)^2}{2}\right)(6/12 - 0)}{0.45\sqrt{6/12 - 0}} = 4.49647.$$

According to Eq. (14.13) the price of the power call is

$$C^p(90; 3/2, 95, 6/12)$$

$$= (90)^{3/2} e^{[(3/2-1)(0.05+(3/2)(0.45)^2/2)-(3/2)(0.015)](6/12-0)} \Phi(4.97376)$$

$$- 95 e^{-0.05(6/12-0)} \Phi(4.49647)$$

$$= \$795.311.$$

Finally using Eq. (14.16) the price of the power put is

$$P^p(90; 3/2, 95, 6/12)$$

$$= 95 e^{-0.05(6/12-0)} \Phi(-4.49647)$$

$$- (90)^{3/2} e^{[(3/2-1)(0.05+(3/2)(0.45)^2/2)-(3/2)(0.015)](6/12-0)} \Phi(-4.97376)$$

$$= \$0.0000285412.$$

Seldom are calls and puts of equal value when their independent variables and parameters are the same, but in the case of Example 14.3 the call has quite a significant price while the put is nearly worthless. There is a Put-Call parity relationship for power options. Note that

$$((S(T))^a - K)^+ - (K - (S(T))^a)^+ = (S(T))^a - K.$$

Taking the time $t$ present value of the risk-neutral expected value of both sides of this equation produces the Put/Call parity formula which follows.

$$C^p(S(t); a, K, T) - P^p(S(t); a, K, T) = F^P_{t,T}((S(t))^a) - Ke^{-r(T-t)}. \quad (14.17)$$

For European power options once again the difference in the premiums for a call and a put equals the difference between the prepaid forward on the underlying security and the present value of the strike price.

Options are more useful to investors when the option positions can be hedged and hedging is more easily accomplished when the Greeks for an option are convenient to calculate. Before launching into calculations of the Greeks for power options a generalization of Lemma 12.1 for power options will be helpful.

**Lemma 14.1.** *For European power calls and puts,*

$$(S(t))^a e^{-\delta^*(T-t)} \varphi(d_1) = K e^{-r(T-t)} \varphi(d_2), \qquad (14.18)$$

*where $\varphi(x)$ is the probability density function for the standard normal random variable.*

The proof of this lemma is similar to the proof of Lemma 12.1 and is left for the reader in Exercise 14.2.7.

Taking the partial derivative with respect to $S$ of both sides of Eq. (14.13) and using Lemma 14.1 produce,

$$\Delta_{C^p} = a(S(t))^{a-1} e^{-\delta^*(T-t)} \Phi(d_1). \qquad (14.19)$$

Performing similar steps on the definition of the European power put option yields,

$$\Delta_{P^p} = -a(S(t))^{a-1} e^{-\delta^*(T-t)} \Phi(-d_1). \qquad (14.20)$$

Differentiating the Delta for the European power call and the Delta for the European power put once again with respect to $S$ reveal the following expressions for the Gamma of a power option.

$$\Gamma_{C^p} = a(S(t))^{a-2} e^{-\delta^*(T-t)} \left( (a-1)\Phi(d_1) + \frac{\varphi(d_1)}{\sigma\sqrt{T-t}} \right), \qquad (14.21)$$

$$\Gamma_{P^p} = a(S(t))^{a-2} e^{-\delta^*(T-t)} \left( (1-a)\Phi(-d_1) + \frac{\varphi(d_1)}{\sigma\sqrt{T-t}} \right). \qquad (14.22)$$

Note that for power options, the Gammas of a call and a put differ from each other. The remaining Greeks for power options are found by computing the appropriate partial derivatives.

**Example 14.4.** A market maker sells 100 units of the European power call option described in Example 14.3. If the market maker wishes to create a Delta-neutral hedging portfolio for the sale, what position in the underlying security should the market maker take? What net cash flow is required to set up the hedging portfolio?

**Solution.** Making use of Eq. (14.19) indicates that for each power call option sold

$$\Delta_{C^p} = \frac{3}{2}(90)^{3/2-1}e^{[(3/2-1)(0.05+(3/2)(0.45)^2/2)-(3/2)(0.015)](6/12-0)}\Phi\,(4.97376)$$

$$= 14.7994,$$

units of the underlying security should be purchased. Thus the market maker should purchase 1479.94 shares. The net cash flow to create the hedging portfolio is

$$B = 100(14.7994(90) - 795.311) = \$53,633.70,$$

which the market maker can borrow at the risk-free rate.

Readers comparing the preceding discussion of power options with information provided by other authors may have noticed there are several similar, though by no means identical, financial instruments which are called power options. For example, some authors [Haug (2007), p. 118] refer to options whose payoffs can be expressed as $((S(T)-K)^+)^a$ or $((K-S(T))^+)^a$ as **symmetric powered options**. Others [McDonald (2013), p. 636] refer to options with payoffs of the form $((S(T))^a - K^a)^+$ or $(K^a - (S(T))^a)^+$ as power options. European call and put options of this latter type have pricing formulas which can be written as follows.

$$C^p(S(t); a, T, K^a) = (S(t))^a e^{-\delta^*(T-t)}\Phi\,(d_1) - K^a e^{-r(T-t)}\Phi\,(d_2) \quad (14.23)$$

$$P^p(S(t); a, T, K^a) = K^a e^{-r(T-t)}\Phi\,(-d_2) - (S(t))^a e^{-\delta^*(T-t)}\Phi\,(-d_1). \quad (14.24)$$

For these types of options,

$$d_1 = \frac{\ln\frac{S(t)}{K} + \left(r - \delta + (a - 1/2)\sigma^2\right)(T - t)}{\sigma\sqrt{T - t}} \quad (14.25)$$

$$d_2 = \frac{\ln\frac{S(t)}{K} + \left(r - \delta - \sigma^2/2\right)(T - t)}{\sigma\sqrt{T - t}}. \quad (14.26)$$

The reader will be asked to derive these formulas in Exercise 14.2.10.

### Exercises

**14.2.1** Starting with the stochastic process in Eq. (10.1) use Itô's Lemma to provide the details in the derivation of Eq. (14.10).

**14.2.2** What is the risk-neutral price of a forward contract for the delivery of $(S(T))^a$ at time $T$?

**14.2.3** Verify that the Put-Call parity relationship of Eq. (14.17) holds for the options described in Example 14.3.

**14.2.4** Find the prices of 1-year, $1-strike European power calls and puts with a power of $1/2$ on a non-dividend-paying stock with current price of $1 and volatility of 20%. The risk-free continuously compounded interest rate is 5%.

**14.2.5** Suppose the current price of a security is $10, the volatility of the security is 30%, and the security pays a continuous dividend with yield 2% per annum. The risk-free interest rate is 4.5% per annum. Find the prices of 3-month, $3-strike European call and put power options on the security with payoffs of $((S(3/12))^{1/2} - 3)^+$ and $(3 - (S(3/12))^{1/2})^+$ respectively.

**14.2.6** Show that the pricing formulas for ordinary European options are special cases of power options when the power $a = 1$.

**14.2.7** Prove Lemma 14.1.

**14.2.8** Suppose after one day the security mentioned in Example 14.4 is worth $91. What is the market maker's one day profit or loss if the hedging portfolio is liquidated?

**14.2.9** Suppose the market maker of Example 14.4 wishes to create a Delta- and Gamma-neutral hedging portfolio by trading ordinary European calls on the security. What positions should the market maker take in the security and the ordinary calls to create a Delta/Gamma-neutral hedging portfolio? What is the net cost to the market maker for setting up this portfolio?

**14.2.10** Following a similar line of reasoning to that which produced the formulas for the European power call and put options, derive the formula for the power option which pays $((S(T))^a - K^a)^+$ (see Eq. (14.23)) and the option which pays $(K^a - (S(T))^a)^+$ (see Eq. (14.24)).

## 14.3 Exchange Options

An **exchange option** is another type of financial instrument which grants an investor the right to exchange one risky asset (such as a stock) for another risky asset (for example, another stock). There are both European style of exercise and American style of exercise exchange options and this section will focus on the pricing and properties of European exchange options. The pricing formula will be developed by generalizing the Black–Scholes option pricing formula for ordinary options. Ordinary European call options give an investor the right to exchange a riskless asset (cash represented by the strike price) for a risky asset (a stock). Ordinary put options perform a similar function. A put option holder may exchange a risky asset (stock) for a riskless asset (cash).

Since two risky assets are at work in pricing exchange options, one of the assets will continue to be denoted $S(t)$ and referred to as the **underlying asset**. The other risky asset will be denoted $K(t)$ and will be called the **strike asset**. The payoff of a European exchange call option with expiry $T$ is $(S(T)-K(T))^+$. Related quantities such as the continuously compounded dividend yield and the volatility of the continuously compounded returns on the assets will be subscripted with either "$S$" or "$K$" to denoted whether they refer to the underlying asset or the strike asset. The symbol $\rho$ will represent the correlation between the continuously compounded returns of the underlying and strike assets. Thus the volatility of the difference in the continuously compounded returns of the underlying and strike assets satisfies the equation,

$$\sigma^2 = \sigma_S^2 - 2\rho\sigma_S\sigma_K + \sigma_K^2. \tag{14.27}$$

To price an exchange call consider a call option whose payoff at expiry is $(S(T)/K(T)-1)^+$. Thus the value of the underlying asset is expressed in units of the strike asset. The "strike price" is 1 and applying the ordinary Black–Scholes call pricing formula in Eq. (10.34) produces

$$\frac{S(t)}{K(t)} e^{-(\delta_S - \delta_K)(T-t)} \Phi\left(d_1\right) - (1)\Phi\left(d_2\right).$$

Since $K(T)(S(T)/K(T) - 1)^+ = (S(T) - K(T))^+$ multiplying the formula by $K(t)e^{-\delta_K(T-t)}$ results in Margrabe's formula [Margrabe (1978)].

$$C^x(S(t), K(t); \sigma, T) = S(t)e^{-\delta_S(T-t)}\Phi\left(d_1\right) - K(t)e^{-\delta_K(T-t)}\Phi\left(d_2\right) \tag{14.28}$$

The values of $d_1$ and $d_2$ can be obtained from Eqs. (10.32) and (10.33).

$$d_1 = \frac{\ln\frac{S(t)}{K(t)} + \left(\delta_K - \delta_S + \frac{\sigma^2}{2}\right)(T-t)}{\sigma\sqrt{T-t}}, \tag{14.29}$$

$$d_2 = \frac{\ln\frac{S(t)}{K(t)} + \left(\delta_K - \delta_S - \frac{\sigma^2}{2}\right)(T-t)}{\sigma\sqrt{T-t}}. \tag{14.30}$$

Note that the risk-free interest rate plays no role in the pricing formula. The value of the European exchange put option whose payoff at expiry is $(K(T) - S(T))^+$ is found similarly to be

$$P^x(S(t), K(t); \sigma, T) = K(t)e^{-\delta_K(T-t)}\Phi\left(-d_2\right) - S(t)e^{-\delta_S(T-t)}\Phi\left(-d_1\right). \tag{14.31}$$

These pricing formulas can be made more general by considering the premium of the option which allows $q_K$ units of the strike asset to be exchanged for $q_S$ units of the underlying asset. In this case, replace $S(t)$ by $q_S S(t)$ and $K(t)$ by $q_K K(t)$ in the formulas above. The same adjustment can be made for exchange puts.

**Example 14.5.** Suppose for security $S$ (the underlying asset) that the current price is \$35/share, the security pays no dividend, and the volatility of the security is 25% per annum. For security $K$ (the strike asset) the current price is \$50/share, the security pays a continuous dividend of 2% per annum, and the volatility is 35%. The correlation between the continuously compounded returns on the two securities is $\rho = 0.4$. Find the premium of an exchange option whose payoff in 6 months will be $(4S(6/12) - 3K(6/12))^+$.

**Solution.** This is an exchange call with $q_S = 4$ and $q_K = 3$.

$$\sigma^2 = (0.25)^2 - 2(0.4)(0.25)(0.35) + (0.35)^2 = 0.115$$

$$\sigma = 0.339116$$

$$d_1 = \frac{\ln\frac{4(35)}{3(50)} + \left(0.02 - 0 + \frac{0.115}{2}\right)(6/12 - 0)}{0.339116\sqrt{6/12 - 0}} = -0.126121$$

$$d_2 = \frac{\ln\frac{4(35)}{3(50)} + \left(0.02 - 0 - \frac{0.115}{2}\right)(6/12 - 0)}{0.339116\sqrt{6/12 - 0}} = -0.365913.$$

Using these calculated input values Eq. (14.28) yields

$$C^x((4)35, (3)50; \sigma, 6/12) = (4)(35)\Phi(d_1) - (3)(5)e^{-0.02(6/12-0)}\Phi(d_2)$$
$$= \$9.92541.$$

Naturally there is a Put-Call parity relationship for exchange options. Since

$$(S(T) - K(T))^+ - (K(T) - S(T))^+ = S(T) - K(T),$$

then

$$C^x(S(t), K(t); \sigma, T) - P^x(S(t), K(t); \sigma, T) = F^P_{t,T}(S(t)) - F^P_{t,T}(K(t)).$$
$$(14.32)$$

A similar formula holds if $S(t)$ is replaced by $q_S S(t)$ and $K(t)$ is replaced by $q_K K(t)$.

Exchange call and put options depend on the prices of the underlying and strike assets. As such, Delta hedging a portfolio containing exchange options requires measuring the response of the option price to changes in both the underlying asset and the strike asset. The partial derivative of an exchange call with respect to the underlying asset can be expressed as

$$\frac{\partial C^x}{\partial S} = e^{-\delta_S(T-t)}\Phi(d_1), \tag{14.33}$$

while the partial derivative of an exchange call with respect to the strike asset can be written as

$$\frac{\partial C^x}{\partial K} = -e^{-\delta_K(T-t)}\Phi(d_2). \tag{14.34}$$

For derivations of these formulas see the exercises at the end of this section. The partial derivatives of an exchange put option with respect to the underlying and strike assets are respectively,

$$\frac{\partial P^x}{\partial S} = -e^{-\delta_S(T-t)}\Phi(-d_1) \tag{14.35}$$

$$\frac{\partial P^x}{\partial K} = e^{-\delta_K(T-t)}\Phi(-d_2). \tag{14.36}$$

**Example 14.6.** Suppose a market maker sells 100 units of the exchange call described in Example 14.5. What position should be market maker take in the strike asset so as to create a Delta-neutral hedging portfolio?

**Solution.** Using Eq. (14.33) the Delta (with respect to the strike asset) of the exchange call is

$$\frac{\partial C^x}{\partial K} = -e^{-0.02(6/12-0)}\Phi(-0.365913) = -0.353661.$$

Since the exchange calls were sold the market maker should take a short position in 35.3661 shares of the strike asset.

The properties of exchange options will be further justified and explored in the following collection of exercises.

### Exercises

**14.3.1** Show that Eqs. (14.28) and (14.31) can be written respectively as:
$$C^x(S(t),K(t);\sigma,T) = F_{t,T}^P(S(t))\Phi(d_1) - F_{t,T}^P(K(t))\Phi(d_2)$$
$$P^x(S(t),K(t);\sigma,T) = F_{t,T}^P(K(t))\Phi(-d_2) - F_{t,T}^P(S(t))\Phi(-d_1),$$
where $F_{t,T}^P$ denotes the prepaid forward price of the asset.

**14.3.2** Suppose $S(t)$ is the time $t$ price of a non-dividend-paying stock with volatility of 30% per annum, while $K(t)$ is the price of another non-dividend-paying stock with volatility of 50% per annum. The current prices of the stocks are $S(0) = 15$ and $K(0) = 20$. The correlation between the continuously compounded returns is $-0.5$. Find the price of an exchange option with payoff in 1 year of $(6K(1) - 8S(1))^+$.

**14.3.3** Suppose $S(t)$ is the time $t$ price of a stock with a continuous dividend rate of 2.5% and with a volatility of 40% per annum, while $K(t)$ is the price of another stock paying a continuous dividend at a rate of 1.3% and with a volatility of 35% per annum. The current prices of the stocks are $S(0) = 17$ and $K(0) = 50$. The correlation between the continuously compounded returns is 0.78. Find the price of an exchange option with payoff in 3 months of $(3S(3/12) - K(3/12))^+$.

**14.3.4** Show that $C^x(S(t), K(t); \sigma, T) = P^x(K(t), S(t); \sigma, T)$ and thus there is no significant difference between exchange puts and calls. The difference is merely that of to which stocks the labels of "underlying" and "strike" are given.

**14.3.5** Suppose a security $S$ has a current price of $20 and pays continuous dividend with yield 1% per annum. Security $K$ does not pay a dividend and has a current price of $5. The option to exchange four shares of security $K$ for one share of security $S$ costs $3. Find the price to exchange five shares of security $S$ for twenty shares of security $K$ in 6 months.

**14.3.6** Show that if the strike asset is a risk-free bond with maturity value $K$, Eqs. (14.28) and (14.31) simplify to the ordinary Black–Scholes call and put prices respectively.

**14.3.7** Show that for European exchange calls and puts,

$$S(t)e^{-\delta_S(T-t)}\varphi(d_1) = K(t)e^{-\delta_K(T-t)}\varphi(d_2)$$

where $\varphi(x)$ is the probability density function for the standard normal random variable.

**14.3.8** Use the identity established in Exercise 14.3.7 to derive the Deltas for exchange calls and puts with respect to the underlying and strike assets.

**14.3.9** Suppose after one day the prices of the underlying and strike assets mentioned in Example 14.5 have changed to $35.50 and $50.25. At time $t = 0$ the market maker set up the Delta-neutral hedging portfolio described in Example 14.6. What is the 1-day profit or loss if the hedging portfolio is liquidated? Assume the risk-free interest rate is 0%.

**14.3.10** Suppose after 1 day the prices of the underlying and strike assets mentioned in Example 14.5 have changed to $35.50 and $50.25. The market maker wishes to rebalance the hedging portfolio so as to keep it Delta-neutral. How many shares of the strike asset should the market maker buy or sell?

**14.3.11** Consider a type of option based on quantities of two securities $S_1(t)$ and $S_2(t)$. Suppose $q_1$ and $q_2$ respectively are the quantities of $S_1(t)$ and $S_2(t)$, and the option pays at time $T$ the $\max\{q_1 S_1(T), q_2 S_2(T)\}$.

Find a formula for the price of this option at time $t < T$. *Hint*: $\max\{x, y\} = (x - y)^+ + y$.

**14.3.12** Consider a type of option based on quantities of two securities $S_1(t)$ and $S_2(t)$. Suppose $q_1$ and $q_2$ respectively are the quantities of $S_1(t)$ and $S_2(t)$, and the option pays at time $T$ the $\min\{q_1 S_1(T), q_2 S_2(T)\}$. Find a formula for the price of this option at time $t < T$. *Hint*: $\min\{x, y\} = x - (x - y)^+$.

**14.3.13** Find a parity relationship for the maximum and minimum paying options of Exercises 14.3.11 and 14.3.12.

## 14.4 Chooser Options

A **chooser option** takes its name the owner's ability to decide after purchase whether the option is a call or a put. Chooser options are also called **as-you-like-it options**. Since a chooser option can become either a call or a put, its value should be the larger of the value of a call or a put. Suppose the underlying asset has value $S(t)$, the strike price of the call or put option is $K$ and the strike time of the call and put is $T > t$, then a chooser option has value,

$$\max\{C^e(S(t); K, T), P^e(S(t); K, T)\}.$$

The chooser option may have the same expiry as the call and put, or the holder of the chooser option may have to decide whether the chooser option becomes a call or a put at a time $t_1 < T$ prior to the expiry of the call and put. If the call and put expire at the same time as the chooser option, the payoff of the chooser is the same as the payoff of a straddle with the same strike price and strike time. Option strategies including the straddle are described in Sec. 6.4. See Fig. 14.3.

If the chooser option must be exercised at time $t_1$ prior to the expiration of the call and put then

$$\max\{C^e(S(t_1); K, T), P^e(S(t_1); K, T)\}$$

$$= C^e(S(t_1); K, T) + (P^e(S(t_1); K, T) - C^e(S(t_1); K, T))^+$$

$$= C^e(S(t_1); K, T) + \left(Ke^{-r(T-t_1)} - S(t_1)e^{-\delta(T-t_1)}\right)^+.$$

This holds due to Put-Call parity Eq. (6.8). Factoring out $e^{-\delta(T-t_1)}$ from the last expression yields

$$\max\{C^e(S(t_1); K, T), P^e(S(t_1); K, T)\}$$

$$= C^e(S(t_1); K, T) + e^{-\delta(T-t_1)}\left(Ke^{-(r-\delta)(T-t_1)} - S(t_1)\right)^+,$$

**payoff**

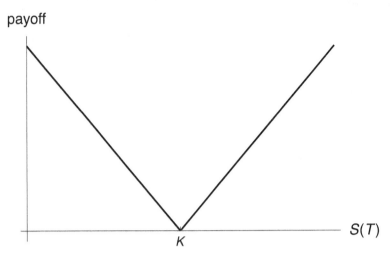

Fig. 14.3 The payoff a chooser option which expires at the same time as the call and put option is the same as that of a straddle with the same strike time and price.

which is the value of a European call with strike time $T$ at time $t_1$ plus $e^{-\delta(T-t_1)}$ units of the payoff of a European put with strike time $t_1$ and strike price $Ke^{-(r-\delta)(T-t_1)}$. Therefore at time $0 \leq t < t_1 < T$ the value of the chooser option can be expressed as

$$V(S(t); t_1, T) = C^e(S(t); K, T) + e^{-\delta(T-t_1)} P^e(S(t); Ke^{-(r-\delta)(T-t_1)}, t_1). \tag{14.37}$$

Note that the European call has strike price $K$ and strike time $T$ while the European put has strike price $Ke^{-(r-\delta)(T-t_1)}$ and strike time $t_1$. Figure 14.4 illustrates the graphs of chooser, European call, and European put options on a stock. For each option the volatility of the stock was 30%, dividend yield 5%, and strike price $100. The risk-free interest rate was set at 10%. The expiration of the options is 2 years hence and the time to choose is 1 year hence. The initial price of the stock is varied to show the relationship between the three option premiums. The chooser is more expensive than the European call or put, but less expensive than a straddle.

The Greeks of a chooser option are straightforward to calculate since the chooser is the sum of a ordinary European call and related put. The Delta of a chooser option is the partial derivative of the option premium with respect to the underlying stock price.

$$\frac{\partial V}{\partial S} = e^{-\delta(T-t)} \left( \Phi\left(d_1\right) - \Phi\left(-d_1^{t_1}\right) \right). \tag{14.38}$$

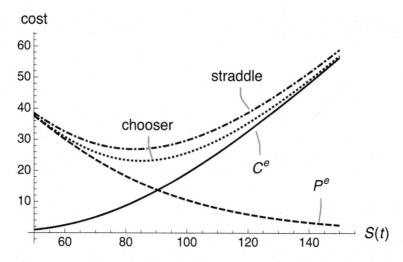

Fig. 14.4    A chooser option is less expensive than a straddle, but more expensive than a call option or a put option, all parameters and variables being the same.

To avoid confusion the expression $d_1^{t_1}$ is introduced to stand in as

$$d_1^{t_1} = \frac{\ln \frac{S}{K} + (r - \delta)(T - t) + \frac{\sigma^2}{2}(t_1 - t)}{\sigma \sqrt{t_1 - t}}. \tag{14.39}$$

Taking the second derivative of the chooser premium with respect to $S$ yields

$$\frac{\partial^2 V}{\partial S^2} = \frac{e^{-\delta(T-t)}}{\sigma S} \left( \frac{\varphi(d_1)}{\sqrt{T - t}} + \frac{\varphi(d_1^{t_1})}{\sqrt{t_1 - t}} \right). \tag{14.40}$$

The value of a chooser option also varies with the time remaining to choose. See Fig. 14.5. This behavior will be further explored in Exercise 14.4.7.

**Example 14.7.** Consider a non-dividend-paying security whose current price is \$50 and whose volatility is 50%. The risk-free interest is 8% continuously compounded. A market maker sells 100 units of a chooser option with 3 months until the choice date and 6 months until the expiry of the put and call options which have a strike price of \$50.

(a) Determine the price of the chooser option.
(b) What position in the underlying security should the market maker take to create a Delta-neutral hedging portfolio?
(c) If after 1 day the price of the underlying security is \$51, what is the market maker's profit/loss if the hedging portfolio is liquidated?

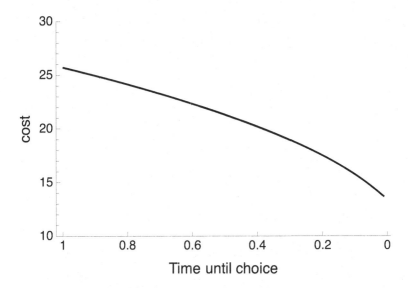

Fig. 14.5   The value of an at-the-money chooser option decreases as the time to choose between a call and a put approaches.

(d) If the market maker wishes to create a Delta/Gamma-neutral portfolio by trading ordinary European calls on the underlying security what positions should the market maker take in the security and the ordinary calls?

**Solution.** Using the parameter and variable values $S(0) = 50$, $K = 50$, $T = 6/12$, $t_1 = 3/12$, $r = 0.08$, $\delta = 0$, and $\sigma = 0.50$ in Eq. (14.37) reveals the chooser premium to be $V(50, 3/12, 6/12) = \$11.86$. The Delta for the chooser as determined by Eq. (14.38) is $\Delta_V = 0.226237$. Thus to create a Delta-neutral hedging portfolio the market maker should take a long position in 22.6237 shares of the security. The net cash flow to create the Delta-neutral hedging portfolio at $t = 0$ is

$$100(11.86 - 0.226237(50)) = \$54.8192,$$

which can be lent at the risk-free rate. If after 1 day the price of the under-lying security is $51, the chooser options will have a value of $12.0669. The market maker will sell the held shares of the security, collect the principal and interest on the funds lent, and re-purchase the chooser options at their new price. The profit/loss will be

$$P/L = 100(0.226237(51) - 12.0669) + 54.8192e^{0.08(1/365)} = \$1.953.$$

If the market maker wishes to create a Delta/Gamma-neutral hedging portfolio by taking positions in the underlying security and ordinary European calls note that the price of an ordinary call with the same parameter and variable values is $C^e(50, 50, 6/12) = \$7.90204$. The Gamma of the chooser option as calculated from the expression in Eq. (14.40) is $\Gamma_V = 0.052284$. The Gamma of the ordinary call is $\Gamma = 0.0216388$. Thus the hedging portfolio is made Gamma-neutral when the market maker takes a long position in

$$100(0.052284/0.216388) = 241.621$$

units of the ordinary call. To make the hedging portfolio simultaneously Delta-neutral, since the Delta of the ordinary call is $\Delta_{C^e} = 0.614059$ the market maker should take a position in $z$ units of the underlying security where

$$241.621(0.614059) - 100(0.226237) + z = 0 \implies z = -125.746.$$

In other words, the market maker should short 125.746 shares of the underlying security.

The type of chooser option described in this section is sometimes referred to as a simple chooser option. There are **complex chooser options** which generalize this scenario. The holder of a complex chooser option may choose at a specified time $t_1$ whether the option is a call option with strike time $T_C$ and strike price $K_C$ or a put option with strike time $T_P$ and strike price $K_P$. At the choice time $t_1$ the value of the complex chooser option is

$$V(S(t_1)) = \max\{C^e(S(t_1); K_C, T_C), P^e(S(t_1); K_P, T_P)\}.$$

However, since the strike prices and strike times of the put and call are not necessarily the same, Put-Call parity relationship cannot be used to decompose the value of the complex chooser option. The value can be determined by a formula. The interested reader should consult [Clewlow and Strickland (1997)].

### Exercises

**14.4.1** The current price of a stock is \$100. The stock pays a continuous dividend at a rate of 2% per annum. The volatility of the continuous return on the stock is 30%. Find the time $t = 0$ price of a chooser option which expires in 2 years with a strike price of \$100. The option holder must decide after 1 year whether the chooser option will be a call or put. The risk-free interest rate is 5% per annum.

**14.4.2** A chooser option has 9 months to expiration and 6 months for the option holder to decide between a put or call. The underlying stock has a price of $75, pays a continuous dividend with yield of 1% per annum, and has volatility of 40%. The strike price of the option is $78 and the continuously compounded risk-free interest rate is 6%. Find the price of the chooser option.

**14.4.3** Compare the price of the chooser option described in Exercise 14.4.2 with the price of an equivalent straddle.

**14.4.4** Show that the chooser option whose value is given in Eq. (14.37) can also be expressed as

$$V(S(t); t_1, T) = P^e(S(t); K, T) + e^{-\delta(T-t_1)} C^e(S(t); K^{-(r-\delta)(T-t_1)}, t_1).$$

**14.4.5** Show that the pricing formula for a chooser option can also be expressed as

$$V(S(t); t_1, T) = C^e(S(t); K, T) + Ke^{-r(T-t)} \Phi\left(-d_2^{t_1}\right) - Se^{-\delta(T-t)} \Phi\left(-d_1^{t_1}\right),$$

where $d_1^{t_1}$ is defined in Eq. (14.39) and $d_2^{t_1} = d_1^{t_1} - \sigma\sqrt{t_1 - t}$.

**14.4.6** Show that for chooser options,

$$S(t)e^{-\delta(T-t)} \varphi\left(d_1^{t_1}\right) = K(t)e^{-r(T-t)} \varphi\left(d_2^{t_1}\right),$$

where $\varphi(x)$ is the probability density function for the standard normal random variable.

**14.4.7** Find an expression for the Theta of a chooser option.

**14.4.8** Find an expression for the Rho of a chooser option.

**14.4.9** Find an expression for the Vega of a chooser option.

**14.4.10** Find the 1-day profit or loss on the Delta/Gamma-neutral portfolio described in Example 14.7.

## 14.5  Forward Start Options

A **forward start option** is a financial instrument which has a payoff at time $\tau$ equal to the value of a $T$-year ordinary option. The strike price of the ordinary option is expressed as a multiple $a$ of the time-$\tau$ value of the underlying stock. If $a = 1$ the ordinary option starts at-the-money. If $a < 1$ the ordinary option starts in-the-money for a forward start call option and out-of-the-money for a forward start put option. If $a > 1$ the ordinary option starts out-of-the-money for a forward start call option and in-the-money for a forward start put option. Employee stock options awarded by a company are sometimes of the forward start type since the employees earn at-the-money options at specific future time points in their careers [Hull (2000), Sec. 24.3].

Forward start options can be priced using a simple modification to the Black–Scholes option pricing formulas for ordinary calls and puts. This discussion begins with the derivation of a pricing formula for a forward start call option. At time $t = \tau$ the value of a forward start call will be the value of a European call option expiring at time $t = \tau + T$,

$$V(S(\tau); \tau) = C^e(S(\tau); aS(\tau), \tau + T) = S(\tau)e^{-\delta T}\Phi(d_1) - aS(\tau)e^{-rT}\Phi(d_2),$$

where

$$d_1 = \frac{\ln\frac{S(\tau)}{aS(\tau)} + (r - \delta + \frac{\sigma^2}{2})T}{\sigma\sqrt{T}} = \frac{-\ln a + (r - \delta + \frac{\sigma^2}{2})T}{\sigma\sqrt{T}} \quad (14.41)$$

$$d_2 = \frac{-\ln a + (r - \delta - \frac{\sigma^2}{2})T}{\sigma\sqrt{T}}. \quad (14.42)$$

Consequently for time $t < \tau$ the price of the forward start call option is

$$C^f(S(t)) = F_{t,\tau}^P(S(t))\left(e^{-\delta T}\Phi(d_1) - ae^{-rT}\Phi(d_2)\right). \quad (14.43)$$

In a similar fashion the price of a forward start put option is expressed as

$$P^f(S(t)) = F_{t,\tau}^P(S(t))\left(ae^{-rT}\Phi(-d_2) - e^{-\delta T}\Phi(-d_1)\right). \quad (14.44)$$

**Example 14.8.** A stock pays dividends continuously at a rate of 1% and has a volatility of 25%. The current price of the stock is \$125. The continuously compounded risk-free rate of interest is 5%. A forward start option on this stock will deliver a 9-month European call in 3 months with a strike price of 110% of the stock price in 3 months. Find the price of this option.

**Solution.** Using the given variable and parameter values,

$$d_1 = \frac{-\ln 1.10 + \left(0.05 - 0.01 + \frac{(0.25)^2}{2}\right)(9/12)}{0.25\sqrt{9/12}} = -0.193402$$

$$d_2 = \frac{-\ln 1.10 + \left(0.05 - 0.01 - \frac{(0.25)^2}{2}\right)(9/12)}{0.25\sqrt{9/12}} = -0.409908$$

$$C^f = 150e^{-0.03(1)}\left(e^{-0.03(2)}\Phi(d_1) - 1.05e^{-0.035(2)}\Phi(d_2)\right) = \$7.34811.$$

The Greeks for forward start options are relatively simple to express. The Delta for a forward start call is the partial derivative of the option value with respect to the underlying stock price. From Eq. (14.43) Delta can be written as

$$\Delta_{Cf} = e^{-\delta(\tau-t)} \left( e^{-\delta T} \Phi\left(d_1\right) - ae^{-rT} \Phi\left(d_2\right) \right). \tag{14.45}$$

For a forward start put the formula for Delta is equally straightforward.

$$\Delta_{Pf} = e^{-\delta(\tau-t)} \left( ae^{-rT} \Phi\left(-d_2\right) - e^{-\delta T} \Phi\left(-d_1\right) \right). \tag{14.46}$$

For both forward start calls and puts, $\Gamma = 0$.

## *Exercises*

**14.5.1** Show that an at-the-money forward start option on a non-dividend paying stock that will start in 1 year and mature in 2 years is worth the same as a 1-year at-the-money option starting today.

**14.5.2** A stock pays dividends continuously at a rate of 3% and has a volatility of 40%. The current price of the stock is $150. The continuously compounded risk-free rate of interest is 3.5%. A forward start call on this stock will expire in 3 years. After 1 year, the strike price of the call will be set to be equal to 105% of the price of the stock at that time. Find the price of this option.

**14.5.3** Find the price of the corresponding forward start put option for the situation described in Exercise 14.5.2.

**14.5.4** Find an expression for the Theta of a forward start option.

**14.5.5** Show that $e^{-\delta T} \varphi\left(d_1\right) = ae^{-rT} \varphi\left(d_2\right)$.

**14.5.6** Find an expression for the Vega of a forward start option.

## 14.6 Compound Options

An option which gives the holder the right to purchase or sell another option in the future is called a **compound option**. Specifically if an option gives an investor the right to purchase a call option, this is known as a call on a call. As such there are four broad types of compound options:

- call on a call (purchase a call),
- call on a put (purchase a put),
- put on a call (sell a call),
- put on a put (sell a put).

When discussing compound options there are also two strike times: (1) the earlier time denoted $t_1$ is the time the option holder may exercise the compound option in order to purchase or sell an ordinary option, and (2) the later time denoted $T$ is the expiry of the underlying ordinary option. There are also two strike prices: $X$ the strike price for the compound option and $K$ the strike of the underlying ordinary option. Throughout this section the current time will be denoted $t$.

As mentioned at the beginning of this chapter, exotic options are a category of options which do not satisfy the Black–Scholes partial differential equation. One of the reasons compound options fail to satisfy Eq. (10.5) is that the values of options are not lognormally distributed even as the returns of the underlying stocks are. In order to price compound options the **binormal distribution** is used. Consider a European call option on a call option. The price of the underlying call option $C^e(S(t), K, T)$ can be determined by the Black–Scholes option pricing formula. At time $t_1$ the payoff of the compound call will be $(C^e(S(t_1), K, T) - X)^+$. The compound option will only be exercised if $C^e(S(t_1), K, T) > X$. A numerical technique such as Newton's method [Burden and Faires (2005)] will determine the stock price $S^*$ such that an ordinary call $C^e(S^*, K, T) = X$ at time $t = t_1$. If the current estimate of $S^*$ is $S_n$ then the next (improved) estimate is

$$S_{n+1} = S_n - \left.\frac{C^e(S, K, T) - X}{\partial C^e / \partial S}\right|_{S=S_n} = S_n - \left.\frac{C^e(S, K, T) - X}{\Delta_{C^e}}\right|_{S=S_n}.$$

(14.47)

Some authors refer to $S^*$ as the **implied stock price**.

**Example 14.9.** A stock paying a continuous dividend at the rate of 1.5% per annum has volatility 40%. The risk-free continuously compounded interest rate is 4%. Find the price of the stock such that a 50-strike European call option with 6-months to expiration has a price of $10.

**Solution.** Use $S_0 = K = 50$ as the initial approximation of the stock price, then repeated using Eq. (14.47) produces the sequence of approximations to $S^*$.

| $n$ | $S_n$ |
|---|---|
| 0 | 50.0000 |
| 1 | 57.2714 |
| 2 | 56.3792 |
| 3 | 56.3683 |
| 4 | 56.3683 |

Thus $S^* \approx \$56.3683$. All iterates beyond the third round to the same four decimal place result.

The pricing formula for compound options can be found as the solution to a system of partial differential equations using Itô's Lemma and mathematical techniques similar to those found in Chapter 10. The interested reader can consult [Geske (1979)] for the details. This section will present a slightly more informal approach to the development of the pricing formula for a call on a call by decomposing the compound European call on a call as the difference between asset-or-nothing calls and cash-or-nothing calls.

Suppose the current price of the stock is $S(t)$, the dividend yield is $\delta$, the volatility is $\sigma$, and the risk-free interest rate is $r$. Suppose an option pays the holder $S(T)$ at time $T$ provided $S(T) > K$ and $S(t_1) > S^*$. Exercise 9.1.5 established that $\rho = \mathbb{C}\mathrm{or}\,(S(t_1), S(T)) = \sqrt{(t_1 - t)/(T - t)}$. The risk-neutral expected value of the payoff of the option is $S(t)e^{-\delta(T-t)}\boldsymbol{\Phi}\,(a_1, d_1; \rho)$ where

$$\boldsymbol{\Phi}\,(x, y; \rho) = \int_{-\infty}^{x} \int_{-\infty}^{y} \frac{1}{2\pi\sqrt{1 - \rho^2}} e^{-(u^2 - 2\rho u u + u^2)/(2(1 - \rho^2))}\, dv\, du,$$

and

$$a_1 = \frac{\ln \frac{S(t)}{S^*} + \left(r - \delta + \frac{\sigma^2}{2}\right)(t_1 - t)}{\sigma\sqrt{t_1 - t}}. \tag{14.48}$$

Now consider an option which pays $K$ provided $S(T) \leq K$ and $S(t_1) > X$. Its risk-neutral expected payoff is $Ke^{-r(T-t)}\boldsymbol{\Phi}\,(a_2, d_2; \rho)$ where

$$a_2 = \frac{\ln \frac{S(t)}{S^*} + \left(r - \delta - \frac{\sigma^2}{2}\right)(t_1 - t)}{\sigma\sqrt{t_1 - t}}. \tag{14.49}$$

The formulas for $d_1$ and $d_2$ are the same as those used in pricing ordinary European options and can be found in Eqs. (10.32) and (10.33). The expressions for $a_1$ and $a_2$ are of a similar form, the only differences being the strike price parameter is replaced by the implied stock price and the strike time is $t_1$ rather than the strike time of the underlying call option. Finally consider a cash-or-nothing call which pays $X$ at time $t_1$ provided $S(t_1) < X$. The risk-neutral expected payoff of this option is $Xe^{-r(t_1-t)}\Phi\,(d_2)$. Combining these formulas creates the pricing formula for a call on a call,

$$\begin{aligned} C^c &= S(t)e^{-\delta(T-t)}\boldsymbol{\Phi}\,(a_1, d_1; \rho) \\ &\quad - Ke^{-r(T-t)}\boldsymbol{\Phi}\,(a_2, d_2; \rho) - Xe^{-r(t_1-t)}\Phi\,(a_2). \end{aligned} \tag{14.50}$$

Using similar reasoning, the formulas for other compound options can be determined. In the case of the call on a put, if the compound option is exercised the owner of the call on a put buys a put and thus receives the value of the underlying put in exchange for the strike price of the compound option. Therefore the aggregate parameters $a_1$, $a_2$, $d_1$, and $d_2$ are negated in the formula expressing the value of $C^p$. For a put on a call, when the compound option is exercised, the holder of the compound option sells the underlying call (*i.e.*, shorts the call) and receives the strike price of the compound option. In this situation $a_1$ and $a_2$ are negated while $d_1$ and $d_2$ are not since the underlying call option is in the money. The correlation $\rho$ is negated as well since as the stock price increases, the underlying call increases in value as well, thus causing a decrease in the value of the put on a call. Lastly for the case of a put on a put, if the compound option is exercised the holder must sell the underlying put. In other words if the put on a put is in the money, expressions $a_1$ and $a_2$ are not negated, but $d_1$ and $d_2$ are negated since the underlying option is a put. Once again the correlation is negated since as the stock price increases the value of the underlying put decreases causing an increase in the price of the put on a put. To summarize, for the call on a put, put on a call, and put on a put respectively the pricing formulas are

$$C^p = Ke^{-r(T-t)}\Phi\left(-a_2, -d_2; \rho\right)$$
$$- S(t)e^{-\delta(T-t)}\Phi\left(-a_1, -d_1; \rho\right) - Xe^{-r(t_1-t)}\Phi\left(-a_2\right) \qquad (14.51)$$
$$P^c = Ke^{-r(T-t)}\Phi\left(-a_2, d_2; -\rho\right)$$
$$- S(t)e^{-\delta(T-t)}\Phi\left(-a_1, d_1; -\rho\right) + Xe^{-r(t_1-t)}\Phi\left(-a_2\right) \qquad (14.52)$$
$$P^p = S(t)e^{-\delta(T-t)}\Phi\left(a_1, -d_1; -\rho\right)$$
$$- Ke^{-r(T-t)}\Phi\left(a_2, -d_2; -\rho\right) + Xe^{-r(t_1-t)}\Phi\left(a_2\right). \qquad (14.53)$$

For the call on a put (Eq. (14.51)) and put on a put (Eq. (14.53)) the implied stock price is the solution to the equation $P^e(S^*, K, T) = X$ when the current time is $t = t_1$.

**Example 14.10.** Suppose a call on a call option gives the owner the right to buy a call option on a stock index 6 months from today. The price of the underlying call will be \$25. The strike price of the underlying call will be \$275 and the expiry of the underlying call will be 1 year from today. The current price of the stock index is \$250, the stock index pays a continuous dividend with a yield of 2.5% per annum, and the volatility is 40%. The risk-free continuously compounded interest rate is 6%. Find the price of the compound option.

**Solution.** Numerically approximating the solution $S^*$ to the equation $C^e(S^*, 275, 1) = 25$ produces $S^* \approx \$260.672$. The correlation $\rho = \sqrt{1/2}$.

Hence,

$$a_1 = \frac{\ln \frac{250}{260.672} + \left(0.06 - 0.025 + \frac{(0.40)^2}{2}\right)\left(\frac{6}{12} - 0\right)}{0.40\sqrt{\frac{6}{12} - 0}} = 0.0555054$$

$$a_2 = \frac{\ln \frac{250}{260.672} + \left(0.06 - 0.025 - \frac{(0.40)^2}{2}\right)\left(\frac{6}{12} - 0\right)}{0.40\sqrt{\frac{6}{12} - 0}} = -0.227337$$

$$d_1 = \frac{\ln \frac{250}{275} + \left(0.06 - 0.025 + \frac{(0.40)^2}{2}\right)(1 - 0)}{0.40\sqrt{1 - 0}} = 0.0492246$$

$$d_2 = \frac{\ln \frac{250}{275} + \left(0.06 - 0.025 - \frac{(0.40)^2}{2}\right)(1 - 0)}{0.40\sqrt{1 - 0}} = -0.350775.$$

$$\Phi\left(a_1, d_1; \rho\right) = 0.396058$$
$$\Phi\left(a_2, d_2; \rho\right) = 0.265458$$
$$\Phi\left(a_2\right) = 0.410081.$$

Thus the premium for this call on a call is

$$C^c = 250e^{-0.025(1-0)}(0.396058) - 275e^{-0.06(1-0)}(0.265458)$$
$$- 25e^{-0.06(6/12-0)}(0.410081)$$
$$= \$17.8711.$$

Once the value of the call on a call option is determined the premium for a put on a call (all parameters and variables remaining the same) can be determined from a Put-Call parity relationship for compound options. Recall from Eq. (6.8) that the value of a call option on a stock plus the present value of the strike price equals the prepaid forward price of the stock plus the value of the put option. This equation is adapted to compound options by treating the underlying option as the underlying security and the time $t$ price of an ordinary call option as the prepaid forward on the security. Thus for the pairing of a call on a call and put on a call on the same underlying call option, the Put-Call parity equation is

$$C^c + Xe^{-r(t_1-t)} = C^e(S(t), K, T) + P^c. \tag{14.54}$$

For compound options in which the underlying option is a put, the Put-Call parity relationship has the form,

$$C^p + Xe^{-r(t_1-t)} = P^e(S(t), K, T) + P^p. \tag{14.55}$$

An Undergraduate Introduction to Financial Mathematics

**Example 14.11.** Suppose a non-dividend-paying security has a current price of \$95 and volatility of 25%. The continuously compounded risk-free interest rate is 4%. A 1-year, \$4-strike call option on a \$100-strike put option that expires 2 years from now has a price of \$11.25. Find the price of a 1-year, \$4-strike put option on a \$100-strike put option that expires 2 years from now.

**Solution.** This is the Put-Call parity relationship between a call on a put and a put on a put. The time $t = 0$ cost of the underlying put is $P^e(95, 100, 2) = \$11.8401$. Using Eq. (14.55),

$$P^p = 11.25 + 4e^{-0.04(1-0)} - 11.8401 = \$3.25307.$$

*Exercises*

**14.6.1** Suppose the current price of a security is \$100, the security pays a continuous dividend at a rate of 2% per annum, and has a volatility of 35%. The risk-free continuously compounded interest rate is 7%. Suppose an investor has a compound call giving them the right to buy for \$10 in 1 year a \$105-strike European put option on the security that will expire in 2 years. At what security price in 1 year will the compound option be exercised?

**14.6.2** Find the premium of a put on a call for the situation described in Example 14.10.

**14.6.3** Suppose a call on a put option gives the owner the right to buy a put option on a stock index 3 months from today. The price of the underlying put will be \$75. The strike price of the underlying put will be \$750 and the expiry of the underlying call will be 9 months from today. The current price of the stock index is \$775, the stock index pays a continuous dividend with a yield of 3% per annum, and the volatility is 35%. The risk-free continuously compounded interest rate is 5%. Find the price of the compound option.

**14.6.4** Suppose a put on a put option gives the owner the right to sell a put option on a stock index 6 months from today. The price of the underlying put will be \$200. The strike price of the underlying put will be \$3,000 and the expiry of the underlying put will be 9 months from today. The current price of the stock index is \$2,900, the stock index pays a continuous dividend with a yield of 1.5% per annum, and the volatility is 45%. The risk-free continuously compounded interest rate is 4%. Find the price of the compound option.

**14.6.5** Suppose a put on a call option gives the owner the right to sell a call option on a stock index 1 year from today. The price of the underlying call will be $15. The strike price of the underlying call will be $225 and the expiry of the underlying put will be 18 months from today. The current price of the stock index is $230, the stock index pays a continuous dividend with a yield of 3.5% per annum, and the volatility is 33%. The risk-free continuously compounded interest rate is 5%. Find the price of the compound option.

**14.6.6** Suppose a call on a call option gives the owner the right to buy a call option on a stock index 3 months from today. The price of the underlying call will be $35. The strike price of the underlying call will be $500 and the expiry of the underlying put will be 1 year from today. The current price of the stock index is $510, the stock index pays a continuous dividend with a yield of 1% per annum, and the volatility is 40%. The risk-free continuously compounded interest rate is 4.5%. Find the price of the compound option.

**14.6.7** Use the Put-Call parity relationship for compound options to find the price of the put on a put corresponding to the call on a put described in Exercise 14.6.3.

**14.6.8** Use the Put-Call parity relationship for compound options to find the price of the call on a put corresponding to the put on a put described in Exercise 14.6.4.

**14.6.9** Show that the limit of the call on a call price as $t \to t_1^-$ is $(C^e(S(t_1), K, T) - X)^+$.

**14.6.10** The value of compound options can be approximated by the ordinary Black–Scholes option pricing formulas provided the volatility of the option is accurately approximated by Eq. (12.16). Justification of this approximation can be found in [Bensoussan *et al.* (1995)]. If the volatility of the underlying stock is $\sigma$, let the option volatility be $\hat{\sigma}$, then for a call on a call

$$C^c \approx C^e(S(t), K, T)\Phi(b_1) - Xe^{-r(T-t)}\Phi(b_2)$$

$$b_1 = \frac{\ln \frac{C^e(S(t),K,T)}{X} + \left(r - \delta + \frac{\hat{\sigma}^2}{2}\right)(t_1 - t)}{\hat{\sigma}\sqrt{t_1 - t}}$$

$$b_2 = \frac{\ln \frac{C^e(S(t),K,T)}{X} + \left(r - \delta - \frac{\hat{\sigma}^2}{2}\right)(t_1 - t)}{\hat{\sigma}\sqrt{t_1 - t}}.$$

Use this approximation to find the value of the call on a call compound option described in Example 14.10.

## 14.7    Asian Options

If the payoff of an option is related to the average price of an underlying
security over a time interval $[0, T]$, the option is an **Asian option**. This
is the first example of a **path dependent** option since the payoff will
strongly depend on the evolution of the price of the asset underlying the
option. There are at least four broad categories of Asian options in addition
to calls and puts. In general an Asian option replaces the strike price $K$ or
the expiry price of the security $S(T)$ of an ordinary option with the average
of of $S(t)$ for $0 < t < T$. When $S(T)$ is replaced by its average, the option
is referred to as an **average price** Asian option. When $K$ is replaced by
the average of $S(t)$, the option is referred to as an **average strike** Asian
option. Averages for $S(t)$ are commonly calculated in two different ways.
If the average is determined as

$$A(T) = \frac{1}{N} \sum_{i=1}^{N} S(t_i), \qquad (14.56)$$

the option is known as an **arithmetic average** Asian option. When the
average is determined by the formula,

$$G(T) = \left( \prod_{i=1}^{N} S(t_i) \right)^{1/N}, \qquad (14.57)$$

the option is called a **geometric average** Asian option. Since $G(T) \leq
A(T)$ (see Exercise 14.7.2) then several inequalities relating prices of Asian
options based on the two notions of the averages exist.

- Since $(G(T) - K)^+ \leq (A(T) - K)^+$ the premium of a geometric average
  price call never exceeds the premium of an arithmetic average price call.
- Since $(K - A(T))^+ \leq (K - G(T))^+$ the premium of an arithmetic average
  price put never exceeds the premium of a geometric average price put.
- Since $(G(T) - S(T))^+ \leq (A(T) - S(T))^+$ the premium of a geometric
  average strike put never exceeds the premium of an arithmetic average
  strike put.
- Since $(S(T) - A(T))^+ \leq (S(T) - G(T))^+$ the premium of an arithmetic
  average strike call never exceeds the premium of a geometric average
  strike call.

Pricing geometric average options is the simpler of the two alternatives, so it will be addressed first. In this derivation the underlying security
has prices which are lognormally distributed with $S(t) \sim \mathcal{LN}(\ln S(0) +$

$(\alpha - \delta - \sigma^2/2)t, \sigma^2 t)$. The interval $[0, T]$ is partitioned into $N$ equal sized subintervals of length $h = T/N$. The expected value and variance of $S(t_i) = S(i\,h)$ are respectively $\mathbb{E}\left(S(i\,h)\right) = S(0)e^{(\alpha - \delta)ih}$ and $\mathbb{V}\mathrm{ar}\left(S(i\,h)\right) = \left(\mathbb{E}\left(S(i\,h)\right)\right)^2 (e^{\sigma^2 t} - 1)$ where $\delta$ is the continuous dividend yield of the underlying security, $\sigma$ is its volatility, and $\alpha$ is its the expected growth rate. The geometric average can be expressed as

$$G(T) = \left(\prod_{i=1}^{N} S(i\,h)\right)^{1/N}$$

$$\ln G(T) = \frac{1}{N} \sum_{i=1}^{N} \ln S(i\,h).$$

According to Eq. (9.1),

$$\ln S(i\,h) = \ln S((i-1)h) + (\alpha - \delta - \sigma^2/2)h + \sigma\sqrt{h}Z,$$

where $Z \sim \mathcal{N}(0,1)$. Therefore for $i = 1, 2, \ldots, N$,

$$\ln S(i\,h) = \ln S(0) + (\alpha - \delta - \sigma^2/2)ih + \sigma\sqrt{h} \sum_{j=1}^{i} Z_j, \qquad (14.58)$$

where $Z_j \sim \mathcal{N}(0,1)$ and $Z_j$ and $Z_k$ are independent for $j \neq k$. Thus the natural logarithm of the geometric average can be written as

$$\ln G(T) = \frac{1}{N} \sum_{i=1}^{N} \left( \ln S(0) + (\alpha - \delta - \sigma^2/2)ih + \sigma\sqrt{h} \sum_{j=1}^{i} Z_j \right)$$

$$= \ln S(0) + \frac{(\alpha - \delta - \frac{\sigma^2}{2})h}{N} \sum_{i=1}^{N} i + \frac{\sigma\sqrt{h}}{N} \sum_{i=1}^{N} \sum_{j=1}^{i} Z_j$$

$$= \ln S(0) + (\alpha - \delta - \frac{\sigma^2}{2}) \frac{(N+1)T}{2N} + \frac{\sigma\sqrt{h}}{N} \sum_{i=1}^{N} (N+1-i)Z_i.$$

The reader will be tasked with justifying some of the steps of this derivation in Exercise 14.7.3. Now the expected value of the logarithm of the geometric mean is

$$\mathbb{E}\left(\ln G(T)\right) = \ln S(0) + (\alpha - \delta - \frac{\sigma^2}{2}) \frac{(N+1)T}{2N} + \frac{\sigma\sqrt{h}}{N} \sum_{i=1}^{N} (N+1-i)\,\mathbb{E}\left(Z_i\right)$$

$$= \ln S(0) + (\alpha - \delta - \frac{\sigma^2}{2}) \frac{(N+1)T}{2N}.$$

The variance of the natural logarithm of the geometric average is

$$\text{Var}\left(\ln G(T)\right) = \text{Var}\left(\frac{\sigma\sqrt{h}}{N}\sum_{i=1}^{N}(N+1-i)Z_i\right) = \frac{\sigma^2 T}{N^3}\sum_{i=1}^{N}(N+1-i)^2$$

$$= \frac{\sigma^2(N+1)(2N+1)T}{6N^2}.$$

Since the sum of normal random variables is normally distributed then $\ln G(T) \sim \mathcal{N}(\ln S(0) + (\alpha - \delta - \frac{\sigma^2}{2})\frac{(N+1)T}{2N}, \frac{\sigma^2(N+1)(2N+1)T}{6N^2})$. Consequently $G(T)$ is lognormally distributed with the same parameter values.

If an investor wished to purchase a prepaid forward on the geometric mean of the security for the interval $[0, T]$, the risk-neutral prepaid forward price would be

$$F_{0,T}^{P}(G(T)) = e^{-rT}\,\mathbb{E}^{*}\left(G(T)\right)$$

$$= S(0)e^{-rT}e^{\left[(r-\delta+\frac{\sigma^2}{2})\frac{N+1}{N}+\frac{\sigma^2(N+1)(2N+1)}{6N^2}\right]\frac{T}{2}}$$

$$= S(0)e^{-\left[r\left(\frac{N-1}{N}\right)+\left(\delta+\frac{\sigma^2}{2}\right)\frac{N+1}{N}-\frac{\sigma^2(N+1)(2N+1)}{6N^2}\right]\frac{T}{2}}.$$

Recall that when the risk-neutral expected value is used, the appreciation rate $\alpha$ of the security is replaced by the risk-free interest rate $r$. For a single security paying a continuous dividend at annual rate $\delta$ the prepaid forward price is $F_{0,T}^{P}(S) = S(0)e^{-\delta T}$, thus the continuous dividend yield of the geometric mean can be expressed as

$$\delta^{*} = \frac{1}{2}\left[r\left(\frac{N-1}{N}\right) + \left(\delta + \frac{\sigma^2}{2}\right)\frac{N+1}{N} - \frac{\sigma^2(N+1)(2N+1)}{6N^2}\right]. \quad (14.59)$$

The volatility of the geometric mean is represented by the quantity

$$\sigma^{*} = \frac{\sigma}{N}\sqrt{\frac{(N+1)(2N+1)}{6}}. \quad (14.60)$$

Perhaps surprisingly, the calculation of the price of an average price option based on the geometric mean price of the underlying security can be calculated simply from the Black–Scholes option pricing formulas upon replacing $\delta$ with $\delta^{*}$ and $\sigma$ with $\sigma^{*}$.

**Example 14.12.** Suppose the current price of a stock is \$100/share, the dividend yield is the stock is continuous at rate 2% per annum, and the volatility of the stock is 25% per annum. The risk-free continuously compounded interest rate is 5%. Find the price of a 1-year at-the-money average price Asian call option for which the average is the geometric average with the stock price observed at the end of each of the next 12 months. Assume it is currently the beginning of the month.

**Solution.** First calculate the dividend yield and volatility for the geometric mean.

$$\delta^* = \frac{1}{2}\left[0.05\left(\frac{11}{12}\right) + \left(0.02 + \frac{(0.25)^2}{2}\right)\frac{13}{12} - \frac{(0.25)^2(13)(25)}{6(12)^2}\right]$$

$$= 0.0389222$$

$$\sigma^* = \frac{0.25}{12}\sqrt{\frac{(12+1)(2(12)+1)}{6}} = 0.153329.$$

Next use Eqs. (10.32) and (10.33) and Eq. (10.34) with $T = 1$ and $K = 100$.

$$d_1 = \frac{\ln\frac{100}{100} + \left(0.05 - 0.0389222 + \frac{(0.153329)^2}{2}\right)(1-0)}{0.153329\sqrt{1-0}} = 0.148913$$

$$d_2 = \frac{\ln\frac{100}{100} + \left(0.05 - 0.0389222 - \frac{(0.153329)^2}{2}\right)(1-0)}{0.153329\sqrt{1-0}} = -0.00441594$$

$$C^{ap,G} = 100e^{-0.0389222(1-0)}\Phi(d_1) - 100e^{-0.05(1-0)}\Phi(d_2) = \$6.39032.$$

When $N = 1$ the Asian options simplify to the corresponding ordinary European options. As $N$ increases the price of an geometric average price Asian call decreases asymptotically to a limiting value which is the continuous geometric average price call. This behavior is illustrated for the average price call described in Example 14.12 in Fig. 14.6. The reader will explore this behavior in Exercise 14.7.7.

It is possible to develop a Put-Call parity formula for average price Asian options. At expiry the difference in the prices of a geometric average price Asian call and put is

$$C^{ap,G}(T) - P^{ap,G}(T) = (G(T) - K)^+ - (K - G(T))^+ = G(T) - K.$$

Taking the present value of the risk-neutral expected values of both sides of this equation yield,

$$C^{ap,G} - P^{ap,G} = F_{0,T}^P(G(T)) - Ke^{-rT}. \tag{14.61}$$

Now consider the case of an geometric average strike Asian option. The geometric average of the security price replaces the strike price in option calculations. This is similar to the situation of exchange options discussed in Sec. 14.3. In the discussion prior to the example, the continuous dividend yield of the geometric average was expressed in Eq. (14.59) as $\delta^*$. Thus $\delta^*$

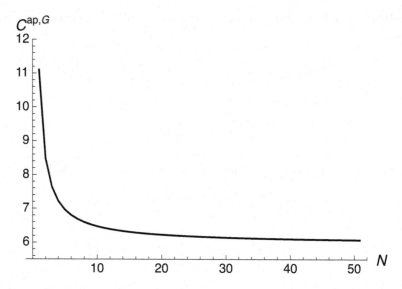

Fig. 14.6 The price of a geometrically average, average price Asian call is a decreasing function of the number of underlying stock prices observed. This is due to the fact that the volatility of the average is a decreasing function of $N$. When $N = 1$ the value value of the Asian option agrees with the value of an ordinary European call, in this case $C^e = \$11.1238$. As $N \to \infty$ the price approaches $C^{ap,G} = \$5.9802$.

plays the role of the risk-free rate in average strike options. The payoff of the average strike Asian option will be the positive part of the difference between the stock price at expiry $S(T)$ and the strike price determined by the average of $S(T)$. Just as in the case of exchange options, the volatility of the difference between these quantities must be determined.

$$\text{Var}\left(\ln S(T) - \ln G(T)\right)$$
$$= \text{Var}\left(\ln S(T)\right) - 2\,\text{Cov}\left(\ln S(T), \ln G(T)\right) + \text{Var}\left(\ln G(T)\right)$$

The term $\text{Var}\left(\ln S(T)\right) = \sigma^2 T$ while the last term on the right-hand side was determined earlier to be

$$\text{Var}\left(\ln G(T)\right) = \frac{\sigma^2(N+1)(2N+1)T}{6N^2}.$$

The covariance term remains to be determined. First consider

$$
\begin{aligned}
\mathbb{C}\mathrm{ov}\,&(\ln S(T), \ln G(T))\\
&= \mathbb{E}\left((\ln S(T) - \mathbb{E}\left(\ln S(T)\right))(\ln G(T) - \mathbb{E}\left(\ln G(T)\right))\right)\\
&= \mathbb{E}\left(\left(\sigma\sqrt{h}\sum_{j=1}^{N}Z_j\right)\left(\frac{\sigma\sqrt{h}}{N}\sum_{i=1}^{N}(N+1-i)Z_i\right)\right)\\
&= \frac{\sigma^2 h}{N}\,\mathbb{E}\left(\left(\sum_{j=1}^{N}Z_j\right)\left(\sum_{i=1}^{N}(N+1-i)Z_i\right)\right)\\
&= \frac{\sigma^2 h}{N}\,\mathbb{E}\left(\sum_{i=1}^{N}\sum_{j=1}^{N}(N+1-i)Z_iZ_j\right).
\end{aligned}
$$

By assumption $Z_i$ and $Z_j$ are independent when $i \neq j$, thus

$$
\mathbb{C}\mathrm{ov}\,(\ln S(T), \ln G(T)) = \frac{\sigma^2 h}{N}\sum_{i=1}^{N}(N+1-i)\,\mathbb{E}\left(Z_i^2\right) = \frac{\sigma^2(N+1)T}{2N}.
$$

The variance or equivalently the square of the volatility of the difference between $\ln S(T)$ and $\ln G(T)$ will be denoted $(\sigma^{**})^2$ and can be expressed as

$$
\begin{aligned}
(\sigma^{**})^2 &= \sigma^2 T - \frac{\sigma^2(N+1)T}{N} + \frac{\sigma^2(N+1)(2N+1)T}{6N^2}\\
&= \frac{\sigma^2(N-1)(2N-1)T}{6N^2}.
\end{aligned}
\tag{14.62}
$$

Thus when pricing a geometric average strike Asian option, the ordinary European option pricing formulas can be used with $\delta^*$ in place of the risk-free interest rate and $\sigma^{**}$ standing in for the volatility of the underlying asset.

**Example 14.13.** For the stock described in Example 14.12 find the price of a geometric average strike put.

**Solution.** In the solution to Example 14.12 $\delta^* = 0.0389222$. Using Eq. (14.62)

$$
\sigma^{**} = \frac{0.25}{12}\sqrt{\frac{(12-1)(2(12)-1)}{6}} = 0.135283.
$$

Next use Eqs. (10.32) and (10.33) and Eq. (10.35) with $T = 1$ and $K = S(0) = 100$.

$$d_1 = \frac{\ln\frac{100}{100} + \left(0.0389222 - 0.02 + \frac{(0.135283)^2}{2}\right)(1-0)}{0.135283\sqrt{1-0}} = 0.207512,$$

$$d_2 = \frac{\ln\frac{100}{100} + \left(0.0389222 - 0.02 - \frac{(0.135283)^2}{2}\right)(1-0)}{0.135283\sqrt{1-0}} = 0.0722294,$$

$$P^{as,G} = 100e^{-0.0389222(1-0)}\Phi(-d_2) - 100e^{-0.02(1-0)}\Phi(-d_1) = \$4.36898.$$

The Put-Call parity relationship between geometric average strike options is

$$C^{as,G} - P^{as,G} = F_{0,T}^P(S) - F_{0,T}^P(G(T)). \tag{14.63}$$

See [Chin *et al.* (2017), Sec. 5.1] for other parity results.

A rigorous discussion of the pricing of arithmetic average Asian options is beyond the scope of this book. Closed-form pricing formulas such as those presented earlier for geometric average options do not exist for arithmetic average options, since the sum of lognormally distributed quantities is not lognormally distributed. Instead the prices of arithmetic average Asian options will be approximated using the binomial tree method introduced in Chapter 7. The arithmetic average of the underlying stock price along each path (not including the initial price) will be calculated to determine the payoff of the option. Then the present value of the risk-neutral expected payoff determines the price of the option as usual.

Consider the following simple example. Suppose the initial price of a non-dividend-paying security is \$100/share. The volatility of the security is 35% and the risk-free interest rate is 7%. The price of a 1-year arithmetic average price call with a strike price of \$105 can be estimated from a binomial tree with time steps of 6 months. The factors by which the stock price increases and decreases at each time step are respectively (according to the formulas derived in Exercise 7.3.1),

$$u = e^{(0.07-0)(1/2)+0.35\sqrt{1/2}} = 1.32643$$

$$d = e^{(0.07-0)(1/2)-0.35\sqrt{1/2}} = 0.808571.$$

The risk-neutral probability of an increase in the stock price is

$$p^* = \frac{1}{1 + e^{0.35\sqrt{1/2}}} = 0.438442.$$

The evolution of stock prices is shown in the following graph.

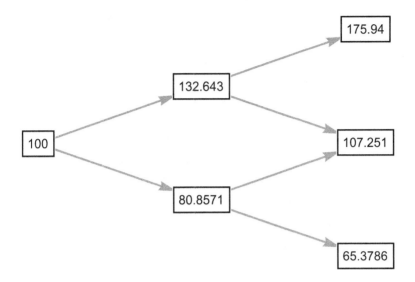

Using the graph of stock prices, the arithmetic average price $A(T)$, the call payoff, and the risk-neutral probability of the payoff are summarized in the following table.

| Path | $A(T)$ | $(A(T) - K)^+$ | Prob. |
|------|--------|----------------|-------|
| $u^2$ | 154.291 | 49.291 | 0.192231 |
| $u\,d$ | 119.947 | 14.947 | 0.246211 |
| $d\,u$ | 94.054 | 0.000 | 0.246211 |
| $d^2$ | 73.118 | 0.000 | 0.315347 |

Therefore the present value of the risk-neutral expected value of the payoff is the value of the arithmetic average price Asian call,

$$C^{ap,A} = e^{-0.07}\left[0.192231(49.291) + 0.246211(14.947) + 0 + 0\right] = \$12.266.$$

This method can be extended to more frequent averaging at the cost of enumerating all the paths from the initial price of the stock to the final prices. This type of work is better handled by a computer algorithm. If the example above is re-worked with arithmetic averaging performed monthly the premium for the arithmetic average price Asian call is approximately $C^{ap,A} = \$8.70628$. Readers interested in the pricing of arithmetic average Asian options should consult Haug *et al.* (2003).

## Exercises

**14.7.1** Jensen's Inequality states that if $f$ is a concave function (a function for which $f'' \leq 0$) and $x_1,\ x_2,\ \ldots,\ x_n \in \mathbb{R}$ and $a_1,\ a_2,\ \ldots,\ a_n \geq 0$ with $a_1 + a_2 + \cdots + a_n = 1$ then

$$a_1 f(x_1) + a_2 f(x_2) + \cdots + a_n f(x_n) \leq f(a_1 x_1 + a_2 x_2 + \cdots + a_n x_n).$$

Prove this inequality using the Fundamental Theorem of Calculus.

**14.7.2** Suppose $\{X_1, X_2, \ldots, X_N\}$ is a set of positive quantities. Use Jensen's Inequality to show that the arithmetic average is always at least as large as the geometric average.

**14.7.3** Show that

$$\sum_{i=1}^{N} \sum_{j=1}^{i} Z_j = \sum_{i=1}^{N} (N + 1 - i) Z_i.$$

**14.7.4** Suppose the returns on security $S(t)$ are lognormally distributed as $\mathcal{LN}((\alpha - \delta + \sigma^2/2)t, \sigma^2 t)$. During the interval $[0, T]$ the returns will be observed at times $T i/N$ with $i = 1, 2, \ldots, N$. Find the expected value of the geometric average for a security.

**14.7.5** Suppose the current price of a stock is \$120/share, the dividend yield of the stock is continuous at a rate 1% per annum, and the volatility of the stock is 35% per annum. The risk-free continuously compounded interest rate is 6%. Find the price of a 1-year, \$115-strike average price Asian put option for which the average is the geometric average with the stock price observed at the end of each of the next 12 months. Assume it is currently the beginning of the month.

**14.7.6** Find the price of the corresponding average price Asian put option for the situation described in Example 14.12. Show that the put and call prices satisfy the Put-Call parity formula in Eq. (14.61).

**14.7.7** Suppose the underlying stock price is observed continuously until expiry. Take the limit as $N \to \infty$ in Eqs. (14.59) and (14.60) and price the option of Example 14.12 under these assumptions.

**14.7.8** Find an expression for the correlation between $\ln S(T)$ and $\ln G(T)$.

**14.7.9** Find the price of the corresponding average strike Asian call option for the situation described in Example 14.13. Show that the put and call prices satisfy the Put-Call parity formula in Eq. (14.63).

**14.7.10** Suppose the underlying stock price is observed continuously until expiry. Take the limit as $N \to \infty$ in Eqs. (14.59) and (14.62) and price the option of Example 14.13 under these assumptions.

**14.7.11** Suppose the initial price of a security is \$125/share. The volatility of the security is 30%, the security pays a dividend at a rate of 2% per year and the risk-free interest rate is 5% per year. Estimate the price of a 1-year arithmetic average strike put from a binomial tree with time steps of 4 months.

**14.7.12** Suppose that $C^{ap,G}$ and $P^{ap,G}$ are the premiums charged for average price Asian options with strike price $K$ and expiry at $T$. Suppose that $C^{as,G}$ and $P^{as,G}$ are the premiums charged for average strike Asian options with expiry at $T$. Finally, suppose that $C^e$ and $P^e$ are the premiums charged for ordinary European options with strike price $K$ and expiry at $T$. Show that $C^{ap,G} + C^{as,G} - C^e = P^{ap,G} + P^{as,G} - P^e$. This problem is adapted from [Hull (2000), Exercise 24.4].

## 14.8 Barrier Options

Another type of path dependent is the **barrier option**. This type of option comes into existence or goes out of existence when the underlying asset crosses a threshold, sometimes called a trigger. The direction of crossing is significant as well. If a barrier option is in existence at expiry, it has payoff properties and values like an ordinary European option. Barrier options are sometimes preferred by investors since barrier options are often less expensive than ordinary European options. This section will explore the pricing and properties of three broad categories of barrier options.

- **Knock-in options** come into existence when the underlying asset crosses the threshold. If the asset price crosses the threshold from above, the knock-in option is called a **down-and-in** option. If the asset price must cross the threshold from below the option is called a **up-and-in** option.

- **Knock-out options** go out of existence when the underlying asset crosses the threshold. If the asset price crosses the threshold from above, the knock-out option is called a **down-and-out** option. If the asset price must cross the threshold from below the option is called a **up-and-out** option.

- **Rebate options** have a fixed payoff depending on whether the barrier is reached. If the barrier must be crossed from above the option is referred to as a **down rebate**, while if the barrier must be crossed from below, the option is called an **up rebate**. To further subdivide this category, the payoff can be made at expiry or at the time the barrier is crossed. In the former case, the option is called a **deferred rebate option**.

Throughout this section the barrier price or trigger price will be denoted $H$ while $K$ and $T$ will continue to denote the strike price and strike time respectively. Of course barrier options can be calls or puts. Consider a knock-out call with an up-and-out barrier $H > K$. If $S(t) < H$ for all $0 \leq t \leq T$ then the option has payoff $(S(T) - K)^+$. If $S(t) \geq H$ for any $0 \leq t \leq T$ the option is knocked-out and has payoff 0. Possible asset price paths and payoff for the knock-out call are illustrated in Fig. 14.7.

The approach taken to price barrier options will rely on building them from components which are cash-or-nothing barrier options and asset-or-nothing barrier options. Ordinary versions of these binary (or digital options) were priced in Sec. 10.4.

A down-and-in, cash-or-nothing call will pay \$1 to the option holder provided $\underline{S} \leq H$ and $\overline{S} > K$ and otherwise pays nothing. Let $C_{di}^{cn}(T)$ be the value of a down-and-in, cash-or-nothing European call at time $T$.

$$C_{di}^{cn}(S(T); H, K, T) = \mathbb{1}_{((\underline{S}(T) \leq H) \cap (S(T) > K))}$$

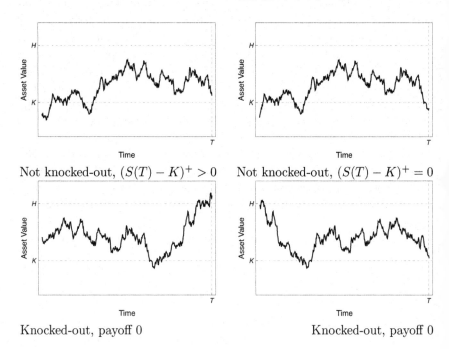

Not knocked-out, $(S(T) - K)^+ > 0$  Not knocked-out, $(S(T) - K)^+ = 0$

Knocked-out, payoff 0  Knocked-out, payoff 0

Fig. 14.7  The evolution of asset prices for an up-and-out call. In the top two diagrams, the barrier option is not knocked-out but finishes in-the-money only in the situation depicted on the left. In the bottom two diagrams, the barrier option is knocked-out and thus pays nothing.

The time $t = 0$ present value of the risk-neutral expected value of this payoff is the time $t = 0$ value of the down-and-in, cash-or-nothing call. There are two cases to consider: (1) $H \leq K$ and (2) $H > K$. To begin, assume $H < S(0) \leq K$.

$$C_{di}^{cn}(S(0); H, K, T)$$
$$= e^{-rT} \, \mathbb{P}^* \left( (\underline{S}(T) \leq H) \cap (S(T) > K) \right)$$
$$= e^{-rT} \, \mathbb{P}^* \left( \left( \ln \frac{\underline{S}(T)}{S(0)} \leq \ln \frac{H}{S(0)} \right) \cap \left( \ln \frac{S(T)}{S(0)} > \ln \frac{K}{S(0)} \right) \right).$$

The underlying stock is lognormally distributed and thus $\ln S(T)/S(0) \sim \mathcal{N}((\alpha - \delta - \frac{\sigma^2}{2})T, \sigma^2 T)$. The risk-neutral probability can be determined from the real-world probability using the *ad hoc* replacement of $\alpha$ with $r$, the risk-free interest rate. Note that by assumption $H < S(0) \leq K$ and hence $\ln(H/S(0)) < 0$ and $\ln(K/S(0)) > \ln(H/S(0))$. Using the result of Exercise 9.6.10,

$$C_{di}^{cn}(S(0); H, K, T)$$
$$= e^{-rT} e^{\frac{2\left(r-\delta-\frac{\sigma^2}{2}\right) \ln \frac{H}{S(0)}}{\sigma^2}} \Phi \left( \frac{2\ln \frac{H}{S(0)} - \ln \frac{K}{S(0)} + \left(r - \delta - \frac{\sigma^2}{2}\right) T}{\sigma\sqrt{T}} \right)$$
$$= e^{-rT} \left( \frac{H}{S(0)} \right)^{\frac{2(r-\delta)}{\sigma^2} - 1} \Phi(d_4),$$

where

$$d_4 = \frac{\ln \frac{H^2}{KS(0)} + \left(r - \delta - \frac{\sigma^2}{2}\right) T}{\sigma\sqrt{T}}. \tag{14.64}$$

Now assume $K < H < S(0)$. The cash-or-nothing, down-and-in call comes into existence as an in-the-money cash-or-nothing call at the time the barrier at $H$ is crossed from above.

$$C_{di}^{cn}(S(0); H, K, T)$$
$$= e^{-rT} [\mathbb{P}^* \left( (\underline{S}(T) \leq K) \cap (S(T) > K) \right)$$
$$\quad + \mathbb{P}^* \left( (K < \underline{S}(T) \leq H) \cap (S(T) > K) \right)]$$
$$= e^{-rT} [\mathbb{P}^* \left( (\underline{S}(T) \leq K) \cap (S(T) > K) \right) + \mathbb{P}^* \left( K < \underline{S}(T) \leq H \right)].$$

The first term in the square brackets can be evaluated using the expression derived in Exercise 9.6.10. The second term can be evaluated using Eq. (9.47) twice.

$$C_{di}^{cn}(S(0); H, K, T)$$

$$= e^{-rT}\left[\mathbb{P}^*\left(\left(\ln\frac{S(T)}{S(0)} \le \ln\frac{K}{S(0)}\right) \cap \left(\ln\frac{S(T)}{S(0)} > \ln\frac{K}{S(0)}\right)\right)\right.$$

$$\left. + \mathbb{P}^*\left(\ln\frac{K}{S(0)} < \ln\frac{S(T)}{S(0)} \le \ln\frac{H}{S(0)}\right)\right]$$

$$= e^{-rT}\left[\left(\frac{K}{S(0)}\right)^{\frac{2(r-\delta)}{\sigma^2}-1}\Phi(-d_2) + \Phi(-d_6) + \left(\frac{H}{S(0)}\right)^{\frac{2(r-\delta)}{\sigma^2}-1}\Phi(d_8)\right.$$

$$\left. - \left(\frac{K}{S(0)}\right)^{\frac{2(r-\delta)}{\sigma^2}-1}\Phi(-d_2) - \Phi(-d_2)\right]$$

$$= e^{-rT}\left[\Phi(d_2) - \Phi(d_6) + \left(\frac{H}{S(0)}\right)^{\frac{2(r-\delta)}{\sigma^2}-1}\Phi(d_8)\right],$$

where

$$d_6 = \frac{\ln\frac{S(0)}{H} + \left(r - \delta - \frac{\sigma^2}{2}\right)T}{\sigma\sqrt{T}} \tag{14.65}$$

$$d_8 = \frac{\ln\frac{H}{S(0)} + \left(r - \delta - \frac{\sigma^2}{2}\right)T}{\sigma\sqrt{T}}. \tag{14.66}$$

Thus the value of a cash-or-nothing, down-and-in call can be expressed as the piecewise defined function,

$$C_{di}^{cn}(S(0); H, K, T)$$

$$= e^{-rT}\begin{cases}\left(\frac{H}{S(0)}\right)^{\frac{2(r-\delta)}{\sigma^2}-1}\Phi(d_4) & \text{if } H \le K, \\ \Phi(d_2) - \Phi(d_6) + \left(\frac{H}{S(0)}\right)^{\frac{2(r-\delta)}{\sigma^2}-1}\Phi(d_8) & \text{if } H > K.\end{cases} \tag{14.67}$$

Since a great deal of work went into the determination of the formula for the cash-or-nothing, down-and-in call it is fortunate that several other results can be determined from Eq. (14.67) with only a small amount of extra effort. Suppose an investor holds a cash-or-nothing, down-and-in call with barrier $H$ and strike $K$ and also possesses a cash-or-nothing. down-and-out call with the same barrier and strike. The latter option will pay the holder \$1 if $\underline{S}(T) > H$ and $S(T) > K$. Therefore by holding the two options the investor receives \$1 so long as $S(T) > K$, in other words, the combination of the two options is equivalent to a cash-or-nothing call with strike price $K$. There the value of a cash-or-nothing, down-and-out call is

$$C_{do}^{cn}(S(0), H, K, T) = C^{cn}(S(0), K, T) - C_{di}^{cn}(S(0), H, K, T). \quad (14.68)$$

Cash down-and-in and down-and-out puts also descend from Eq. (14.67). Suppose the strike price $K$ approaches 0, then the cash-or-nothing, down-and-in call pays \$1 at expiry provided the barrier $H$ has not been crossed. This type of option is called a **deferred down rebate** and will be denoted $C_{dr}^{cn}(S(0), H, T)$. The reader will show in Exercise 14.8.4 that

$$C_{dr}^{cn}(S(0); H, T) = e^{-rT} \left[ 1 - \Phi(d_6) + \left( \frac{H}{S(0)} \right)^{\frac{2(r-\delta)}{\sigma^2} - 1} \Phi(d_8) \right]. \quad (14.69)$$

A cash-or-nothing, down-and-in put option will pay the holder \$1 if $S(T) < K$ and $\underline{S}(T) < H$. If an investor also holds a cash-or-nothing, down-and-in call option with the same barrier and strike, then provided the barrier is crossed from above during the life of the option, the investor receives \$1, the same payoff as the deferred down rebate. Thus the cash-or-nothing, down-and-in put can be priced as

$$P_{di}^{cn}(S(0); H, K, T) = C_{dr}^{cn}(S(0); H, T) - C_{di}^{cn}(S(0); H, K, T). \quad (14.70)$$

Finally suppose an investor holds the cash-or-nothing, down-and-in put described above and a cash-or-nothing, down-and-out put with the same barrier and strike. The second option will pay \$1 provided $\underline{S}(T) > H$ and $S(T) < K$. The portfolio of these two options pays \$1 if $S(T) < K$ whether the barrier is reached or not, thus the portfolio can be considered a synthetic cash-or-nothing put. Consequently

$$P_{do}^{cn}(S(0); H, K, T) = P^{cn}(S(0); K, T) - P_{di}^{cn}(S(0); H, K, T). \quad (14.71)$$

The next set of cash-or-nothing barrier options will assume the current price of the underlying asset is below the barrier. A cash-or-nothing, up-and-in put option will pay \$1 provided $\overline{S}(T) \geq H$ and $S(T) < K$, thus

$$\begin{aligned}
P_{ui}^{cn}&(S(0); H, K, T) \\
&= e^{-rT} \mathbb{P}^* \left( (\overline{S}(T) \geq H) \cap (S(T) < K) \right) \\
&= e^{-rT} \mathbb{P}^* \left( \left( \ln \frac{\overline{S}(T)}{S(0)} \geq \ln \frac{H}{S(0)} \right) \cap \left( \ln \frac{S(T)}{S(0)} < \ln \frac{K}{S(0)} \right) \right).
\end{aligned}$$

Suppose $H \geq K$ then $\ln \frac{H}{S(0)} \geq \ln \frac{K}{S(0)}$ and according to the result of Exercise 9.6.9,

$$P_{ui}^{cn}(S(0); H, K, T)$$

$$= e^{-rT} e^{\frac{2\left(r - \delta - \frac{\sigma^2}{2}\right) \ln \frac{H}{S(0)}}{\sigma^2}} \Phi\left(\frac{\ln \frac{K}{S(0)} - 2\ln \frac{H}{S(0)} - \left(r - \delta - \frac{\sigma^2}{2}\right)T}{\sigma\sqrt{T}}\right)$$

$$= e^{-rT} \left(\frac{H}{S(0)}\right)^{\frac{2(r-\delta)}{\sigma^2} - 1} \Phi(-d_4).$$

If $H < K$ then $\ln \frac{H}{S(0)} < \ln \frac{K}{S(0)}$ and the risk-neutral probability,

$$\mathbb{P}^*\left((\overline{S}(T) \geq H) \cap (S(T) < K)\right)$$
$$= \mathbb{P}^*\left((\overline{S}(T) \geq H) \cap (S(T) \leq H)\right) + \mathbb{P}^*\left((\overline{S}(T) \geq H) \cap (H < S(T) < K)\right)$$
$$= \mathbb{P}^*\left((\overline{S}(T) \geq H) \cap (S(T) \leq H)\right) + \mathbb{P}^*\left(H < S(T) < K\right).$$

This implies

$$P_{ui}^{cn}(S(0); H, K, T) = e^{-rT}\left[\Phi(-d_2) - \Phi(-d_6) + \left(\frac{H}{S(0)}\right)^{\frac{2(r-\delta)}{\sigma^2} - 1} \Phi(-d_8)\right].$$

Combining the two expressions for the cash-or-nothing, up-and-in put into a piecewise defined expression yields,

$$P_{ui}^{cn}(S(0); H, K, T)$$

$$= e^{-rT} \begin{cases} \left(\frac{H}{S(0)}\right)^{\frac{2(r-\delta)}{\sigma^2} - 1} \Phi(-d_4) & \text{if } H \geq K, \\ \Phi(-d_2) - \Phi(-d_6) + \left(\frac{H}{S(0)}\right)^{\frac{2(r-\delta)}{\sigma^2} - 1} \Phi(-d_8) & \text{if } H < K. \end{cases}$$

$$(14.72)$$

An up-and-in cash put and an up-and-out cash put together are a synthetic cash-or-nothing put. Hence the premium of a cash-or-nothing, up-and-out put is

$$P_{uo}^{cn}(S(0); H, K, T) = P^{cn}(S(0); K, T) - P_{ui}^{cn}(S(0); H, K, T). \quad (14.73)$$

A financial derivative which pays \$1 at expiry provided the barrier at $H$ is crossed from below during the lifetime of the option is called a **deferred up rebate** option. The premium for a deferred up rebate option can be expressed as

$$P_{ur}^{cn}(S(0); H, T) = e^{-rT}\left[\Phi(d_6) + \left(\frac{H}{S(0)}\right)^{\frac{2(r-\delta)}{\sigma^2} - 1} \Phi(-d_8)\right]. \quad (14.74)$$

A portfolio consisting of a cash-or-nothing, up-and-in put and a cash-or-nothing, up-and-in call will have a payoff of \$1 if the barrier is crossed from below. Therefore the portfolio is a synthetic deferred up rebate option. This enables the price of the cash-or-nothing, up-and-in call to be determined.

$$C_{ui}^{cn}(S(0); H, K, T) = P_{ur}^{cn}(S(0); H, T) - P_{ui}^{cn}(S(0); H, K, T). \quad (14.75)$$

Finally a cash-or-nothing, up-and-in call together with a cash-or-nothing, up-and-out call is a synthetic cash-or-nothing call option. Therefore the price of a cash-or-nothing, up-and-out call is

$$C_{uo}^{cn}(S(0); H, K, T) = C^{cn}(S(0); K, T) - C_{ui}^{cn}(S(0); H, K, T). \quad (14.76)$$

Half of the building blocks needed to price barrier options are in place. The remaining half are asset-or-nothing barrier options. In many texts, the prices of asset-or-nothing barrier options are derived quickly from the cash-or-nothing barrier pricing formulas using a technique known as **change of numeraire**. While this may be convenient it assumes more background in probability and measure theory than the technique used here, which relies only on calculus. Consider an asset-or-nothing, down-and-in call option with barrier $H$, strike $K$, and expiry $T$. This option will provide the owner with a share of the underlying worth $S(T)$ provided $S(T) > K$ and $\underline{S}(T) \leq H$. The price of the option will be the time $t = 0$ present value of the risk-neutral conditional expected value of $S(T)$ assuming $S(T) > K$ and $\underline{S}(T) \leq H$. Since the present value can be determined in a later step of the derivation, consider

$$\mathbb{E}^* \left( S(T) \mid (\underline{S}(T) \leq H) \cap (S(T) > K) \right) = \mathbb{E}^* \left( S(T) \mathbb{1}_{((\underline{S}(T) \leq H) \cap (S(T) > K))} \right).$$

Divide the stock, barrier, and strike prices by $S(0)$, then

$$\mathbb{E}^* \left( S(T) \mid (\underline{S}(T) \leq H) \cap (S(T) > K) \right)$$
$$= S(0) \, \mathbb{E}^* \left( e^{\ln \frac{S(T)}{S(0)}} \mathbb{1}_{\left( \left( \ln \frac{S(T)}{S(0)} \leq \ln \frac{H}{S(0)} \right) \cap \left( \ln \frac{S(T)}{S(0)} > \ln \frac{K}{S(0)} \right) \right)} \right).$$

If $a = \ln(H/S(0))$, $b = \ln(K/S(0))$, $X = \ln(\underline{S}(T)/S(0))$, and $Y = \ln S(T)/S(0)$, then

$$\mathbb{E}^* \left( S(T) \mid (\underline{S}(T) \leq H) \cap (S(T) > K) \right) = S(0) \, \mathbb{E}^* \left( e^Y \mathbb{1}_{((X \leq a) \cap (Y > b))} \right).$$

The joint probability distribution function for $X$ and $Y$ is given in Eq. (9.45) (with $X = \underline{M}(t)$ and $Y = W(t)$). To price the asset-or-nothing, down-and-in call assume $S(0) > H$, else the barrier option is trivially an ordinary asset-or-nothing call. As was the case for the cash-or-nothing barrier options, there are two cases to consider. The first considered is $H \leq K$, which implies $a < 0$ and $b \geq a$.

$$\mathbb{E}^* \left( S(T) \,|\, (\underline{S}(T) \leq H) \cap (S(T) > K) \right)$$

$$= S(0) \int_b^\infty \int_{-\infty}^a \frac{2(y - 2x)}{\sigma^2 T \sqrt{2\pi\sigma^2 T}} e^y e^{\frac{-(y-2x)^2 + 2\mu y T - \mu^2 T^2}{2\sigma^2 T}} \, dx \, dy$$

$$= S(0) \int_b^\infty \int_{-\infty}^a \frac{2(y - 2x)}{\sigma^2 T \sqrt{2\pi\sigma^2 T}} e^{\frac{-(y-2x)^2 + 2(\mu+\sigma^2)y T - \mu^2 T^2}{2\sigma^2 T}} \, dx \, dy$$

$$= S(0) \int_b^\infty \frac{1}{\sqrt{2\pi\sigma^2 T}} e^{\frac{2(\mu+\sigma^2)y T - \mu^2 T^2}{2\sigma^2 T}} \int_{\frac{y-2a}{\sigma\sqrt{T}}}^\infty z e^{-\frac{z^2}{2}} \, dz \, dy$$

$$= S(0) \int_b^\infty \frac{1}{\sqrt{2\pi\sigma^2 T}} e^{\frac{2(\mu+\sigma^2)y T - \mu^2 T^2}{2\sigma^2 T}} e^{-\frac{(y-2a)^2}{2\sigma^2 T}} \, dy.$$

Complete the square in variable $y$ and integrate by substitution.

$$\mathbb{E}^* \left( S(T) \,|\, (\underline{S}(T) \leq H) \cap (S(T) > K) \right)$$

$$= S(0) e^{\frac{4a(\mu+\sigma^2)T + (2\mu+\sigma^2)\sigma^2 T^2}{2\sigma^2 T}} \int_{\frac{b-2a-(\mu+\sigma^2)T}{\sigma\sqrt{T}}}^\infty \frac{1}{\sqrt{2\pi}} e^{-\frac{z^2}{2}} \, dz$$

$$= S(0) e^{\frac{4a(\mu+\sigma^2)T + (2\mu+\sigma^2)\sigma^2 T^2}{2\sigma^2 T}} \Phi\left( \frac{-b + 2a + (\mu+\sigma^2)T}{\sigma\sqrt{T}} \right)$$

$$= S(0) e^{(r-\delta)T} \left( \frac{H}{S} \right)^{\frac{2(r-\delta)}{\sigma^2}+1} \Phi(d_3), \tag{14.77}$$

where

$$d_3 = \frac{\ln \frac{H^2}{KS(0)} + \left( r - \delta + \frac{\sigma^2}{2} \right) T}{\sigma\sqrt{T}}. \tag{14.78}$$

Taking the time $t = 0$ present value of the risk-neutral expected value gives

$$C_{di}^{an}(S(0); H, K, T) = S(0) e^{-\delta T} \left( \frac{H}{S(0)} \right)^{\frac{2(r-\delta)}{\sigma^2}+1} \Phi(d_3),$$

when $H \leq K$. Now consider the situation in which $K < H < S(0)$. The events for which

$$(\underline{S}(T) \leq H) \cap (S(T) > K)$$

$$= [(\underline{S}(T) \leq K) \cap (S(T) > K)] \cup [(K < \underline{S}(T) \leq H) \cap (S(T) > K)]$$

$$= [(\underline{S}(T) \leq K) \cap (S(T) > K)] \cup (K < \underline{S}(T) \leq H).$$

The risk-neutral conditional expected value of $S(T)$ can be expressed as the sum of two quantities.

$$\mathbb{E}^* \left( S(T) \,|\, (\underline{S}(T) \leq H) \cap (S(T) > K) \right)$$
$$= \mathbb{E}^* \left( S(T) \,|\, (\underline{S}(T) \leq K) \cap (S(T) > K) \right) + \mathbb{E}^* \left( S(T) \,|\, K < \underline{S}(T) \leq H \right)$$
$$= \mathbb{E}^* \left( S(T) \mathbb{1}_{((\underline{S}(T) \leq K) \cap (S(T) > K))} \right) + \mathbb{E}^* \left( S(T) \mathbb{1}_{(K < \underline{S}(T) \leq H)} \right).$$

The first of the two expected values can be determined from earlier work by replacing $K$ with $H$ in Eq (14.78) and Eq (14.78). Therefore

$$\mathbb{E}^* \left( S(T) \mathbb{1}_{((\underline{S}(T) \leq K) \cap (S(T) > K))} \right) = S(0)e^{(r-\delta)T} \left( \frac{H}{S} \right)^{\frac{2(r-\delta)}{\sigma^2}+1} \Phi(d_7),$$
$$(14.79)$$

where

$$d_7 = \frac{\ln \frac{H}{S(0)} + \left( r - \delta + \frac{\sigma^2}{2} \right) T}{\sigma\sqrt{T}}. \tag{14.80}$$

The remaining integral can be thought of as the risk-neutral expected payoff of an option which gives the holder a share of stock if $S(T) > K$ (which occurs with probability $\Phi(d_1)$) and $\underline{S}(T) \leq H$. Note that $\underline{S}(T) > H$ with probability $\Phi(d_5)$ where

$$d_5 = \frac{\ln \frac{S(0)}{H} + \left( r - \delta + \frac{\sigma^2}{2} \right) T}{\sigma\sqrt{T}}. \tag{14.81}$$

Hence when $K < H < S(0)$

$$C_{di}^{an}(S(0); H, K, T)$$
$$= S(0)e^{-\delta T} \left( \Phi(d_1) - \Phi(d_5) + \left( \frac{H}{S(0)} \right)^{\frac{2(r-\delta)}{\sigma^2}+1} \Phi(d_7) \right).$$

Combining the two cases into a piecewise-defined function yields the pricing formula for an asset-or-nothing, down-and-in call.

$$C_{di}^{an}(S(0); H, K, T)$$
$$= S(0)e^{-\delta T} \begin{cases} \left( \frac{H}{S(0)} \right)^{\frac{2(r-\delta)}{\sigma^2}+1} \Phi(d_3) & \text{if } H \leq K, \\ \Phi(d_1) - \Phi(d_5) + \left( \frac{H}{S(0)} \right)^{\frac{2(r-\delta)}{\sigma^2}+1} \Phi(d_7) & \text{if } H > K. \end{cases}$$
$$(14.82)$$

The asset-or-nothing, down-and-in call and the cash-or-nothing, down-and-in call pricing formulas are closely related. In Exercise 14.8.11 the reader will show that

$$C_{di}^{an}(S(0); H, K, T) = S(0)e^{(r-\delta)T} C_{di}^{cn}(S(0); H, K, T) \Big|_{\delta \mapsto \delta - \sigma^2}.$$

This is no mere coincidence. All of the asset-or-nothing barrier options can be derived from the corresponding cash-or-nothing barrier options by replacing $\delta$ in the cash-or-nothing barrier option formula with $\delta - \sigma^2$ and then multiplying by $F_{0,T}(S)$, the price of the forward on the underlying stock. The remaining asset-or-nothing barrier options have prices given by the formulas below.

$$C_{do}^{an}(S(0); H, K, T) \tag{14.83}$$

$$= S(0)e^{-\delta T} \begin{cases} \Phi(d_1) - \left(\frac{H}{S(0)}\right)^{\frac{2(r-\delta)}{\sigma^2}+1} \Phi(d_3) \text{ if } H \leq K, \\ \Phi(d_5) - \left(\frac{H}{S(0)}\right)^{\frac{2(r-\delta)}{\sigma^2}+1} \Phi(d_7) \text{ if } H > K. \end{cases}$$

$$P_{di}^{an}(S(0); H, K, T) \tag{14.84}$$

$$= S(0)e^{-\delta T} \begin{cases} \Phi(-d_5) + \left(\frac{H}{S(0)}\right)^{\frac{2(r-\delta)}{\sigma^2}+1} (\Phi(-d_3) - \Phi(-d_7)) \text{ if } H \leq K, \\ \Phi(-d_1) \hspace{6.5cm} \text{ if } H > K. \end{cases}$$

$$P_{do}^{an}(S(0); H, K, T) \tag{14.85}$$

$$= S(0)e^{-\delta T} \begin{cases} \Phi(d_6) - \Phi(d_2) - \left(\frac{H}{S(0)}\right)^{\frac{2(r-\delta)}{\sigma^2}+1} (\Phi(d_8) - \Phi(d_4)) \text{ if } H \leq K, \\ 0 \hspace{8cm} \text{ if } H > K \end{cases}$$

$$P_{ui}^{an}(S(0); H, K, T) \tag{14.86}$$

$$= S(0)e^{-\delta T} \begin{cases} \left(\frac{H}{S(0)}\right)^{\frac{2(r-\delta)}{\sigma^2}+1} \Phi(-d_3) \hspace{4cm} \text{ if } H \geq K, \\ \Phi(-d_1) - \Phi(-d_5) + \left(\frac{H}{S(0)}\right)^{\frac{2(r-\delta)}{\sigma^2}+1} \Phi(-d_7) \text{ if } H < K. \end{cases}$$

$$P_{uo}^{an}(S(0); H, K, T) \tag{14.87}$$

$$= S(0)e^{-\delta T} \begin{cases} \Phi(-d_1) - \left(\frac{H}{S(0)}\right)^{\frac{2(r-\delta)}{\sigma^2}+1} \Phi(-d_3) \text{ if } H \geq K, \\ \Phi(-d_5) - \left(\frac{H}{S(0)}\right)^{\frac{2(r-\delta)}{\sigma^2}+1} \Phi(-d_7) \text{ if } H < K. \end{cases}$$

$$C_{ui}^{an}(S(0); H, K, T) \tag{14.88}$$

$$= S(0)e^{-\delta T} \begin{cases} \Phi(d_5) + \left(\frac{H}{S(0)}\right)^{\frac{2(r-\delta)}{\sigma^2}+1} (\Phi(d_3) - \Phi(d_7)) \text{ if } H \geq K, \\ \Phi(d_1) \hspace{6.5cm} \text{ if } H < K. \end{cases}$$

$$C_{uo}^{an}(S(0); H, K, T) \tag{14.89}$$

$$= S(0)e^{-\delta T} \begin{cases} \Phi(d_1) - \Phi(d_5) + \left(\frac{H}{S(0)}\right)^{\frac{2(r-\delta)}{\sigma^2}+1} (\Phi(d_7) - \Phi(d_3)) \text{ if } H \geq K \\ 0 \hspace{8cm} \text{ if } H < K \end{cases}$$

Now that all of the components of the barrier options have been derived, the eight European style, single barrier options can be formulated.

$$C_{di}^e(S; H, K, T) = C_{di}^{an}(S; H, K, T) - KC_{di}^{cn}(S; H, K, T) \qquad (14.90)$$

$$C_{do}^e(S; H, K, T) = C_{do}^{an}(S; H, K, T) - KC_{do}^{cn}(S; H, K, T) \qquad (14.91)$$

$$C_{ui}^e(S; H, K, T) = C_{ui}^{an}(S; H, K, T) - KC_{ui}^{cn}(S; H, K, T) \qquad (14.92)$$

$$C_{uo}^e(S; H, K, T) = C_{uo}^{an}(S; H, K, T) - KC_{uo}^{cn}(S; H, K, T) \qquad (14.93)$$

$$P_{di}^e(S; H, K, T) = KP_{di}^{cn}(S; H, K, T) - P_{di}^{an}(S; H, K, T) \qquad (14.94)$$

$$P_{do}^e(S; H, K, T) = KP_{do}^{cn}(S; H, K, T) - P_{do}^{an}(S; H, K, T) \qquad (14.95)$$

$$P_{di}^e(S; H, K, T) = KP_{ui}^{cn}(S; H, K, T) - P_{ui}^{an}(S; H, K, T) \qquad (14.96)$$

$$P_{uo}^e(S; H, K, T) = KP_{uo}^{cn}(S; H, K, T) - P_{uo}^{an}(S; H, K, T). \qquad (14.97)$$

**Example 14.14.** Suppose the current price of a stock is \$55/share, the stock has a volatility of 35%, and pays a continuous dividend with a yield of 2% per annum. The continuously compounded, risk-free interest rate is 6%. Find the price of a \$57-strike up-and-out European call with a barrier of \$60 and a expiry of 6 months.

**Solution.** Using the values $S(0) = 55$, $K = 57$, $H = 60$, $T = 6/12$, $r = 0.06$, $\sigma = 0.35$, and $\delta = 0.02$ then

$$d_1 = 0.060233 \qquad d_3 = 0.763391 \qquad d_5 = -0.147023 \qquad d_7 = 0.556135$$
$$d_2 = -0.187254 \qquad d_4 = 0.515904 \qquad d_6 = -0.394511 \qquad d_8 = 0.308648,$$

and $\left(\dfrac{H}{S}\right)^{\frac{2(r-\delta)}{\sigma^2}-1} = 0.970263$ and $\left(\dfrac{H}{S}\right)^{\frac{2(r-\delta)}{\sigma^2}+1} = 1.15469$. Making use of Eq. (14.93) then

$$C_{uo}^e = \$0.00533071.$$

**Example 14.15.** Suppose the current price of a stock is \$120/share, the stock has a volatility of 25%, and pays a continuous dividend with a yield of 1% per annum. The continuously compounded, risk-free interest rate is 4%. Find the price of a \$115-strike up-and-in European put with a barrier of \$125 and a expiry of 9 months.

**Solution.** Using the values $S(0) = 120$, $K = 115$, $H = 125$, $T = 9/12$, $r = 0.04$, $\sigma = 0.25$, and $\delta = 0.01$ then

$$d_3 = 0.785848$$
$$d_4 = 0.569342,$$

and $\left(\dfrac{H}{S}\right)^{\frac{2(r-\delta)}{\sigma^2}-1} = 0.998368$ and $\left(\dfrac{H}{S}\right)^{\frac{2(r-\delta)}{\sigma^2}+1} = 1.08330.$ Making use of Eq. (14.96) then

$$P^e_{ui} = \$3.83917.$$

A portfolio consisting of a knock-in barrier option and a knock-out barrier option with the same barrier is equivalent to an ordinary option since the knock-in and knock-out features of the option effectively remove the barrier. Hence the following four parity relations hold for barrier options.

$$C_{di}(S; H, K, T) + C_{do}(S; H, K, T) = C(S, K, T) \qquad (14.98)$$
$$C_{ui}(S; H, K, T) + C_{uo}(S; H, K, T) = C(S, K, T) \qquad (14.99)$$
$$P_{di}(S; H, K, T) + P_{do}(S; H, K, T) = P(S, K, T) \qquad (14.100)$$
$$P_{ui}(S; H, K, T) + P_{uo}(S; H, K, T) = P(S, K, T). \qquad (14.101)$$

These parity relations hold for barrier options with both European and American style of exercise.

## Exercises

**14.8.1** Find the prices of the following barrier options.

(a) A down-and-out put with $H > K$.
(b) A down-and-in put with $H > K$.
(c) An up-and-out call with $H < K$.
(d) An up-and-in call with $H < K$.

**14.8.2** Suppose the following closing prices for Google stock (GOOG) were observed weekly during 2020.

| Date | Price ($) | Date | Price ($) |
|---|---|---|---|
| 01/03 | 1,360.66 | 02/14 | 1,520.74 |
| 01/10 | 1,429.73 | 02/21 | 1,485.11 |
| 01/17 | 1,480.39 | 02/28 | 1,339.33 |
| 01/24 | 1,466.39 | 03/06 | 1,298.41 |
| 01/31 | 1,434.23 | 03/13 | 1,219.73 |
| 02/07 | 1,479.23 | 03/20 | 1,072.32 |

An up-and-in barrier put was issued on 01/03 with a strike price of $1,500 and an expiry of 03/20. What is the payoff of this option?

**14.8.3** Show that a cash-or-nothing, down-and-out call can be priced as

$$C_{do}^{cn}(S(0); H, K, T) = e^{-rT} \begin{cases} \Phi(d_2) - \left(\frac{H}{S(0)}\right)^{\frac{2(r-\delta)}{\sigma^2}-1} \Phi(d_4) \text{ if } H \leq K, \\ \Phi(d_6) - \left(\frac{H}{S(0)}\right)^{\frac{2(r-\delta)}{\sigma^2}-1} \Phi(d_8) \text{ if } H > K. \end{cases}$$

**14.8.4** Take the limit as $K \to 0^+$ of the expression for a cash-or-nothing, down-and-in call to find the value of a deferred down rebate option as given in Eq. (14.69).

**14.8.5** Show that a cash-or-nothing, down-and-in put can be priced as

$$P_{di}^{cn}(S(0); H, K, T)$$
$$= e^{-rT} \begin{cases} \Phi(-d_6) + \left(\frac{H}{S(0)}\right)^{\frac{2(r-\delta)}{\sigma^2}-1} (\Phi(-d_4) - \Phi(-d_8)) \text{ if } H \leq K, \\ \Phi(-d_2) \qquad\qquad\qquad\qquad\qquad\qquad \text{ if } H > K. \end{cases}$$

**14.8.6** Show that a cash-or-nothing, down-and-out put can be priced as

$$P_{do}^{cn}(S(0); H, K, T)$$
$$= e^{-rT} \begin{cases} \Phi(d_6) - \Phi(d_2) - \left(\frac{H}{S(0)}\right)^{\frac{2(r-\delta)}{\sigma^2}-1} (\Phi(d_8) - \Phi(d_4)) \text{ if } H \leq K, \\ 0 \qquad\qquad\qquad\qquad\qquad\qquad\qquad\qquad \text{ if } H > K. \end{cases}$$

**14.8.7** Show that a cash-or-nothing, up-and-out put can be priced as

$$P_{uo}^{cn}(S(0); H, K, T)$$
$$= e^{-rT} \begin{cases} \Phi(-d_2) - \left(\frac{H}{S(0)}\right)^{\frac{2(r-\delta)}{\sigma^2}-1} \Phi(-d_4) \text{ if } H \geq K, \\ \Phi(-d_6) - \left(\frac{H}{S(0)}\right)^{\frac{2(r-\delta)}{\sigma^2}-1} \Phi(-d_8) \text{ if } H < K. \end{cases}$$

**14.8.8** Take the limit as $K \to \infty$ of the expression for a cash-or-nothing, up-and-in put to find the value of a deferred down rebate option as given in Eq. (14.74).

**14.8.9** Show that a cash-or-nothing, up-and-in call can be priced as

$$C_{ui}^{cn}(S(0); H, K, T)$$
$$= e^{-rT} \begin{cases} \Phi(d_6) + \left(\frac{H}{S(0)}\right)^{\frac{2(r-\delta)}{\sigma^2}-1} (\Phi(d_4) - \Phi(d_8)) \text{ if } H \geq K, \\ \Phi(d_2) \qquad\qquad\qquad\qquad\qquad\qquad \text{ if } H < K. \end{cases}$$

**14.8.10** Show that a cash-or-nothing, up-and-out call can be priced as

$$C_{uo}^{cn}(S(0); H, K, T)$$

$$= e^{-rT} \begin{cases} \Phi(d_2) - \Phi(d_6) + \left(\frac{H}{S(0)}\right)^{\frac{2(r-\delta)}{\sigma^2}-1} (\Phi(d_8) - \Phi(d_4)) & \text{if } H \geq K, \\ 0 & \text{if } H < K. \end{cases}$$

**14.8.11** Show that the price of the asset-or-nothing, down-and-in call can be determined by replacing $\delta$ with $\delta - \sigma^2$ in the formula for the cash-or-nothing, down-and-in call and then multiplying by the forward price of the underlying stock.

$$C_{di}^{an}(S(0); H, K, T) = S(0)e^{(r-\delta)T}C_{di}^{cn}(S(0); H, K, T)\Big|_{\delta \mapsto \delta - \sigma^2}.$$

**14.8.12** Suppose the current price of a stock is \$75/share, the stock has a volatility of 45%, and pays a continuous dividend with a yield of 1% per annum. The continuously compounded, risk-free interest rate is 8%. Find the price of a \$70-strike down-and-out European put with a barrier of \$65 and a expiry of 3 months.

**14.8.13** Suppose the current price of a stock is \$250/share, the stock has a volatility of 40%, and pays a continuous dividend with a yield of 1.5% per annum. The continuously compounded, risk-free interest rate is 6.5%. Find the price of a \$260-strike up-and-in European call with a barrier of \$275 and a expiry of 4 months.

**14.8.14** The current price of a stock is \$105. The volatility of the stock is 27% per annum and the stock pays a continuous dividend of 2.75% per annum. The continuously compounded, risk-free interest rate is 6%. A 6-month, \$110-strike up-and-out call with a barrier of \$120 has a price of \$0.13928. Find the price of a 6-month, \$110-strike up-and-in call with a barrier of \$120.

There are many more types of exotic options. This chapter has introduced some of the techniques that can develop pricing formulas for novel exotic options. These techniques have included binomial trees, all-or-nothing type binary options, the general idea of the present value of risk-neutral expected value of the option payoff, and parity relations involving exotic options and ordinary options. Other techniques available to practitioners include **Monte Carlo simulation** and the solution of partial differential equations using either closed form expressions or numerical methods.

# Appendix A

# Linear Algebra Primer

In this chapter, a brief introduction to matrices, vectors, and matrix algebra is given in support of the discussion of linear programming covered in Chapter 3. The coverage of linear algebra topics will be selective and is intended to make the notation and operations of linear programming more easily understood and documented.

## A.1  Matrices and their Properties

A **linear system of equations** with $m$ equations and $n$ unknowns can be written as

$$a_{11}x_1 + a_{12}x_2 + \cdots + a_{1n}x_n = b_1$$
$$a_{21}x_1 + a_{22}x_2 + \cdots + a_{2n}x_n = b_2$$
$$\vdots$$
$$a_{m1}x_1 + a_{m2}x_2 + \cdots + a_{mn}x_n = b_m.$$

The expressions $a_{ij}$ for $i = 1, 2, \ldots, m$ and $j = 1, 2, \ldots, n$ are called the **coefficients** of the linear system and the expressions $x_j$ are called its **unknowns**. The linear system of equations above can be expressed in terms of matrices. A **matrix** is a rectangular array of numbers (real or complex). An $m \times n$ matrix contains $m\,n$ entries arranged in $m$ rows and $n$ columns. Matrices are often denoted with capital letters such as $A$, $B$, $C$, *etc.* The entries of a matrix associated with a particular linear system are the coefficients of the linear system. For example, a matrix, denoted as $A$, could be associated with the linear system above and defined as

$$A = \begin{bmatrix} a_{11} & a_{12} & \cdots & a_{1n} \\ a_{21} & a_{22} & \cdots & a_{2n} \\ \vdots & \vdots & & \vdots \\ a_{m1} & a_{m2} & \cdots & a_{mn} \end{bmatrix},$$

and the linear system of equations can be written as $A\mathbf{x} = \mathbf{b}$ where $\mathbf{x}$ is an $n \times 1$ matrix with entries $x_1, x_2, \ldots, x_n$ and $\mathbf{b}$ is an $m \times 1$ matrix with entries $b_1, b_2, \ldots, b_m$. An $n \times 1$ matrix is sometimes identified with an $n$-vector. For convenience, set $(A)_{ij} = a_{ij}$, the entry of matrix $A$ in the $i$th row and the $j$th column. A **column vector** or simply **vector** is an $m \times 1$ matrix. In the interest of saving printed space a vector written out in component form will often be denoted as $\mathbf{x} = \langle x_1, x_2, \ldots, x_m \rangle$.

Arithmetic operations for matrices are defined in the natural way. Two matrices $A$ and $B$ are **equal**, denoted as $A = B$, if they have the same number of rows and columns as each other (henceforth referred to as having the same size) and if $(A)_{ij} = (B)_{ij}$ for all $i$ and $j$. If matrix $A$ and matrix $B$ have the same size, the **sum** of $A$ and $B$, denoted as $A + B$, is a matrix in which $(A + B)_{ij} = (A)_{ij} + (B)_{ij}$ for all $i$ and $j$. Matrix addition can be thought of as being carried out component-wise. If $c$ is a scalar (a real or complex number) and $A$ is a matrix, the **scalar product** of $c$ and $A$ is denoted as $cA$ and is defined as a matrix for which $(cA)_{ij} = c(A)_{ij}$ for all $i$ and $j$. Matrix subtraction can be defined as $A - B = A + (-1)B$ for matrices of the same size. The **transpose** of an $m \times n$ matrix $A$ is an $n \times m$ matrix, denoted as $A^T$, where $(A^T)_{ij} = (A)_{ji}$ for all $i$ and $j$.

The operation of **matrix multiplication** requires more care to define. If $A$ is an $m \times p$ matrix and if $B$ is a $p \times n$ matrix, the **product** is denoted as $AB$ and is an $m \times n$ matrix where

$$(AB)_{ij} = \sum_{k=1}^{p} (A)_{ik}(B)_{kj},$$

for $i = 1, 2, \ldots, m$ and $j = 1, 2, \ldots, n$.

The **inner product** of two vectors $\mathbf{x}$ and $\mathbf{y}$ is denoted $\mathbf{x} \cdot \mathbf{y}$ and is defined as a special case of matrix multiplication.

$$\mathbf{x} \cdot \mathbf{y} = \mathbf{x}^T \mathbf{y} = \sum_{k=1}^{n} x_k y_k.$$

The inner product of two vectors is a $1 \times 1$ matrix, in other words a scalar (real or complex number).

The following theorem presents many of the properties of matrices and matrix operations. Many of these properties are matrix analogues of properties of the real numbers.

**Theorem A.1.** *Let $A$, $B$, and $C$ be matrices of the appropriate size so that the following operations are well-defined and let $a$ and $b$ be scalars, then*

   *(i)* $A + B = B + A$ *(commutative property of matrix addition),*
  *(ii)* $A + (B+C) = (A+B) + C$ *(associative property of matrix addition),*
 *(iii)* $A(BC) = (AB)C$ *(associative property of matrix multiplication),*
 *(iv)* $A(B + C) = AB + AC$ *(left distributive law),*
  *(v)* $(B + C)A = BA + CA$ *(right distributive law),*
 *(vi)* $a(B + C) = aB + aC$,
*(vii)* $(a + b)C = aC + bC$,
*(viii)* $a(bC) = (ab)C$,
 *(ix)* $a(BC) = (aB)C = B(aC)$,
  *(x)* $(AB)^T = B^T A^T$,
 *(xi)* $(aB)^T = aB^T$,
*(xii)* $(A^T)^T = A$.

Each of these statements can be proved using the relevant definitions given so far.

Often matrices have special organization or structure. A **square matrix of order** $n$ is a matrix with $n$ rows and $n$ columns. Matrix powers can be defined for square matrices. If $A$ is an $n \times n$ matrix and $k \in \mathbb{N}$ then the $k$th power of $A$, denoted as $A^k$, is defined as

$$A^k = \underbrace{A A \cdots A}_{k \text{ factors}}.$$

A matrix for which all entries are 0 will be called a **zero matrix** and denoted simply as 0. The **diagonal** of a square matrix $A$ of order $n$ is the ordered sequence of entries $(a_{11}, a_{22}, \ldots, a_{nn})$. A **lower triangular matrix** $L$ is a matrix for which $(L)_{ij} = 0$ for each $i = 1, 2, \ldots, n$ and $j = i + 1, i + 2, \ldots, n$. An **upper triangular matrix** $U$ is a matrix for which $(U)_{ij} = 0$ for each $j = 1, 2, \ldots, n$ and $i = j + 1, j + 2, \ldots, n$. A square matrix $A$ for which $A = A^T$ is said to be a **symmetric matrix**. A square matrix of order $n$ is called a **banded matrix** if there exist integers $r$ and $s$ with $1 < r, s < n$ for which $(A)_{ij} = 0$ when $r \leq j - i$ or $s \leq i - j$. The

**band width** is $r + s - 1$. Banded matrices of band width 3 are sometimes called **tridiagonal matrices** and have the form

$$A = \begin{bmatrix} a_{11} & a_{12} & 0 & \cdots & 0 & 0 & 0 \\ a_{21} & a_{22} & a_{23} & \cdots & 0 & 0 & 0 \\ \vdots & \vdots & \vdots & & \vdots & & \vdots \\ 0 & 0 & 0 & \cdots & a_{n-1,n-2} & a_{n-1,n-1} & a_{n-1,n} \\ 0 & 0 & 0 & \cdots & 0 & a_{n,n-1} & a_{n,n} \end{bmatrix}.$$

Banded matrices of band width 1 are called **diagonal matrices** and may have non-zero entries only on their diagonals. A diagonal matrix with 1's along its diagonal is known as the **identity matrix** and is denoted as $I$ or $I_n$ when the order of the matrix is significant.

**Theorem A.2.** *If $A$ is an $m \times n$ matrix then $A I_n = A = I_m A$.*

*Proof.* Consider entry

$$(A I_n)_{ij} = \sum_{k=1}^{n} (A)_{ik} (I_n)_{kj} = (A)_{ij}.$$

Since $(I_n)_{kj} = 1$ if and only if $k = j$ and is zero otherwise. Thus $A I_n = A$. Similarly,

$$(I_m A)_{ij} = \sum_{k=1}^{n} (I_m)_{ik} (A)_{kj} = (A)_{ij}.$$

Since $(I_m)_{ik} = 1$ if and only if $i = k$ and is zero otherwise. Thus $I_m A = A$. $\square$

A set of vectors $\{\mathbf{x}_1, \mathbf{x}_2, \ldots, \mathbf{x}_n\}$ is said to be a **linearly independent set** if the equation,

$$c_1 \mathbf{x}_1 + c_2 \mathbf{x}_2 + \cdots + c_n \mathbf{x}_n = \mathbf{0},$$

implies the scalars $c_1 = c_2 = \cdots = c_n = 0$. When dealing with vectors with $n$ components, a useful set of linearly independent vectors is $\{\mathbf{e}_1, \mathbf{e}_2, \ldots, \mathbf{e}_n\}$ where $\mathbf{e}_i = \langle 0, \ldots, 0, 1, 0, \ldots, 0 \rangle$ and the 1 appears in the $i$th position for $1 \leq i \leq n$. Any set of $n$ linearly independent vectors $\{\mathbf{x}_1, \mathbf{x}_2, \ldots, \mathbf{x}_n\}$ in $\mathbb{R}^n$ forms a **basis** for $\mathbb{R}^n$ and hence if $\mathbf{x} \in \mathbb{R}^n$ there exists a unique set of scalars $c_1, c_2, \ldots, c_n$ such that

$$c_1 \mathbf{x}_1 + c_2 \mathbf{x}_2 + \cdots + c_n \mathbf{x}_n = \mathbf{x}.$$

A related, but stronger property of a set of vectors is known as **orthogonality**. Vectors $\mathbf{v}_1$ and $\mathbf{v}_2$ are orthogonal if $\mathbf{v}_1^T \mathbf{v}_2 = \mathbf{v}_2^T \mathbf{v}_1 = 0$.

If $\{\mathbf{v}_1, \mathbf{v}_2, \ldots, \mathbf{v}_n\}$ is a set of non-zero, mutually orthogonal vectors, the set is also linearly independent. To see this, suppose there exist scalars $c_1$, $c_2$, $\ldots$, $c_n$ such that

$$c_1\mathbf{v}_1 + c_2\mathbf{v}_2 + \cdots + c_n\mathbf{v}_n = \mathbf{0}.$$

Then for each $i = 1, 2, \ldots, n$,

$$\mathbf{v}_i^T(c_1\mathbf{v}_1 + c_2\mathbf{v}_2 + \cdots + c_n\mathbf{v}_n) = c_1\mathbf{v}_i^T\mathbf{v}_1 + c_2\mathbf{v}_i^T\mathbf{v}_2 + \cdots + c_n\mathbf{v}_i^T\mathbf{v}_n$$
$$= c_1(0) + c_2(0) + \cdots + c_i\mathbf{v}_i^T\mathbf{v}_i + \cdots + c_n(0)$$
$$= c_i\|\mathbf{v}_i\|_2^2 = 0.$$

This implies $c_i = 0$ for $i = 1, 2, \ldots, n$ and hence the vectors are linearly independent.

If $A$ and $B$ are $n \times n$ matrices for which $AB = BA = I_n$ then $A$ is said to be an **invertible** matrix and $B$ is the **inverse** of $A$. In this case matrix $B$ is often denoted as $A^{-1}$. Likewise $A$ is the inverse of matrix $B$. If matrix $A$ has no inverse then $A$ is said to be a **singular** matrix.

**Theorem A.3.** *The inverse of $A$, if it exists, is unique.*

*Proof.* Suppose there exist two matrices $B$ and $C$ for which $AB = BA = I$ and $AC = CA = I$, then $AB - AC = A(B - C) = 0$ and thus

$$BA(B - C) = I(B - C) = B - C = 0 \iff B = C.$$

Consequently the inverse of matrix $A$ is unique. □

**Theorem A.4.** *If $A$ and $B$ are $n \times n$ invertible matrices and $k \neq 0$ is a scalar, then*

*(i)* $(AB)^{-1} = B^{-1}A^{-1}$,
*(ii)* $(A^{-1})^{-1} = A$,
*(iii)* $(A^m)^{-1} = (A^{-1})^m = A^{-m}$ *for* $m = 0, 1, 2, \ldots$,
*(iv)* $(kA)^{-1} = (1/k)A^{-1}$,
*(v)* $(A^T)^{-1} = (A^{-1})^T$.

*Proof.* Let $A$ and $B$ be invertible $n \times n$ matrices, then

$$(AB)(B^{-1}A^{-1}) = A(BB^{-1})A^{-1} = AI_nA^{-1} = AA^{-1} = I_n.$$

Similarly $(B^{-1}A^{-1})(AB) = I_n$ and thus $AB$ is invertible with inverse $B^{-1}A^{-1}$.

By assumption $A$ is invertible, so

$$A^{-1}A = I_n = AA^{-1}$$

which implies $A^{-1}$ is invertible with inverse $A$. Thus $(A^{-1})^{-1} = A$.

Suppose $m = 0$ then $A^0 = I_n = A^{-0}$ which is an invertible $n \times n$ matrix. If $m = 1$ then by assumption $(A^1)^{-1} = A^{-1} = A^{-(1)}$. Now suppose $(A^k)^{-1} = (A^{-1})^k = A^{-k}$ for some natural number $k$.

$$A^{k+1}(A^{-1})^{k+1} = A(A^k A^{-k})A^{-1} = A I_n A^{-1} = A A^{-1} = I_n$$

Similarly $(A^{-1})^{k+1}A^{k+1} = I_n$ which implies $A^{k+1}$ is invertible with inverse $(A^{-1})^{k+1}$. Define $A^{-(k+1)}$ to be $(A^{-1})^{k+1}$. Thus by mathematical induction, $(A^m)^{-1} = (A^{-1})^m = A^{-m}$ for $m = 0, 1, 2, \ldots$.

Let $k$ be a non-zero scalar.

$$(k\,A)\left(\frac{1}{k}A^{-1}\right) = \left(k\,\frac{1}{k}\right)(A\,A^{-1}) = (1)I_n = I_n$$

Likewise $(1/k)A^{-1}(k\,A) = I_n$ and thus $k\,A$ is invertible with inverse $(1/k)A^{-1}$.

Finally consider,

$$A^T(A^{-1})^T = (A^{-1}A)^T = (I_n)^T = I_n.$$

Similarly $(A^{-1})^T A^T = I_n$ so $A^T$ is invertible with inverse $(A^{-1})^T$.  □

Invertible matrices play a fundamental role in solving systems of linear equations.

**Theorem A.5.** *The following statements are equivalent for any square matrix $A$ of order $n$.*

*(i) Matrix $A$ is invertible.*

*(ii) The equation $A\mathbf{x} = \mathbf{b}$ has a unique solution for any column vector $\mathbf{b}$ with $n$ components.*

*(iii) The columns of $A$ are linearly independent.*

*Proof.* Suppose $A$ is invertible and define $\mathbf{x} = A^{-1}\mathbf{b}$, then

$$A\mathbf{x} = A(A^{-1}\mathbf{b}) = (A\,A^{-1})\mathbf{b} = I_n\mathbf{b} = \mathbf{b},$$

which establishes the existence of a solution. Now suppose $A\mathbf{x} = \mathbf{b}$ and $A\mathbf{y} = \mathbf{b}$, then

$$\mathbf{x} = (A^{-1}A)\mathbf{x} = A^{-1}(A\mathbf{x}) = A^{-1}\mathbf{b} = A^{-1}(A\mathbf{y}) = (A^{-1}A)\mathbf{y} = \mathbf{y},$$

which establishes the uniqueness of the solution.

Let the columns of matrix $A$ be the vectors $\mathbf{a}_1, \mathbf{a}_2, \ldots, \mathbf{a}_n$ and suppose

$$\begin{aligned}
\mathbf{0} &= c_1\mathbf{a}_1 + c_2\mathbf{a}_2 + \cdots + c_n\mathbf{a}_n \\
&= c_1 A\mathbf{e}_1 + c_2 A\mathbf{e}_2 + \cdots + c_n A\mathbf{e}_n \\
&= A(c_1\mathbf{e}_1 + c_2\mathbf{e}_2 + \cdots + c_n\mathbf{e}_n).
\end{aligned}$$

If the equation $A\mathbf{x} = \mathbf{b}$ has a unique solution for every vector $\mathbf{b}$, then the equation above has the unique solution

$$c_1\mathbf{e}_1 + c_2\mathbf{e}_2 + \cdots + c_n\mathbf{e}_n = \mathbf{0}.$$

Since the vectors $\mathbf{e}_1, \mathbf{e}_2, \ldots, \mathbf{e}_n$ are linearly independent, then $c_1 = c_2 = \cdots = c_n = 0$. Hence the columns of matrix $A$ are linearly independent.

Now suppose the columns of matrix $A$ are linearly independent. For each canonical basis vector $\mathbf{e}_i$ there exists a unique set of scalars $b_{1i}, b_{2i}, \ldots, b_{ni}$ such that

$$b_{1i}\mathbf{a}_1 + b_{2i}\mathbf{a}_2 + \cdots + b_{ni}\mathbf{a}_n = \mathbf{e}_i.$$

Define matrix $B$ as

$$B = \begin{bmatrix} b_{11} & b_{12} & \cdots & b_{1n} \\ b_{21} & b_{22} & \cdots & b_{2n} \\ \vdots & \vdots & \ddots & \vdots \\ b_{n1} & b_{n2} & \cdots & b_{nn} \end{bmatrix},$$

where the $i$th column of $B$ consists of the coefficients of the column representation of $\mathbf{e}_i$. Then $AB = I_n$ and $B^T A^T = I_n$. Suppose $A^T\mathbf{x} = \mathbf{0}$, then $B^T A^T\mathbf{x} = \mathbf{0}$ and consequently $\mathbf{x} = \mathbf{0}$. As a result $A^T\mathbf{x} = \mathbf{0}$ if and only if $\mathbf{x} = \mathbf{0}$. Thus the only linear combination of the columns of $A^T$ equal to the zero vector is the trivial linear combination. This implies the columns of $A^T$, and hence the rows of $A$, are linearly independent. Using the argument above there exists an $n \times n$ matrix $\hat{B}$ such that $\hat{B} A = I_n$. Now consider

$$\hat{B} = \hat{B}(AB) = (\hat{B} A) B = B,$$

which implies $AB = BA = I_n$ and $A$ is invertible. □

The reader should note that not all the familiar operations and properties of the real numbers have analogues found among the properties of matrices listed in Thm. A.1. For example there is no commutative property of matrix multiplication, no zero factor property, and no multiplicative inverse for all non-zero matrices. In fact, these properties do not hold in general for matrices.

## A.2 Vector and Matrix Norms

Two vital concepts used throughout mathematical analysis are the notions of norm (or length) and distance. If $\mathbf{x}$ is a vector of $n$ components, define

$$\|\mathbf{x}\|_2 = \left(x_1^2 + x_2^2 + \cdots + x_n^2\right)^{1/2},$$

which is sometimes called the $l_2$-**norm** of a vector. Many readers may know the $l_2$-norm as the Euclidean length of a vector. Another useful norm is the $l_\infty$-**norm** defined as

$$\|\mathbf{x}\|_\infty = \max_{1\le i\le n} |x_i|.$$

The choice of which vector norm to use is one of convenience since the two norms are equivalent in the sense that

$$\|\mathbf{x}\|_\infty \le \|\mathbf{x}\|_2 \le \sqrt{n}\|\mathbf{x}\|_\infty, \tag{A.1}$$

for all $\mathbf{x} \in \mathbb{R}^n$. There are other equivalent norms, but their use is not necessary to establish the results mentioned in this textbook.

The distance between two vectors $\mathbf{u}$ and $\mathbf{v}$ is defined as $\|\mathbf{u}-\mathbf{v}\|_\infty$ (if using the $l_\infty$-norm) or as $\|\mathbf{u} - \mathbf{v}\|_2$ (if using the $l_2$-norm). Thus the notion of a sequence of vectors $\{\mathbf{u}^{(k)}\}_{k=0}^\infty$ **converging** to a vector $\mathbf{u}$ can be made more precise using the concept of the distance between vectors. The sequence $\mathbf{u}^{(k)} \to \mathbf{u}$ as $k \to \infty$ provided for all $\epsilon > 0$ there exists a natural number $N$ such that

$$\|\mathbf{u}^{(k)} - \mathbf{u}\| < \epsilon \text{ for all } k \ge N.$$

The norm used above can be either the $l_2$-norm or the $l_\infty$-norm.

**Theorem A.6.** *The sequence of vectors $\{\mathbf{u}^{(k)}\}_{k=0}^\infty$ converges to $\mathbf{u} \in \mathbb{R}^n$ with respect to the $l_\infty$-norm (or $l_2$-norm) if and only if*

$$\lim_{k\to\infty} u_i^{(k)} = u_i \text{ for each } i = 1, 2, \ldots, n.$$

*Proof.* Suppose $\mathbf{u}^{(k)} \to \mathbf{u}$ with respect to the $l_\infty$-norm. By definition, given any $\epsilon > 0$ there exists $N \in \mathbb{N}$ such that if $k \ge N$ it is true that

$$\|\mathbf{u}^{(k)} - \mathbf{u}\|_\infty = \max_{1\le i\le n} |u_i^{(k)} - u_i| < \epsilon.$$

Therefore for each $i = 1, 2, \ldots, n$ it is the case that $|u_i^{(k)} - u_i| < \epsilon$ which implies $u_i^{(k)} \to u_i$ as a sequence of scalars.

Now suppose that for each $i = 1, 2, \ldots, n$ it is true that $u_i^{(k)} \to u_i$ as a sequence of scalars. By definition given $\epsilon > 0$ there exists for each $i = 1, 2, \ldots, n$ a natural number $N_i$ such that $|u_i^{(k)} - u_i| < \epsilon$ for all $k \ge N_i$. Define $N = \max_{1\le i\le n} N_i$ then if $k \ge N$,

$$\max_{1\le i\le n} |u_i^{(k)} - u_i| = \|\mathbf{u}^{(k)} - \mathbf{u}\|_\infty < \epsilon.$$

This implies that $\mathbf{u}^{(k)} \to \mathbf{u}$ with respect to the $l_\infty$-norm. The case using $l_2$-norm can be proved similarly.   $\square$

The vector norms defined above induce norms on matrices in the following manner. If $A$ is an $m \times n$ matrix and $\mathbf{x} \in \mathbb{R}^n$ with $\|\mathbf{x}\|_\infty = 1$ then the $l_\infty$-norm of $A$ denoted as $\|A\|_\infty$ is defined as

$$\|A\|_\infty = \max_{\|\mathbf{x}\|_\infty = 1} \|A\mathbf{x}\|_\infty.$$

Similarly the $l_2$-norm of $A$ is defined as

$$\|A\|_2 = \max_{\|\mathbf{x}\|_2 = 1} \|A\mathbf{x}\|_2.$$

Since the geometrical interpretation of norm is length of a vector, the norm of a matrix can be thought of as the maximum factor by which the length of any unit vector is multiplied when the unit vector is multiplied by the matrix. Suppose $\|\mathbf{x}\|_\infty = 1$, then $\|A\mathbf{x}\|_\infty \leq \|A\|_\infty = \|A\|_\infty \|\mathbf{x}\|_\infty$. If $\mathbf{x} \neq \mathbf{0}$, then $\|\mathbf{x}\|_\infty > 0$ and hence $\mathbf{z} = \mathbf{x}/\|\mathbf{x}\|_\infty$ is a unit vector (with respect to the $l_\infty$-norm). Similar to the previous calculation,

$$\|A\mathbf{z}\|_\infty \leq \|A\|_\infty \|\mathbf{z}\|_\infty$$
$$\|\mathbf{x}\|_\infty \|A\mathbf{z}\|_\infty \leq \|\mathbf{x}\|_\infty \|A\|_\infty \|\mathbf{z}\|_\infty$$
$$\|A\|\mathbf{x}\|_\infty \mathbf{z}\|_\infty \leq \|A\|_\infty \|\|\mathbf{x}\|_\infty \mathbf{z}\|_\infty$$
$$\|A\mathbf{x}\|_\infty \leq \|A\|_\infty \|\mathbf{x}\|_\infty.$$

The final inequality is trivially true when $\mathbf{x} = \mathbf{0}$.

**Theorem A.7.** $\|A\|_\infty = \max_{\mathbf{x} \neq 0} \dfrac{\|A\mathbf{x}\|_\infty}{\|\mathbf{x}\|_\infty}.$

*Proof.* Let $\mathbf{x} \in \mathbb{R}^n$ with $\mathbf{x} \neq \mathbf{0}$, then the vector $\mathbf{z} = \mathbf{x}/\|\mathbf{x}\|_\infty$ is a unit vector with respect to the $l_\infty$-norm. Hence $\|A\mathbf{z}\|_\infty \leq \|A\|_\infty$ which implies

$$\max_{\mathbf{x} \neq 0} \frac{\|A\mathbf{x}\|_\infty}{\|\mathbf{x}\|_\infty} = \max_{\mathbf{x} \neq 0} \left\| A\left(\frac{\mathbf{x}}{\|\mathbf{x}\|_\infty}\right) \right\|_\infty \leq \|A\|_\infty.$$

Note that

$$\|A\|_\infty = \max_{\|\mathbf{x}\|_\infty = 1} \|A\mathbf{x}\|_\infty = \max_{\|\mathbf{x}\|_\infty = 1} \frac{\|A\mathbf{x}\|_\infty}{\|\mathbf{x}\|_\infty} \leq \max_{\mathbf{x} \neq 0} \frac{\|A\mathbf{x}\|_\infty}{\|\mathbf{x}\|_\infty}.$$

Thus the theorem is proved. □

A similar theorem holds for the $l_2$-norm of matrices.

## A.3   Separating Hyperplanes and Convex Cones

The strong version of the Duality Theorem (Thm. 3.3) relies upon Farkas'
Lemma and its corollary introduced in this section. The proof of Farkas'
Lemma in turn depends on the Separating Hyperplane Theorem, of which
there are several versions. A simple version of this theorem is introduced
and proved in this section. More general versions of this theorem can be
stated and proved, but the one included here is sufficient for the purposes
of this text. Before introducing these theorems, several terms should be
defined. A subset $A$ of $\mathbb{R}^n$ is **bounded** provided there exists a real number
$M > 0$ such that $\|\mathbf{x}\| \leq M$ for all $\mathbf{x} \in A$. A subset $B$ of $\mathbb{R}^n$ is **closed** if for
all convergent sequences $\{\mathbf{x}^{(k)}\}_{k=0}^\infty$ of vectors drawn from $B$ the limit of the
sequence belongs to $B$. Set $B$ is closed if when $\mathbf{x}^{(k)} \in B$ for all $k = 0, 1, 2, \ldots$
and $\lim_{k\to\infty} \mathbf{x}^{(k)} = \mathbf{x}$ exists, then $\mathbf{x} \in B$ as well.

The notion of a plane in three dimensions is extended to higher dimen-
sions by defining a **hyperplane**. Given a fixed vector $\mathbf{a} \in \mathbb{R}^n$ and a real
scalar $b$, the set $H$ defined as

$$H = \{\mathbf{x} \in \mathbb{R}^n \mid \mathbf{a}^T\mathbf{x} = b\},$$

is called a hyperplane. Just as for lines in two dimensions and planes in
three dimensions, hyperplanes separate $\mathbb{R}^n$ into two **halfspaces**. Theses
halfspaces can be described using the inequalities $\mathbf{a}^T\mathbf{z} \leq b$ and $\mathbf{a}^T\mathbf{z} \geq b$. An
illustration of a special case of the following theorem is shown in Fig. A.1.

**Theorem A.8 (Separating Hyperplane).** *Suppose $C$ and $D$ are two
closed convex sets in $\mathbb{R}^n$ and suppose $D$ is bounded. If $C \cap D = \emptyset$ then there
exists $\mathbf{a} \in \mathbb{R}^n$ with $\mathbf{a} \neq \mathbf{0}$ and there exists $b \in \mathbb{R}$ such that $\mathbf{a}^T\mathbf{x} > b$ for all
$\mathbf{x} \in D$ and $\mathbf{a}^T\mathbf{x} < b$ for all $\mathbf{x} \in C$.*

*Proof.* The distance between the closed sets $C$ and $D$ is the minimum
of $\|\mathbf{u} - \mathbf{v}\|$ taken over all $\mathbf{u} \in C$ and $\mathbf{v} \in D$. The distance between $C$
and $D$ is strictly positive, for otherwise suppose there exists a sequence
$\{\mathbf{u}^{(n)}\}_{n=1}^\infty \subset C$ and a sequence $\{\mathbf{v}^{(n)}\}_{n=1}^\infty \subset D$ such that $\|\mathbf{u}^{(n)} - \mathbf{v}^{(n)}\| \to 0$
as $n \to \infty$. Since $D$ is closed and bounded the sequence $\{\mathbf{v}^{(n)}\}_{n=1}^\infty$ possesses
a convergent subsequence $\{\mathbf{v}^{(n_j)}\}_{j=1}^\infty$. Assume $\mathbf{v}^{(n_j)} \to \mathbf{v} \in D$. By the
triangle inequality,

$$\|\mathbf{u}^{(n_j)} - \mathbf{v}\| \leq \|\mathbf{u}^{(n_j)} - \mathbf{v}^{(n_j)}\| + \|\mathbf{v}^{(n_j)} - \mathbf{v}\| \to 0,$$

as $j \to \infty$. Therefore the sequence $\mathbf{u}^{(n_j)} \to \mathbf{v}$. Since $C$ is a closed set, $\mathbf{v} \in C$
which contradicts $C \cap D = \emptyset$.

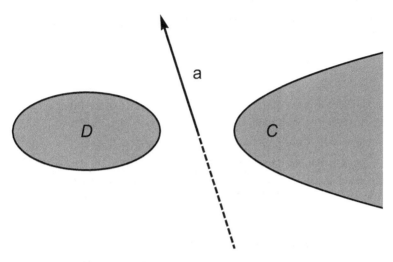

Fig. A.1   Non-intersecting closed, convex sets in $\mathbb{R}^n$ (at least one of which is bounded) can be separated by a hyperplane. The case for $n = 2$ is illustrated here.

Now let

$$m = \inf_{\substack{\mathbf{u} \in C \\ \mathbf{v} \in D}} \|\mathbf{u} - \mathbf{v}\|,$$

and let $\{\mathbf{u}^{(n)}\}_{n=1}^{\infty}$ and $\{\mathbf{v}^{(n)}\}_{n=1}^{\infty}$ be sequences in $C$ and $D$ respectively such that

$$\lim_{n \to \infty} \|\mathbf{u}^{(n)} - \mathbf{v}^{(n)}\| = m.$$

Since $D$ is closed and bounded there exists a convergent subsequence of $\{\mathbf{v}^{(n)}\}_{n=1}^{\infty}$, call it $\{\mathbf{v}^{(n_j)}\}_{j=1}^{\infty}$ such that $\mathbf{v}^{(n_j)} \to \mathbf{v} \in D$ and

$$\lim_{j \to \infty} \|\mathbf{u}^{(n_j)} - \mathbf{v}\| = m.$$

Hence the subsequence $\{\mathbf{u}^{(n_j)}\}_{j=1}^{\infty}$ is bounded and thus has a convergent subsequence. Without loss of generality assume the subsequence $\mathbf{u}^{(n_j)} \to \mathbf{u} \in C$. Since $C$ is a closed set, $\mathbf{u} \in C$ and thus $m = \|\mathbf{u} - \mathbf{v}\|$. Consequently the minimum distance between sets $C$ and $D$ is attained at the vectors $\mathbf{u} \in C$ and $\mathbf{v} \in D$.

Define the vector $\mathbf{a} = \mathbf{v} - \mathbf{u}$ and the scalar $b = (\|\mathbf{v}\|^2 - \|\mathbf{u}\|^2)/2$. The expression $\mathbf{a}^T \mathbf{x} > b$ for all $\mathbf{x} \in D$. To establish this, suppose there exists vector $\mathbf{d} \in D$ for which $\mathbf{a}^T \mathbf{d} \leq b$. Define the function $g(\mathbf{x}) = \|\mathbf{x} - \mathbf{u}\|^2$. Recall that the gradient of a scalar-valued multivariable function is a vector-valued function pointing in the direction of greatest increase. Note that

$\nabla g(\mathbf{x}) = 2(\mathbf{x} - \mathbf{u})$. Consider the inner product of $\nabla g(\mathbf{v})$ and $\mathbf{d} - \mathbf{v}$.

$$\begin{aligned}(\nabla g(\mathbf{v}))^T(\mathbf{d} - \mathbf{v}) &= 2\mathbf{a}^T\mathbf{d} - 2\|\mathbf{v}\|^2 + 2\mathbf{u}^T\mathbf{v}\\ &\leq \|\mathbf{v}\|^2 - \|\mathbf{u}\|^2 - 2\|\mathbf{v}\|^2 + 2\mathbf{u}^T\mathbf{v}\\ &= -\|\mathbf{u} - \mathbf{v}\|^2 < 0.\end{aligned}$$

Thus the vector $\mathbf{d} - \mathbf{v}$ points in a direction of decrease for function $g$. Hence there exists $0 < \lambda < 1$ such that

$$g(\mathbf{v} + \lambda(\mathbf{d} - \mathbf{v})) < g(\mathbf{v}) \iff \|\lambda\mathbf{d} + (1 - \lambda)\mathbf{v} - \mathbf{u}\|^2 < \|\mathbf{v} - \mathbf{u}\|^2.$$

Since $D$ is a convex set, the vector $\lambda\mathbf{d} + (1 - \lambda)\mathbf{v} \in D$ and hence the minimum distance between $C$ and $D$ is not attained at $\mathbf{u} \in C$ and $\mathbf{v} \in D$. This is a contradiction and therefore $g(\mathbf{x}) > 0$ for all $\mathbf{x} \in D$. In a similar fashion it can be shown that $g(\mathbf{x}) < 0$ for all $\mathbf{x} \in C$. $\qquad\square$

In addition to convex sets there is a need to study **convex cones**. A set $X \subset \mathbb{R}^n$ is called a convex cone if for all $\mathbf{x}, \mathbf{y} \in X$ and all $\alpha, \beta \geq 0$ then $\alpha\mathbf{x} + \beta\mathbf{y} \in X$. For a set $X \subset \mathbb{R}^n$ containing the origin, the **polar cone** denoted $X^+$ is defined as

$$X^+ = \{\mathbf{a} \in \mathbb{R}^n \,|\, \mathbf{a}^T\mathbf{x} \geq 0 \text{ for all } \mathbf{x} \in X\}. \tag{A.2}$$

**Example A.1.** The non-negative orthant is defined as $\mathbb{R}^n_+ = \{\langle x_1, x_2, \ldots, x_n\rangle \,|\, x_i \geq 0 \text{ for } i = 1, 2, \ldots, n\}$. Show that $(\mathbb{R}^n_+)^+ = \mathbb{R}^n_+$.

**Solution.** By Eq. (A.2) $(\mathbb{R}^n_+)^+ = \{\mathbf{a} \in \mathbb{R}^n \,|\, \mathbf{a}^T\mathbf{x} \geq 0 \text{ for all } \mathbf{x} \in \mathbb{R}^n_+\}$. Let $\mathbf{a} \in (\mathbb{R}^n_+)^+$ then for $i \in \{1, 2, \ldots, n\}$ it is the case that $\mathbf{a}^T\mathbf{e}_i = a_i \geq 0$. Thus $\mathbf{a} \geq \mathbf{0}$ and consequently $(\mathbb{R}^n_+)^+ \subset \mathbb{R}^n_+$. Now let $\mathbf{a}, \mathbf{x} \in \mathbb{R}^n_+$, then

$$\mathbf{a}^T\mathbf{x} = \sum_{i=1}^n a_i x_i \geq 0.$$

Since $a_i \geq 0$ and $x_i \geq 0$ for $i = 1, 2, \ldots, n$. Thus $\mathbb{R}^n_+ \subset (\mathbb{R}^n_+)^+$ and finally $(\mathbb{R}^n_+)^+ = \mathbb{R}^n_+$.

The next several theorems about convex cones and polar cones are taken from the discussion in [Karlin (1959), Sec. B.3].

**Theorem A.9.** *Let $X$ and $Y$ be any subsets of $\mathbb{R}^n$ containing the origin, then*

   *(i) $X^+$ is a closed convex cone,*
   *(ii) if $X \subset Y$ then $Y^+ \subset X^+$,*
   *(iii) $X \subset X^{++}$,*
   *(iv) $X^+ = X^{+++}$,*
   *(v) $(X + Y)^+ = X^+ \cap Y^+$.*

*Proof.*

(i) Let $\mathbf{a}, \mathbf{b} \in X^+$ and $\alpha, \beta \geq 0$, then

$$(\alpha\mathbf{a} + \beta\mathbf{b})^T\mathbf{x} = \alpha\mathbf{a}^T\mathbf{x} + \beta\mathbf{b}^T\mathbf{x} \geq 0,$$

which implies $\alpha\mathbf{a} + \beta\mathbf{b} \in X^+$ and thus $X^+$ is a convex cone. Now suppose $\{\mathbf{a}^{(k)}\}_{k=1}^\infty$ is a sequence of vectors in $X^+$ and suppose $\lim_{k\to\infty} \mathbf{a}^{(k)} = \mathbf{a}$. By assumption $(\mathbf{a}^{(k)})^T\mathbf{x} \geq 0$ for all $\mathbf{x} \in X$ and thus

$$\lim_{k\to\infty} (\mathbf{a}^{(k)})^T\mathbf{x} = \left(\lim_{k\to\infty} \mathbf{a}^{(k)}\right)^T \mathbf{x} = \mathbf{a}^T\mathbf{x} \geq 0,$$

for all $\mathbf{x} \in X$. Consequently $\mathbf{a} \in X^+$ and $X^+$ is closed.

(ii) Suppose $X \subset Y$ and let $\mathbf{b} \in Y^+$. By Eq. (A.2) $\mathbf{b}^T\mathbf{y} \geq 0$ for all $\mathbf{y} \in Y$. Since if $\mathbf{x} \in X$ then $\mathbf{x} \in Y$ then $\mathbf{b}^T\mathbf{x} \geq 0$ which implies $\mathbf{b} \in X^+$. Therefore $Y^+ \subset X^+$.

(iii) Let $\mathbf{x} \in X$ then for all $\mathbf{a} \in X^+$ it is the case that $\mathbf{a}^T\mathbf{x} = \mathbf{x}^T\mathbf{a} \geq 0$ and hence by Eq. (A.2) $\mathbf{x} \in X^{++}$. Therefore $X \subset X^{++}$.

(iv) By (iii) $X^+ \subset X^{+++}$. Applying (ii) to the result in (iii) then $X^{+++} \subset X^+$ and consequently $X^+ = X^{+++}$.

(v) Let $\mathbf{a} \in X^+ \cap Y^+$ then $\mathbf{a}^T\mathbf{x} \geq 0$ for all $\mathbf{x} \in X$ and $\mathbf{a}^T\mathbf{y} \geq 0$ for all $\mathbf{y} \in Y$. Therefore

$$0 \leq \mathbf{a}^T\mathbf{x} + \mathbf{a}^T\mathbf{y} = \mathbf{a}^T(\mathbf{x} + \mathbf{y}),$$

which implies $\mathbf{a} \in (X + Y)^+$. Now suppose $\mathbf{a} \in (X + Y)^+$ then $\mathbf{a}^T\mathbf{x} = \mathbf{a}^T(\mathbf{x} + \mathbf{0}) \geq 0$ for all $\mathbf{x} \in X$ which implies $\mathbf{a} \in X^+$. In a similar manner it can be shown that $\mathbf{a} \in Y^+$. Therefore $\mathbf{a} \in X^+ \cap Y^+$ and consequently $(X + Y)^+ = X^+ \cap Y^+$.

$\square$

**Theorem A.10.** *Let $X$ and $Y$ be closed convex cones, then*

*(i)* $X = X^{++}$,
*(ii)* $(X + Y)^+ = X^+ \cap Y^+$,
*(iii)* $(X \cap Y)^+ = \overline{X^+ + Y^+}$.

*Proof.*

(i) Suppose $\mathbf{y} \notin X$ then by Thm. A.8 there exists $\mathbf{a}$ such that $\mathbf{a}^T\mathbf{x} > \mathbf{a}^T\mathbf{y}$ for all $\mathbf{x} \in X$. By assumption $X$ is a convex cone which implies $\mathbf{a}^T\mathbf{x} \geq 0 > \mathbf{a}^T\mathbf{y}$ for all $\mathbf{x} \in X$. Thus according to Eq. (A.2), $\mathbf{a} \in X^+$ and $\mathbf{y} \notin X^{++}$. Combining this result with Thm. A.9(iii) demonstrates $X = X^{++}$.

(ii) This is a consequence of (v).

(iii) By (i) and Thm. A.9(v) then

$$(X \cap Y)^+ = (X^{++} \cap Y^{++})^+ = (X^+ + Y^+)^{++} = \overline{X^+ + Y^+}. \qquad \square$$

**Theorem A.11.** *Let $X$ be a closed convex cone for which $X \cap \mathbb{R}_+^n = \{\mathbf{0}\}$. There exists a vector $\mathbf{a} > \mathbf{0}$ such that $\mathbf{a}^T \mathbf{x} \leq 0$ for all $\mathbf{x} \in X$.*

*Proof.* Suppose that $\mathbf{a} \geq \mathbf{0}$ and $\mathbf{a}^T \mathbf{x} \leq 0$ for all $\mathbf{x} \in X$ imply that some component of vector $\mathbf{a}$ is zero. By Eq. (A.2) the polar cone $(-X)^+$ consists of

$$\{\mathbf{a} \in \mathbb{R}^n \,|\, \mathbf{a}^T \mathbf{x} \geq 0 \text{ for all } \mathbf{x} \in -X\} = \{\mathbf{a} \in \mathbb{R}^n \,|\, \mathbf{a}^T \mathbf{x} \leq 0 \text{ for all } \mathbf{x} \in X\}.$$

Consider the intersection of the polar cones, $(-X)^+ \cap (\mathbb{R}_+^n)^+$ which is a closed set since the intersection of two closed sets is closed and by Thm. A.9(i).

$$(-X)^+ \cap (\mathbb{R}_+^n)^+ \subset H = \{\mathbf{h} \in \mathbb{R}_+^n \,|\, h_i = 0\}.$$

According to Thm. A.9(v) $(-X)^+ \cap (\mathbb{R}_+^n)^+ = (-X + \mathbb{R}_+^n)^+$ and thus by (iii) of the same theorem, $H^+ \subset (-X + \mathbb{R}_+^n)^{++}$. The set $-X + \mathbb{R}_+^n$ is a convex cone since for any $\alpha, \beta \geq 0$ and $\mathbf{x}_1 + \mathbf{y}_1, \mathbf{x}_2 + \mathbf{y}_2 \in -X + \mathbb{R}_+^n$ then

$$\alpha(\mathbf{x}_1 + \mathbf{y}_1) + \beta(\mathbf{x}_2 + \mathbf{y}_2) = (\alpha \mathbf{x}_1 + \beta \mathbf{x}_2) + (\alpha \mathbf{y}_1 + \beta \mathbf{y}_2) \in -X + \mathbb{R}_+^n.$$

Since $-X$ and $\mathbb{R}_+^n$ are convex cones. Using [Hestenes (1975), Lemma 3.4, p. 197] then $-X + \mathbb{R}_+^n$ is closed. Therefore by Thm. A.10(i) $H^+ \subset -X + \mathbb{R}_+^n$. For any real number $a$ the vector $a\mathbf{e}_i \in H^+$ and thus there exist vectors $\mathbf{x} \in X$ and $y \in \mathbb{R}_+^n$ such that $a\mathbf{e}_i = -\mathbf{x} + \mathbf{y}$ which is equivalent to $\mathbf{x} = \mathbf{y} - a\mathbf{e}_i$. Thus when $a < 0$ then $\mathbf{x} \geq \mathbf{0}$ and $x_i > 0$. This implies $\mathbf{x} \in \mathbb{R}_+^n$ which contradicts the assumption that $X \cap \mathbb{R}_+^n = \{\mathbf{0}\}$. Therefore $\mathbf{a} > \mathbf{0}$. $\qquad \square$

Another separating hyperplane theorem can be stated with slightly weaker assumptions and a correspondingly weaker conclusion. This theorem is stated below but not proved. Interested readers can find the proof in [Boyd and Vandenberghe (2004), Sec. 2.5].

**Theorem A.12.** *Let $A$ and $B$ be two disjoint, non-empty, convex subsets of $\mathbb{R}^n$.*

   (i) *There exists a vector $\mathbf{y} \in \mathbb{R}^n$ and a real number $b$ such that $\mathbf{y}^T \mathbf{x} \geq b$ for all $\mathbf{x} \in A$ and $\mathbf{y}^T \mathbf{x} \leq b$ for all $\mathbf{x} \in B$.*

   (ii) *If $A$ is open there exists a vector $\mathbf{y} \in \mathbb{R}^n$ and a real number $b$ such that $\mathbf{y}^T \mathbf{x} > b$ for all $\mathbf{x} \in A$ and $\mathbf{y}^T \mathbf{x} \leq b$ for all $\mathbf{x} \in B$.*

   (iii) *If both $A$ and $B$ are open there exists a vector $\mathbf{y} \in \mathbb{R}^n$ and a real number $b$ such that $\mathbf{y}^T \mathbf{x} > b$ for all $\mathbf{x} \in A$ and $\mathbf{y}^T \mathbf{x} < b$ for all $\mathbf{x} \in B$.*

## A.4 Farkas' Lemma

**Lemma A.1 (Farkas').** *Let $A$ be an $m \times n$ matrix with real entries and let $\mathbf{b}$ be a vector with $m$ components. Exactly one of the following two statements is true.*

*(i) There exists $\mathbf{x} \in \mathbb{R}^n$ with $\mathbf{x} \geq \mathbf{0}$ such that $A\mathbf{x} = \mathbf{b}$.*
*(ii) There exists $\mathbf{y} \in \mathbb{R}^m$ such that $\mathbf{b}^T\mathbf{y} > 0$ and $A^T\mathbf{y} \leq \mathbf{0}$.*

*Proof.* Suppose (i) holds and suppose there exists $\mathbf{y} \in \mathbb{R}^m$ such that $A^T\mathbf{y} \leq \mathbf{0}$. Since $\mathbf{x} \geq \mathbf{0}$ then

$$\mathbf{x}^T A^T \mathbf{y} = (A\mathbf{x})^T\mathbf{y} = \mathbf{b}^T\mathbf{y} \leq 0,$$

which implies (ii) does not hold.

Define the set

$$C = \{\mathbf{z} \mid A\mathbf{x} = \mathbf{z} \text{ for some } \mathbf{x} \geq \mathbf{0}\}.$$

If $\mathbf{u}, \mathbf{v} \in C$ and if $\lambda \in [0,1]$ then there exist $\mathbf{s} \geq \mathbf{0}$ and $\mathbf{t} \geq \mathbf{0}$ such that $A\mathbf{s} = \mathbf{u}$ and $A\mathbf{t} = \mathbf{v}$. Note that $\lambda\mathbf{s} + (1 - \lambda)\mathbf{t} \geq \mathbf{0}$ which implies

$$A\left(\lambda\mathbf{s} + (1 - \lambda)\mathbf{t}\right) = \lambda A\mathbf{s} + (1 - \lambda)A\mathbf{t} = \lambda\mathbf{u} + (1 - \lambda)\mathbf{v} \in C.$$

Hence $C$ is a convex set. Showing that $C$ is a closed set is more difficult. A detailed proof of the closed-ness of $C$ can be found in [Ciarlet (1989), pp. 332–334]. The justification given here is based on Ciarlet's approach. Denote the $i$th column of matrix $A$ as the vector $\mathbf{a}_i$ for $i = 1, 2, \dots, n$. As a first case, assume the columns of matrix $A$ are linearly independent. Let $\{\mathbf{z}^{(k)}\}_{k=1}^{\infty}$ be a sequence of vectors in $C$. By construction there exists a sequence $\{\mathbf{x}^{(k)}\}_{k=1}^{\infty}$ with $\mathbf{x}^{(k)} \geq \mathbf{0}$ for which

$$A\mathbf{x}^{(k)} = \sum_{i=1}^{n} x_i^{(k)}\mathbf{a}_i = \mathbf{z}^{(k)}.$$

If $\lim_{k\to\infty} \mathbf{z}^{(k)} = \mathbf{z} \in \mathbb{R}^m$ then the sequence $\{\mathbf{z}^{(k)}\}_{k=1}^{\infty}$ converges to $\mathbf{z}$ in the column space of $A$. Convergence in the column space of $A$ is equivalent to the convergence of the coordinates $\{\mathbf{x}^{(k)}\}_{k=1}^{\infty}$. Hence,

$$\lim_{k\to\infty} \mathbf{z}^{(k)} = \lim_{k\to\infty} \sum_{i=1}^{n} x_i^{(k)}\mathbf{a}_i = \sum_{i=1}^{n} \left(\lim_{k\to\infty} x_i^{(k)}\right) \mathbf{a}_i = \sum_{i=1}^{n} x_i\mathbf{a}_i = \mathbf{z}.$$

Since $\{x_i^{(k)}\}_{k=1}^{\infty}$ is a non-negative, convergent sequence, then $\lim_{k\to\infty} x_i^{(k)} = x_i \geq 0$ for $i = 1, 2, \dots, n$ which implies $\mathbf{z} \in C$ and thus $C$ is closed. If the columns of $A$ are not linearly independent, the columns must be partitioned into a finite number of sets of linearly independent vectors. Each set of linearly independent column vectors will correspond to a closed set using the proof just given. Since the finite union of closed sets is closed, then again $C$ is closed.

If (i) does not hold, then there does not exist $\mathbf{x} \geq 0$ for which $A\mathbf{x} = \mathbf{b}$ which implies $\mathbf{b} \notin C$. By Thm. A.8 there exists $\mathbf{y} \in \mathbb{R}^m$ with $\mathbf{y} \neq \mathbf{0}$ and $r \in \mathbb{R}$ such that $\mathbf{y}^T\mathbf{z} \leq r$ for all $\mathbf{z} \in C$ and $\mathbf{y}^T\mathbf{b} > r$. Note that $\mathbf{z} = \mathbf{0} \in C$ and therefore $r \geq 0$. If $r > 0$ then the inequality $\mathbf{y}^T\mathbf{b} > 0$ also holds since $\mathbf{y}^T\mathbf{b} > r$. The inner product $\mathbf{y}^T\mathbf{z}$ is bounded above by 0, since if there exists $\mathbf{z} \in C$ for which $\mathbf{y}^T\mathbf{z} > 0$ then $\mathbf{y}^T(\lambda \mathbf{z})$ becomes arbitrarily large as $\lambda \to \infty$ contradicting the inequality $\mathbf{y}^T\mathbf{z} \leq r$. Consequently $\mathbf{y}^T\mathbf{z} \leq 0$ for all $\mathbf{z} \in C$ and $\mathbf{y}^T\mathbf{b} > 0$. Finally since $\mathbf{a}_i = A\mathbf{e}_i$ for $i = 1, 2, \ldots, n$ then $A^T\mathbf{y} \leq \mathbf{0}$ which shows (ii) holds.                    □

An independent proof of Farkas' Lemma can be found in [Dax (1997)]. Farkas' Lemma has a geometric interpretation. Either vector $\mathbf{b}$ is in the vector space formed by the non-negative linear combinations of the columns of matrix $A$ or there exists a hyperplane such that the vector space lies on one side of the hyperplane and vector $\mathbf{b}$ lies on the other. An illustration of Farkas' Lemma in two dimensions is shown in Fig. A.2.

Due to the variety of formats in which linear equations and inequalities can be written, there are several versions and variations on Farkas' Lemma.

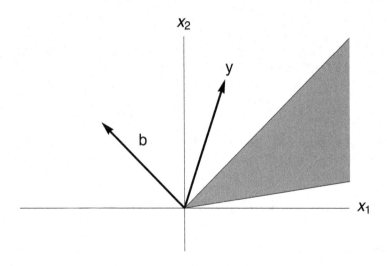

Fig. A.2   In the two-dimensional plane the case of (ii) for Farkas' Lemma is shown. The vector $\mathbf{b}$, which does not lie in the space of non-negative linear combinations of the columns of a matrix $A$, can be separated from the vector space by a hyperplane.

One useful restatement of Farkas' Lemma called the Farkas' Variant below is used in the proof of the Duality Theorem (Thm. 3.3) in Chapter 3.

**Corollary A.1 (Farkas' Variant).** *Let $A$ be an $m \times n$ matrix with real entries and let $\mathbf{b}$ be a vector with $m$ real components. Exactly one of the following two statements is true.*

*(i) There exists $\mathbf{x} \in \mathbb{R}^n$ such that $A\mathbf{x} \leq \mathbf{b}$.*
*(ii) There exists $\mathbf{y} \in \mathbb{R}^m$ with $\mathbf{y} \geq \mathbf{0}$ such that $\mathbf{b}^T\mathbf{y} = -1$ and $A^T\mathbf{y} = \mathbf{0}$.*

*Proof.* Suppose (i) holds, then there exists $\mathbf{x} \in \mathbb{R}^n$ such that $A\mathbf{x} \leq \mathbf{b}$. If (ii) holds as well there exists $\mathbf{y} \in \mathbb{R}^m$ such that $A^T\mathbf{y} = \mathbf{0}$ and $\mathbf{y}^T\mathbf{b} < 0$. Therefore,

$$0 = \mathbf{x}^T A^T \mathbf{y} = (A\mathbf{x})^T\mathbf{y} \leq \mathbf{b}^T\mathbf{y} = \mathbf{y}^T\mathbf{b} = -1,$$

since $\mathbf{y} \geq \mathbf{0}$. This is a contradiction and consequently (ii) cannot hold if (i) holds.

Suppose (ii) does not hold, then for no $\mathbf{y} \in \mathbb{R}^m$ with $\mathbf{y} \geq \mathbf{0}$ is $\mathbf{b}^T\mathbf{y} = -1$ and $A^T\mathbf{y} = \mathbf{0}$. This is equivalent to the statement that the equation

$$\begin{bmatrix} A^T \\ \mathbf{b}^T \end{bmatrix} \mathbf{y} = \begin{bmatrix} \mathbf{0} \\ -1 \end{bmatrix},$$

has no solution with $\mathbf{y} \geq \mathbf{0}$. Applying Lemma A.1 (where the equation above states that (i) of the lemma does not hold) then there exists $\mathbf{x} \in \mathbb{R}^{n+1}$ such that

$$\begin{bmatrix} \mathbf{0}^T \big| -1 \end{bmatrix} \mathbf{x} = -x_{n+1} > 0,$$

and

$$\begin{bmatrix} A \big| \mathbf{b} \end{bmatrix} \mathbf{x} \leq \mathbf{0} \iff A \begin{bmatrix} x_1 \\ x_2 \\ \vdots \\ x_n \end{bmatrix} + x_{n+1}\mathbf{b} \leq \mathbf{0} \iff A \begin{bmatrix} x_1 \\ x_2 \\ \vdots \\ x_n \end{bmatrix} \leq -x_{n+1}\mathbf{b}.$$

Define the vector $z \in \mathbb{R}^n$ as

$$\mathbf{z} = \frac{-1}{x_{n+1}} \begin{bmatrix} x_1 \\ x_2 \\ \vdots \\ x_n \end{bmatrix},$$

then $A\mathbf{z} \leq \mathbf{b}$. Hence (i) holds. $\square$

Statement (ii) is also logically equivalent to the statement that there exists $\mathbf{y} \in \mathbb{R}^m$ with $\mathbf{y} \geq \mathbf{0}$ such that $\mathbf{b}^T\mathbf{y} < 0$ and $A\mathbf{y} = \mathbf{0}$. The proof of this equivalence is left to the reader.

# Bibliography

Barrow, J. D. (2008). *One Hundred Essential Things You Didn't Know You Didn't Know: Math Explains Your World*, W. W. Norton and Company, Inc., New York, USA.

Bensoussan, A., Crouhy, M., and Galai, D. (1995). Black–Scholes approximation of warrant prices, *Journal of Advances in Futures and Options Research*, **8**, pp. 1–14.

Bleecker, D. and Csordas, G. (1996). *Basic Partial Differential Equations*, International Press, Cambridge, Massachusetts, USA.

Boyce, W. E. and DiPrima, R. C. (2001). *Elementary Differential Equations and Boundary Value Problems*, 7th edition, John Wiley and Sons, Inc., New York, USA.

Bradley, S. P., Hax, A. C., and Magnanti, T. L. (1977). *Applied Mathematical Programming*, Addison-Wesley Publishing Company, Reading, MA, USA.

Broverman, S. A. (2004). *Mathematics of Investment and Credit*, 3rd edition, ACTEX Publications, Winsted, CT, USA.

Buchanan, J. R. and Shao, Z. (2018). *A First Course in Partial Differential Equations*, World Scientific Publishing Company, Hackensack, NJ, USA.

Burden, R. L. and Faires, J. D. (2005). *Numerical Analysis*, 8th edition, Thomson Brooks/Cole, Belmont, CA, USA.

Boyd, S. P. and Vandenberghe, L. (2004). *Convex Optimization*, Cambridge University Press, Cambridge, UK.

Chance, D. M. (2008). A synthesis of binomial option pricing models for lognormally distributed assets, *Journal of Applied Finance*, **18**, pp. 38–56.

Chawla, M. M. (2006). Accurate computation of the Greeks, *International Journal of Applied Mathematics*, **7**(4), pp. 379–388.

Chawla, M. M. and Evans, D. J. (2005). High-accuracy finite difference methods for the valuation of options, *International Journal of Computer Mathematics*, **82**(9), pp. 1157–1165.

Chin, E., Nel, D., and Olafsson, S. (2017). *Problems and Solutions in Mathematical Finance: Equity Derivatives*, Volume 2, John Wiley & Sons Inc., West Sussex, UK.

Churchill, R. V., Brown, J. W., and Verhey, R. F. (1976). *Complex Variables and Applications*, 3rd edition, McGraw-Hill Book Company, New York, USA.

Ciarlet, P. G. (1989). *Introduction to Numerical Linear Algebra and Optimization*, Cambridge University Press, Cambridge, UK.

Clewlow, L. and Strickland C. (1997). *Exotic Options: The State of the Art*, International Thomson Business Press, London, UK.

Cornuejols, G. and Tütüncü, R. (2007). *Optimization Methods in Finance*, Cambridge University Press, Cambridge, UK.

Courant, R. and Robbins, H. (1969). *What is Mathematics?*, Oxford University Press, Oxford, UK.

Cox, J. C., Ross, S., and Rubinstein, M. (1979), Option pricing: A simplified approach, *Journal of Financial Economics*, **7**, pp. 229–264.

Cuthbertson, K. and Nitzsche, D. (2004). *Quantitative Financial Economics: Stocks, Bonds, and Foreign Exchange*, 2nd edition, John Wiley & Sons Inc., West Sussex, UK.

Dax, A. (1997). An elementary proof of Farkas' Lemma, *SIAM Review*, **39**(3), pp. 503–507.

Dean, J., Corvin, J., and Ewell, D. (2001). List of *Playboy*'s playmates of the month with data sheet stats, Available at http://www3.sympatico.ca/jimdean/pmstats.txt.

DeGroot, M. H. (1975). *Probability and Statistics, Behavioral Science: Quantitative Methods*, Addison-Wesley Publishing Company, Reading, MA, USA.

Durrett, R. (1996). *Stochastic Calculus: A Practical Introduction*, CRC Press, Boca Raton, FL, USA.

Dybvig, P. H. and Ross, S. A. (2008). "Arbitrage", in S. N. Durlauf and L. E. Blume, editors, *The New Palgrave Dictionary of Economics*, Palgrave Macmillan, pp. 188–197.

Franklin, J. (1980). *Methods of Mathematical Economics*, Springer-Verlag, New York, NY, USA.

Gale, D., Kuhn, H. W., and Tucker, A. W. (1951). Linear programming and the theory of games, in T. C. Koopmans, editor, *Activity Analysis of Production and Allocation*, Wiley, New York, NY, USA, pp. 317–329.

Gale, D. (1960). *The Theory of Linear Economic Models*, McGraw-Hill, New York, NY, USA.

Gard, T. C. (1988). *Introduction to Stochastic Differential Equations*, Marcel Dekker, New York, USA.

Gardner, M. (October 1959). "Mathematical Games" column, *Scientific American*, pp. 180–182.

Garman, M. B. and Kohlhagan, S. W. (1983). Foreign currency option values, *Journal of International Money and Finance*, **2**(3), pp. 231–237.

Geske, R. (1979). The valuation of compound options, *Journal of Financial Economics*, **7**(1), pp. 63–81.

Goldstein, L. J., Lay, D. C., and Schneider, D. I. (1999). *Brief Calculus and Its Applications*, 8th edition, Prentice Hall, Upper Saddle River, NJ, USA.

Golub, G. H. and Van Loan, C. F. (1989). *Matrix Computations*, 2nd edition, Johns Hopkins University Press, Baltimore, Maryland, USA.

González-Díaz, J., García-Jurado, I., and Fiestras-Janiero, M. G. (2010). *An Introductory Course on Mathematical Game Theory*, Volume 115 of Graduate Studies in Mathematics, American Mathematical Society, Providence, Rhode Island, USA.

Guo, W. and Su, T. (2006). Option put-call parity relations when the underlying security pays dividends, *International Journal of Business and Economics*, **5**, pp. 225–230.

Greenberg, M. D. (1998). *Advanced Engineering Mathematics*, 2nd edition, Prentice-Hall, Inc., Upper Saddle River, New Jersey, USA.

Grimmett, G. R. and Stirzaker, D. R. (1982). *Probability and Random Processes*, Oxford University Press, London, UK.

Haug, E. G., Haug, J., and Margrabe, W. (2003). Asian Pyramid Power, *Wilmott Magazine*.

Haug, E. G. (2007). *The Complete Guide to Option Pricing Formulas*, 2nd edition, McGraw-Hill, New York, NY, USA.

Hestenes, M. R. (1975). *Optimization Theory: The Finite Dimensional Case*, John Wiley & Sons, New York, USA.

Hull, J. C. (2000). *Options, Futures, and Other Derivatives*, Prentice-Hall, Inc., Upper Saddle River, New Jersey, USA.

Jeffrey, A. (2002). *Advanced Engineering Mathematics*, Harcourt Academic Press, Burlington, Massachusetts, USA.

Karatzas, I. and Shreve, S. E. (1991). *Brownian Motion and Stochastic Calculus*, Springer-Verlag, New York, NY, USA.

Karlin, S. (1959). *Mathematical Methods and Theory in Games, Programming, and Economics*, Addison-Wesley, Reading, MA, USA.

Kijima, M. (2003). *Stochastic Processes with Applications to Finance*, Chapman & Hall/CRC, Boca Raton, FL, USA.

Lawler, G. F. (2006). *Introduction to Stochastic Processes*, 2nd edition, Chapman & Hall/CRC, Boca Raton, FL, USA.

Luenberger, D. G. (1998). *Investment Science*, Oxford University Press, New York, USA.

Margrabe, W. (1978). The value of an option to exchange one asset for another, *Journal of Finance*, **33**(1), pp. 177–186.

Markowitz, H. (1952). Portfolio selection, *Journal of Finance*, **7**, pp. 77–91.

Marsden, J. E. and Hoffman, M. J. (1987). *Basic Complex Analysis*, W. H. Freeman and Company, New York, NY, USA.

McDonald, R. L. (2013). *Derivatives Markets*, 3rd edition, Pearson Education Inc., Boston, MA, USA.

Mehta, N., Thomasson, L., and Barret, P. M. (2010). The machines that ate the market, *Bloomberg Businessweek*, **4180**, pp. 48–55.

Mikosch, T. (1998). *Elementary Stochastic Calculus With Finance in View*, Vol. 6 of *Advanced Series on Statistical Science & Applied Probability*, World Scientific Publishing, River Edge, New Jersey, USA.

Neftci, S. N. (2000). *An Introduction to the Mathematics of Financial Derivatives*, 2nd edition, Academic Press, San Diego, CA, USA.

Noble, D. and Daniel, J. W. (1988). *Applied Linear Algebra*, 3rd edition, Prentice Hall, Englewood Cliffs, NJ, USA.

Øksendal, B. (2003). *Stochastic Differential Equations: An Introduction with Applications*, 6th edition, Springer-Verlag, Berlin, Germany.

OpenStax College. (2018). *Calculus Volume 1*, OpenStax College. Retrieved from https://openstax.org/details/books/calculus-volume-1.

OpenStax College. (2018). *Calculus Volume 3*, OpenStax College. Retrieved from https://openstax.org/details/books/calculus-volume-3.

Proschan, M. A. (2008). The normal approximation to the binomial, *The American Statistician*, **62**(1), pp. 62–63.

Redner, S. (2001). *A Guide to First Passage Processes*, Cambridge University Press, Cambridge, UK.

Ross, S. M. (1999). *An Introduction to Mathematical Finance: Options and Other Topics*, Cambridge University Press, Cambridge, UK.

Ross, S. M. (2003). *Introduction to Probability Models*, 8th edition, Academic Press, San Diego, CA, USA.

Ross, S. M. (2006). *A First Course in Probability*, 7th edition, Prentice Hall, Inc., Upper Saddle River, NJ, USA.

Selvin, S. (1975). A problem in probability, *American Statistician*, **29**(1), p. 67.

Seydel, R. (2002). *Tools for Computational Finance*, Springer-Verlag, New York, NY, USA.

Sharpe, W. (1964). Capital asset prices: A Theory of Market Equilibrium under Conditions of Risk, *Journal of Finance*, **19**, pp. 425–442.

Sharpe, W. (1966). Mutual fund performance, *Journal of Business*, **39**, pp. 119–138.

Shodor Education Foundation, Inc., "Graphing and Interpreting Bivariate Data", 20 May 2008, http://www.shodor.org/interactivate/discussions/Graphing Data/.

Shreve, S. E. (2004). *Stochastic Calculus for Finance I*, Springer-Verlag, New York, NY, USA.

Shreve, S. E. (2004). *Stochastic Calculus for Finance II*, Springer-Verlag, New York, NY, USA.

Smith, R. T. and Minton, R. B. (2002). *Calculus*, 2nd edition, McGraw-Hill, Boston, MA, USA.

Steele, J. M. (2001). *Stochastic Calculus and Financial Applications*, Volume 45 in Applications of Mathematics, Springer-Verlag, New York, NY, USA.

Stewart, J. (1999). *Calculus*, 4th edition, Brooks/Cole Publishing Company, Pacific Grove, CA, USA.

Strang, G. (1986). *Introduction to Applied Mathematics*, Wellesley-Cambridge Press, Wellesley, MA, USA.

Sullivan, M. (2018). *Fundamentals of Statistics*, 5th edition, Pearson Education, Boston, MA, USA.

Taylor, A. E. and Mann, W. R. (1983). *Advanced Calculus*, 3rd edition, John Wiley & Sons, Inc., New York, NY, USA.

Tobin, J. (1958). Liquidity preference in behavior toward risk, *Review of Economic Studies*, **25**, pp. 65–86.

vos Savant, M. (February 1990). "Ask Marilyn" column, *Parade Magazine*, p. 12.

Williams, A. C. (1970). Complementarity theorems for linear programming, *SIAM Review*, **12**(1), pp. 135–137.

Wilmott, P.(2006). *Paul Wilmott on Quantitative Finance*, 2nd edition, John Wiley & Sons, Inc., Hoboken, NJ, USA.

Wilmott, P., Howison, S., and Dewynne, J. (1995). *The Mathematics of Financial Derivatives: A Student Introduction*, Cambridge University Press, Cambridge, UK.

Winston, W. L. (1994). *Operations Research: Applications and Algorithms*, International Thompson Publishing, Belmont, CA, USA.

# Index

**S**

saddle point, 91
sample variance, 200
sample space, 24
scalar product, 418
Schwarz Inequality, 62
secant lines, 127
security market line, 116
self-financing, 358
    portfolio, 358
sensitivity, 323
separating hyperplane, 426
set
    convex, 75
set theory, 25
settlement date, 136
Sharpe Ratio, 109
short position, 135
simple interest, 2
simplex method, 75
singular matrix, 421
slack variables, 75
spot contract, 139
spot rate, 17
spread, *see also* standard
    deviation, 182
    bear, 183
    bull, 183
    butterfly, 185
square matrix, 419
standard bivariate normal
    distribution, 246
standard deviation, 23, 57,
    201, 224
standard form, 75
standard linear program, 75
standard normal random
    variable, 231
stationarity, 261
stochastic
    calculus, 249, 260
    differential equation, 314
    integral, 266, 268
        differential form, 268
    process, 260, 288

straddle, 184, 380
strangle, 184
strike
    asset, 376
    price, 163
    time, 163
stub quotes, 97
symmetric form, 74
symmetric linear program, 74
symmetric matrix, 419
symmetric powered options, 374
synthetic forward, 155
systematic risk, 102

**T**

Taylor remainder, 273
Taylor's Theorem, 273
Theta, 329
threshold, 242
transaction costs, 137
transpose of a matrix, 418
Treasury Bills, 165
tridiagonal matrix, 420

**U**

uncorrelated random variable, 61
underlying asset, 376
uniform random variable, 213
United States Treasury, 19
    Bonds, 141
up rebate, 403
up-and-in option, 403
up-and-out option, 403
upper triangular matrix, 419
utility, 125
    decreasing, 127
utility function, 125–126, 238

**V**

value-weighted average, 344
vanilla options, 287

Printed in the United States
by Baker & Taylor Publisher Services